SCHAUM'S OUTLINE OF

THEORY AND PROBLEMS

OF

GENERAL TOPOLOGY

•

BY

SEYMOUR LIPSCHUTZ, Ph.D.

Professor of Mathematics
Temple University

•

McGraw-Hill
New York St. Louis San Francisco Auckland Bogotá
Caracas Lisbon London Madrid Mexico City Milan
Montreal New Delhi San Juan Singapore
Sydney Tokyo Toronto

ISBN 07-037988-2

19 20 21 22 23 24 25 26 27 28 29 30 BAW BAW 9 9 8 7 6

McGraw-Hill

*A Division of The **McGraw·Hill** Companies*

Preface

General topology, also called point set topology, has recently become an essential part of the mathematical background of both graduate and undergraduate students. This book is designed to be used either as a textbook for a formal course in topology or as a supplement to all current standard texts. It should also be of considerable value as a source and reference book for those who require a comprehensive and rigorous introduction to the subject.

Each chapter begins with clear statements of pertinent definitions, principles and theorems together with illustrative and other descriptive material. This is followed by graded sets of solved and supplementary problems. The solved problems serve to illustrate and amplify the theory, bring into sharp focus those fine points without which the student continually feels himself on unsafe ground, and provide the repetition of basic principles so vital to effective learning. Numerous proofs of theorems are included among the solved problems. The supplementary problems serve as a complete review of the material of each chapter.

Topics covered include the basic properties of topological, metric and normed spaces, the separation axioms, compactness, the product topology, and connectedness. Theorems proven include Urysohn's lemma and metrization theorem, Tychonoff's product theorem and Baire's category theorem. The last chapter, on function spaces, investigates the topologies of pointwise, uniform and compact convergence. In addition, the first three chapters present the required concepts of set theory, the fourth chapter treats of the topology of the line and plane, and the appendix gives the basic principles of the real numbers.

More material is included here than can be covered in most first courses. This has been done to make the book more flexible, to provide a more useful book of reference, and to stimulate further interest in the subject.

I wish to thank many of my friends and colleagues, especially Dr. Joan Landman, for invaluable suggestions and critical review of the manuscript. I also wish to express my gratitude to the staff of the Schaum Publishing Company, particularly to Jeffrey Albert and Alan Hopenwasser, for their helpful cooperation.

SEYMOUR LIPSCHUTZ

Temple University
May, 1965

CONTENTS

CONTENTS

Chapter 1

Sets and Relations

SETS, ELEMENTS

The concept *set* appears in all branches of mathematics. Intuitively, a set is any well-defined list or collection of objects, and will be denoted by capital letters $A, B, X, Y, \ldots$. The objects comprising the set are called its *elements* or *members* and will be denoted by lower case letters $a, b, x, y, \ldots$. The statement *"p is an element of A"* or, equivalently, *"p belongs to A"* is written

$$p \in A$$

The negation of $p \in A$ is written $p \notin A$.

There are essentially two ways to specify a particular set. One way, if it is possible, is by actually listing its members. For example,

$$A = \{a, e, i, o, u\}$$

denotes the set A whose elements are the letters a, e, i, o and u. Note that the elements are separated by commas and enclosed in braces { }. The other way is by stating those properties which characterize the elements in the set. For example,

$$B = \{x : x \text{ is an integer, } x > 0\}$$

which reads "B is the set of x such that x is an integer and x is greater than zero," denotes the set B whose elements are the positive integers. A letter, usually x, is used to denote an arbitrary member of the set; the colon is read as 'such that' and the comma as 'and'.

Example 1.1: The set B above can also be written as $B = \{1, 2, 3, \ldots\}$. Note that $-6 \notin B$, $3 \in B$ and $\pi \notin B$.

Example 1.2: Intervals on the real line, defined below, appear very often in mathematics. Here a and b are real numbers with $a < b$.

Open interval from a to b $\quad = (a, b) = \{x : a < x < b\}$

Closed interval from a to b $\quad = [a, b] = \{x : a \leqq x \leqq b\}$

Open-closed interval from a to b $= (a, b] = \{x : a < x \leqq b\}$

Closed-open interval from a to b $= [a, b) = \{x : a \leqq x < b\}$

The open-closed and closed-open intervals are also called *half-open* intervals.

Two sets A and B are *equal*, written $A = B$, if they consist of the same elements, i.e. if each member of A belongs to B and each member of B belongs to A. The negation of $A = B$ is written $A \neq B$.

Example 1.3: Let $E = \{x : x^2 - 3x + 2 = 0\}$, $F = \{2, 1\}$ and $G = \{1, 2, 2, 1\}$. Then $E = F = G$. Observe that a set does not depend on the way in which its elements are displayed. A set remains the same if its elements are repeated or rearranged.

Sets can be *finite* or *infinite*. A set is finite if it consists of n different elements, where n is some positive integer; otherwise a set is infinite. In particular, a set which consists of exactly one element is called a *singleton set*.

1

SUBSETS, SUPERSETS

A set A is a *subset* of a set B or, equivalently, B is a *superset* of A, written

$$A \subset B \quad \text{or} \quad B \supset A$$

iff each element in A also belongs to B; that is, if $x \in A$ implies $x \in B$. We also say that A *is contained in* B or B *contains* A. The negation of $A \subset B$ is written $A \not\subset B$ or $B \not\supset A$ and states that there is an $x \in A$ such that $x \notin B$.

> **Example 2.1:**   Consider the sets
>
> $$A = \{1, 3, 5, 7, \ldots\}, \quad B = \{5, 10, 15, 20, \ldots\}$$
>
> $$C = \{x : x \text{ is prime}, x > 2\} = \{3, 5, 7, 11, \ldots\}$$
>
> Then $C \subset A$ since every prime number greater than 2 is odd. On the other hand, $B \not\subset A$ since $10 \in B$ but $10 \notin A$.

> **Example 2.2:**   We will let **N** denote the set of positive integers, **Z** denote the set of integers, **Q** denote the set of rational numbers and **R** denote the set of real numbers. Accordingly,
>
> $$\mathbf{N} \subset \mathbf{Z} \subset \mathbf{Q} \subset \mathbf{R}$$

Observe that $A \subset B$ does not exclude the possibility that $A = B$. In fact, we are able to restate the definition of equality of sets as follows:

> **Definition:**   Two sets A and B are equal if and only if $A \subset B$ and $B \subset A$.

In the case that $A \subset B$ but $A \neq B$, we say that A is a *proper subset* of B or B contains A properly. The reader should be warned that some authors use the symbol $\subseteq$ for a subset and the symbol $\subset$ only for a proper subset.

Our first theorem follows from the preceding definitions.

Theorem 1.1:   Let A, B and C be any sets. Then (i) $A \subset A$; (ii) if $A \subset B$ and $B \subset A$ then $A = B$; and (iii) if $A \subset B$ and $B \subset C$ then $A \subset C$.

UNIVERSAL AND NULL SETS

In any application of the theory of sets, all sets under investigation are subsets of a fixed set. We call this set the *universal set* or *universe of discourse* and denote it in this chapter by U. It is also convenient to introduce the concept of the *empty* or *null set*, that is, a set which contains no elements. This set, denoted by $\emptyset$, is considered finite and a subset of every other set. Thus, for any set A, $\emptyset \subset A \subset U$.

> **Example 3.1:**   In plane geometry, the universal set consists of all the points in the plane.

> **Example 3.2:**   Let $A = \{x : x^2 = 4, x \text{ is odd}\}$. Then A is empty, i.e. $A = \emptyset$.

> **Example 3.3:**   Let $B = \{\emptyset\}$. Then $B \neq \emptyset$ for B contains one element.

CLASSES, COLLECTIONS, FAMILIES AND SPACES

Frequently, the members of a set are sets themselves. For example, each line in a set of lines is a set of points. To help clarify these situations, we use the words *"class"*, *"collection"* and *"family"* synonymously with set. Usually we use class for a set of sets, and *collection* or *family* for a set of classes. The words *subclass*, *subcollection* and *subfamily* have meanings analogous to subset.

> **Example 4.1:**   The members of the class $\{\{2, 3\}, \{2\}, \{5, 6\}\}$ are the sets $\{2, 3\}$, $\{2\}$ and $\{5, 6\}$.

Example 4.2:   Consider any set A. The *power set* of A, denoted by $\mathcal{P}(A)$ or 2^A, is the class of all subsets of A. In particular, if $A = \{a, b, c\}$, then

$$\mathcal{P}(A) = \{A, \{a, b\}, \{a, c\}, \{b, c\}, \{a\}, \{b\}, \{c\}, \emptyset\}$$

In general, if A is finite, say A has n elements, then $\mathcal{P}(A)$ will have 2^n elements.

The word *space* shall mean a non-empty set which possesses some type of mathematical structure, e.g. vector space, metric space or topological space. In such a situation, we will call the elements in a space *points*.

SET OPERATIONS

The *union* of two sets A and B, denoted by $A \cup B$, is the set of all elements which belong to A or B, i.e.,

$$A \cup B = \{x : x \in A \text{ or } x \in B\}$$

Here "or" is used in the sense of "and/or".

The *intersection* of two sets A and B, denoted by $A \cap B$, is the set of elements which belong to both A and B, i.e.,

$$A \cap B = \{x : x \in A \text{ and } x \in B\}$$

If $A \cap B = \emptyset$, that is, if A and B do not have any elements in common, then A and B are said to be *disjoint* or *non-intersecting*. A class $\mathcal{A}$ of sets is called a *disjoint class of sets* if each pair of distinct sets in $\mathcal{A}$ is disjoint.

The *relative complement* of a set B with respect to a set A or, simply the *difference* of A and B, denoted by $A \setminus B$, is the set of elements which belong to A but which do not belong to B. In other words,

$$A \setminus B = \{x : x \in A, \ x \notin B\}$$

Observe that $A \setminus B$ and B are disjoint, i.e. $(A \setminus B) \cap B = \emptyset$.

The *absolute complement* or, simply, *complement* of a set A, denoted by A^c, is the set of elements which do not belong to A, i.e.,

$$A^c = \{x : x \in U, \ x \notin A\}$$

In other words, A^c is the difference of the universal set U and A.

Example 5.1:   The following diagrams, called Venn diagrams, illustrate the above set operations. Here sets are represented by simple plane areas and U, the universal set, by the area in the entire rectangle.

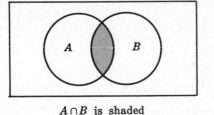

$A \cap B$ is shaded

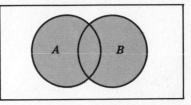

$A \cup B$ is shaded

$A \setminus B$ is shaded

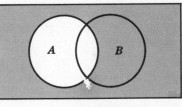

A^c is shaded

Sets under the above operations satisfy various laws or identities which are listed in the table below (Table 1).   In fact, we state

Theorem 1.2:   Sets satisfy the laws in Table 1.

LAWS OF THE ALGEBRA OF SETS	
Idempotent Laws	
1a. $A \cup A = A$	1b. $A \cap A = A$
Associative Laws	
2a. $(A \cup B) \cup C = A \cup (B \cup C)$	2b. $(A \cap B) \cap C = A \cap (B \cap C)$
Commutative Laws	
3a. $A \cup B = B \cup A$	3b. $A \cap B = B \cap A$
Distributive Laws	
4a. $A \cup (B \cap C) = (A \cup B) \cap (A \cup C)$	4b. $A \cap (B \cup C) = (A \cap B) \cup (A \cap C)$
Identity Laws	
5a. $A \cup \emptyset = A$	5b. $A \cap U = A$
6a. $A \cup U = U$	6b. $A \cap \emptyset = \emptyset$
Complement Laws	
7a. $A \cup A^c = U$	7b. $A \cap A^c = \emptyset$
8a. $(A^c)^c = A$	8b. $U^c = \emptyset, \ \emptyset^c = U$
De Morgan's Laws	
9a. $(A \cup B)^c = A^c \cap B^c$	9b. $(A \cap B)^c = A^c \cup B^c$

Table 1

Remark: Each of the above laws follows from an analogous logical law.   For example,

$$A \cap B \ = \ \{x : x \in A \text{ and } x \in B\} \ = \ \{x : x \in B \text{ and } x \in A\} \ = \ B \cap A$$

Here we use the fact that the composite statement "p and q", written $p \wedge q$, is logically equivalent to the composite statement "q and p", i.e. $q \wedge p$.

The relationship between set inclusion and the above set operations follows.

Theorem 1.3:   Each of the following conditions is equivalent to $A \subset B$:

(i) $A \cap B = A$	(iii) $B^c \subset A^c$	(v) $B \cup A^c = U$
(ii) $A \cup B = B$	(iv) $A \cap B^c = \emptyset$	

PRODUCT SETS

Let A and B be two sets.   The *product set* of A and B, written $A \times B$, consists of all ordered pairs $\langle a, b \rangle$ where $a \in A$ and $b \in B$, i.e.,

$$A \times B \ = \ \{\langle a, b \rangle : a \in A, \ b \in B\}$$

The product of a set with itself, say $A \times A$, will be denoted by A^2.

Example 6.1:   The reader is familiar with the Cartesian plane $\mathbf{R}^2 = \mathbf{R} \times \mathbf{R}$ (Fig. 1-1 below).   Here each point P represents an ordered pair $\langle a, b \rangle$ of real numbers and vice versa.

Example 6.2:   Let $A = \{1, 2, 3\}$ and $B = \{a, b\}$.   Then

$$A \times B \ = \ \{\langle 1, a \rangle, \ \langle 1, b \rangle, \ \langle 2, a \rangle, \ \langle 2, b \rangle, \ \langle 3, a \rangle, \ \langle 3, b \rangle\}$$

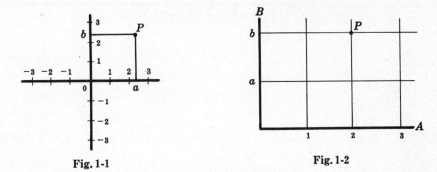

Fig. 1-1   Fig. 1-2

Since A and B do not contain many elements, it is possible to represent $A \times B$ by a coordinate diagram as shown in Fig. 1-2 above. Here the vertical lines through the points of A and the horizontal lines through the points of B meet in 6 points which represent $A \times B$ in the obvious way. The point P is the ordered pair $\langle 2, b \rangle$. In general, if a set A has s elements and a set B has t elements, then $A \times B$ has s times t elements.

Remark: The notion "ordered pair" $\langle a, b \rangle$ is defined rigorously by $\langle a, b \rangle \equiv \{\{a\}, \{a, b\}\}$. From this definition, the "order" property may be proven:

$$\langle a, b \rangle = \langle c, d \rangle \quad \text{implies} \quad a = c \text{ and } b = d$$

The concept of product set can be extended to any finite number of sets in a natural way. The product set of the sets $A_1, \ldots, A_m$, denoted by

$$A_1 \times A_2 \times \cdots \times A_m \quad \text{or} \quad \prod_{i=1}^{m} A_i$$

consists of all m-tuples $\langle a_1, a_2, \ldots, a_m \rangle$ where $a_i \in A_i$ for each i.

RELATIONS

A *binary relation* (or *relation*) R from a set A to a set B assigns to each pair $\langle a, b \rangle$ in $A \times B$ exactly one of the following statements:

(i) "a is related to b", written $a \, R \, b$

(ii) "a is not related to b", written $a \, \not{R} \, b$

A relation from a set A to the same set A is called a *relation in A*.

Example 7.1: Set inclusion is a relation in any class of sets. For, given any pair of sets A and B, either $A \subset B$ or $A \not\subset B$.

Observe that any relation R from a set A to a set B uniquely defines a subset R^* of $A \times B$ as follows:

$$R^* = \{\langle a, b \rangle : a \, R \, b\}$$

On the other hand, any subset R^* of $A \times B$ defines a relation R from A to B as follows:

$$a \, R \, b \quad \text{iff} \quad \langle a, b \rangle \in R^*$$

In view of the correspondence between relations R from A to B and subsets of $A \times B$, we redefine a relation by

Definition: A relation R from A to B is a subset of $A \times B$.

The *domain* of a relation R from A to B is the set of first coordinates of the pairs in R and its *range* is the set of second coordinates, i.e.,

$$\text{domain of } R = \{a : \langle a, b \rangle \in R\}, \quad \text{range of } R = \{b : \langle a, b \rangle \in R\}$$

The *inverse of R*, denoted by R^{-1}, is the relation from B to A defined by

$$R^{-1} = \{\langle b, a \rangle : \langle a, b \rangle \in R\}$$

Note that R^{-1} can be obtained by reversing the pairs in R.

> **Example 7.2:**   Consider the relation
> $$R \;=\; \{\,\langle 1,2\rangle,\ \langle 1,3\rangle,\ \langle 2,3\rangle\,\}$$
> in $A = \{1,2,3\}$. Then the domain of $R = \{1,2\}$, the range of $R = \{2,3\}$, and
> $$R^{-1} \;=\; \{\,\langle 2,1\rangle,\ \langle 3,1\rangle,\ \langle 3,2\rangle\,\}$$
> Observe that R and R^{-1} are identical, respectively, to the relations $<$ and $>$ in A, i.e.,
> $$\langle a,b\rangle \in R \ \text{ iff } \ a < b \qquad \text{and} \qquad \langle a,b\rangle \in R^{-1} \ \text{ iff } \ a > b$$

The *identity relation* in any set A, denoted by Δ or Δ_A, is the set of all pairs in $A \times A$ with equal coordinates, i.e.,
$$\Delta_A \;=\; \{\,\langle a,a\rangle : a \in A\,\}$$

The identity relation is also called the *diagonal* by virtue of its position in a coordinate diagram of $A \times A$.

EQUIVALENCE RELATIONS

A relation R in a set A, i.e. a subset R of $A \times A$, is termed an *equivalence relation* iff it satisfies the following axioms:

[E₁]   For every $a \in A$, $\langle a,a\rangle \in R$.

[E₂]   If $\langle a,b\rangle \in R$, then $\langle b,a\rangle \in R$.

[E₃]   If $\langle a,b\rangle \in R$ and $\langle b,c\rangle \in R$, then $\langle a,c\rangle \in R$.

In general, a relation is said to be *reflexive* iff it satisfies **[E₁]**, *symmetric* iff it satisfies **[E₂]** and *transitive* iff it satisfies **[E₃]**. Accordingly, a relation R is an equivalence relation iff it is reflexive, symmetric and transitive.

> **Example 8.1:**   Consider the relation $\subset$, i.e. set inclusion. Recall, by Theorem 1.1, that $A \subset A$ for every set A, and
> $$\text{if } A \subset B \text{ and } B \subset C \text{ then } A \subset C$$
> Hence $\subset$ is both reflexive and transitive. On the other hand,
> $$A \subset B \text{ and } A \neq B \quad \text{implies} \quad B \not\subset A$$
> Accordingly, $\subset$ is not symmetric and hence is not an equivalence relation.

> **Example 8.2:**   In Euclidian geometry, similarity of triangles is an equivalence relation. For if α, β and γ are any triangles then: (i) α is similar to itself; (ii) if α is similar to β, then β is similar to α; and (iii) if α is similar to β and β is similar to γ then α is similar to γ.

If R is an equivalence relation in A, then the *equivalence class* of any element $a \in A$, denoted by $[a]$, is the set of elements to which a is related:
$$[a] \;=\; \{x : \langle a,x\rangle \in R\}$$

The collection of equivalence classes of A, denoted by A/R, is called the *quotient* of A by R:
$$A/R \;=\; \{[a] : a \in A\}$$

The quotient set A/R possesses the following properties:

Theorem 1.4:   Let R be an equivalence relation in A and let $[a]$ be the equivalence class of $a \in A$. Then:

> (i)   For every $a \in A$, $a \in [a]$.
>
> (ii) $[a] = [b]$ if and only if $\langle a,b\rangle \in R$.
>
> (iii)   If $[a] \neq [b]$, then $[a]$ and $[b]$ are disjoint.

A class $\mathcal{A}$ of non-empty subsets of A is called a *partition* of A iff (1) each $a \in A$ belongs to some member of $\mathcal{A}$ and (2) the members of $\mathcal{A}$ are pair-wise disjoint. Accordingly, the previous theorem implies the following *fundamental theorem of equivalence relations*:

Theorem 1.5: Let R be an equivalence relation in A. Then the quotient set A/R is a partition of A.

Example 8.3: Let R_5 be the relation in $\mathbf{Z}$, the set of integers, defined by

$$x \equiv y \pmod 5$$

which reads "x is congruent to y modulo 5" and which means "$x - y$ is divisible by 5". Then R_5 is an equivalence relation in $\mathbf{Z}$. There are exactly five distinct equivalence classes in $\mathbf{Z}/R_5$:

$$E_0 = \{\ldots, -10, -5, 0, 5, 10, \ldots\} = \cdots = [-10] = [-5] = [0] = [5] = \cdots$$
$$E_1 = \{\ldots, -9, -4, 1, 6, 11, \ldots\} = \cdots = [-9] = [-4] = [1] = [6] = \cdots$$
$$E_2 = \{\ldots, -8, -3, 2, 7, 12, \ldots\} = \cdots = [-8] = [-3] = [2] = [7] = \cdots$$
$$E_3 = \{\ldots, -7, -2, 3, 8, 13, \ldots\} = \cdots = [-7] = [-2] = [3] = [8] = \cdots$$
$$E_4 = \{\ldots, -6, -1, 4, 9, 14, \ldots\} = \cdots = [-6] = [-1] = [4] = [9] = \cdots$$

Observe that each integer x, which is uniquely expressible in the form $x = 5q + r$ where $0 \leq r < 5$, is a member of the equivalence class E_r where r is the remainder. Note that the equivalence classes are pairwise disjoint and that $\mathbf{Z} = E_0 \cup E_1 \cup E_2 \cup E_3 \cup E_4$.

COMPOSITION OF RELATIONS

Let U be a relation from A to B and let V be a relation from B to C, i.e. $U \subset A \times B$ and $V \subset B \times C$. Then the relation from A to C which consists of all ordered pairs $\langle a, c \rangle \in A \times C$ such that, for some $b \in B$,

$$\langle a, b \rangle \in U \quad \text{and} \quad \langle b, c \rangle \in V$$

is called the *composition* of U and V and is denoted by $V \circ U$. (The reader should be warned that some authors denote this relation by $U \circ V$.)

It is convenient to introduce some more symbols:

$\exists$, there exists   s.t., such that $\forall$, for all $\Rightarrow$, implies

We may then write:

$$V \circ U = \{\langle x, y \rangle : x \in A, \, y \in C; \, \exists b \in B \text{ s.t. } \langle x, b \rangle \in U, \langle b, y \rangle \in V\}$$

Example 9.1: Let $A = \{1, 2, 3, 4\}$, $B = \{x, y, z, w\}$ and $C = \{5, 6, 7, 8\}$, and let

$$U = \{\langle 1, x \rangle, \langle 1, y \rangle, \langle 2, x \rangle, \langle 3, w \rangle, \langle 4, w \rangle\} \quad \text{and} \quad V = \{\langle y, 5 \rangle, \langle y, 6 \rangle, \langle z, 8 \rangle, \langle w, 7 \rangle\}$$

That is, U is a relation from A to B and V is a relation from B to C. We may illustrate U and V as follows:

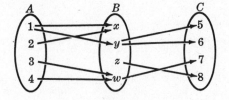

Accordingly,
$$\langle 1, 5 \rangle \in V \circ U \quad \text{since} \quad y \in B \text{ and } \langle 1, y \rangle \in U, \, \langle y, 5 \rangle \in V$$
$$\langle 1, 6 \rangle \in V \circ U \quad \text{since} \quad y \in B \text{ and } \langle 1, y \rangle \in U, \, \langle y, 6 \rangle \in V$$
$$\langle 3, 7 \rangle \in V \circ U \quad \text{since} \quad w \in B \text{ and } \langle 3, w \rangle \in U, \, \langle w, 7 \rangle \in V$$
$$\langle 4, 7 \rangle \in V \circ U \quad \text{since} \quad w \in B \text{ and } \langle 4, w \rangle \in U, \, \langle w, 7 \rangle \in V$$

No other ordered pairs belong to $V \circ U$, that is,

$$V \circ U = \{\langle 1, 5 \rangle, \langle 1, 6 \rangle, \langle 3, 7 \rangle, \langle 4, 7 \rangle\}$$

Observe that $V \circ U$ consists precisely of those pairs $\langle x, y \rangle$ for which there exists, in the above diagram, a "path" from $x \in A$ to $y \in C$ composed of two arrows, one following the other.

Example 9.2:   Let U and V be the relations in **R** defined by

$$U = \{\langle x, y \rangle : x^2 + y^2 = 1\} \quad \text{and} \quad V = \{\langle y, z \rangle : 2y + 3z = 4\}$$

Then the relation $V \circ U$, the composition of U and V, can be found by eliminating y from the two equations $x^2 + y^2 = 1$ and $2y + 3z = 4$. In other words,

$$V \circ U = \{\langle x, z \rangle : 4x^2 + 9z^2 - 24z + 12 = 0\}$$

Example 9.3:   Let **N** denote the set of positive integers, and let R denote the relation $<$ in **N**, i.e. $\langle a, b \rangle \in R$ iff $a < b$. Hence $\langle a, b \rangle \in R^{-1}$ iff $a > b$. Then

$$
\begin{aligned}
R \circ R^{-1} &= \{\langle x, y \rangle : x, y \in \mathbf{N}; \ \exists b \in \mathbf{N} \ \text{s.t.} \ \langle x, b \rangle \in R^{-1}, \langle b, y \rangle \in R\} \\
&= \{\langle x, y \rangle : x, y \in \mathbf{N}; \ \exists b \in \mathbf{N} \ \text{s.t.} \ b < x, \ b < y\} \\
&= (\mathbf{N} \backslash \{1\}) \times (\mathbf{N} \backslash \{1\}) \quad = \quad \{\langle x, y \rangle : x, y \in \mathbf{N}; \ x, y \neq 1\}
\end{aligned}
$$

and

$$
\begin{aligned}
R^{-1} \circ R &= \{\langle x, y \rangle : x, y \in \mathbf{N}; \ \exists b \in \mathbf{N} \ \text{s.t.} \ \langle x, b \rangle \in R, \langle b, y \rangle \in R^{-1}\} \\
&= \{\langle x, y \rangle : x, y \in \mathbf{N}; \ \exists b \in \mathbf{N} \ \text{s.t.} \ b > x, \ b > y\} \\
&= \mathbf{N} \times \mathbf{N}
\end{aligned}
$$

Note that $R \circ R^{-1} \neq R^{-1} \circ R$.

Solved Problems

SETS, ELEMENTS, SUBSETS

1. Let $A = \{x : 3x = 6\}$. Does $A = 2$?

Solution:

A is the set which consists of the single element 2, i.e. $A = \{2\}$. The number 2 belongs to A; it does not equal A. There is a basic difference between an element p and the singleton set $\{p\}$.

2. Determine which of the following sets are equal: $\emptyset$, $\{0\}$, $\{\emptyset\}$.

Solution:

Each is different from the other. The set $\{0\}$ contains one element, the number zero. The set $\emptyset$ contains no elements; it is the null set. The set $\{\emptyset\}$ also contains one element, the null set.

3. Determine whether or not each of the following sets is the null set:

(i) $X = \{x : x^2 = 9, 2x = 4\}$, (ii) $Y = \{x : x \neq x\}$, (iii) $Z = \{x : x + 8 = 8\}$.

Solution:

(i)   There is no number which satisfies both $x^2 = 9$ and $2x = 4$; hence $X = \emptyset$.

(ii)   We assume that any object is itself, so Y is empty. In fact, some texts define the null set by $\emptyset \equiv \{x : x \neq x\}$.

(iii) The number zero satisfies $x + 8 = 8$; hence $Z = \{0\}$. Accordingly, $Z \neq \emptyset$.

4. Prove that $A = \{2, 3, 4, 5\}$ is not a subset of $B = \{x : x \text{ is even}\}$.

Solution:

It is necessary to show that at least one member of A does not belong to B. Since $3 \in A$ and $3 \notin B$, A is not a subset of B.

5. Prove Theorem 1.1 (iii):   If $A \subset B$ and $B \subset C$ then $A \subset C$.

   Solution:

   We must show that each element in A also belongs to C. Let $x \in A$. Now $A \subset B$ implies $x \in B$. But $B \subset C$, so $x \in C$. We have therefore shown that $x \in A$ implies $x \in C$, or $A \subset C$.

6. Prove: If A is a subset of the null set $\emptyset$, then $A = \emptyset$.

   Solution:

   The null set $\emptyset$ is a subset of every set; in particular, $\emptyset \subset A$. But, by hypothesis, $A \subset \emptyset$; hence, by Definition 1.1, $A = \emptyset$.

7. Find the power set $\mathcal{P}(S)$ of the set $S = \{1, 2, 3\}$.

   Solution:

   Recall that the power set $\mathcal{P}(S)$ of S is the class of all subsets of S. The subsets of S are $\{1, 2, 3\}$, $\{1, 2\}$, $\{1, 3\}$, $\{2, 3\}$, $\{1\}$, $\{2\}$, $\{3\}$ and the empty set $\emptyset$. Hence

$$\mathcal{P}(S) = \{S, \{1,3\}, \{2,3\}, \{1,2\}, \{1\}, \{2\}, \{3\}, \emptyset\}$$

   Note that there are $2^3 = 8$ subsets of S.

8. Find the power set $\mathcal{P}(S)$ of $S = \{3, \{1, 4\}\}$.

   Solution:

   Note first that S contains two elements, 3 and the set $\{1, 4\}$. Therefore $\mathcal{P}(S)$ contains $2^2 = 4$ elements: S itself, the empty set $\emptyset$, the singleton set $\{3\}$ containing 3 and the singleton set $\{\{1, 4\}\}$ containing the set $\{1, 4\}$. In other words,

$$\mathcal{P}(S) = \{S, \{3\}, \{\{1,4\}\}, \emptyset\}$$

SET OPERATIONS

9. Let $U = \{1, 2, \ldots, 8, 9\}$, $A = \{1, 2, 3, 4\}$, $B = \{2, 4, 6, 8\}$ and $C = \{3, 4, 5, 6\}$. Find: (i) A^c,   (ii) $(A \cap C)^c$,   (iii) $B \setminus C$,   (iv) $(A \cup B)^c$.

   Solution:

   (i) A^c consists of the elements in U that are not in A; hence $A^c = \{5, 6, 7, 8, 9\}$.

   (ii) $A \cap C$ consists of the elements in both A and C; hence

$$A \cap C = \{3, 4\} \quad \text{and} \quad (A \cap C)^c = \{1, 2, 5, 6, 7, 8, 9\}$$

   (iii) $B \setminus C$ consists of the elements in B which are not in C; hence $B \setminus C = \{2, 8\}$.

   (iv) $A \cup B$ consists of the elements in A or B (or both); hence

$$A \cup B = \{1, 2, 3, 4, 6, 8\} \quad \text{and} \quad (A \cup B)^c = \{5, 7, 9\}$$

10. Prove: $(A \setminus B) \cap B = \emptyset$.

   Solution:
$$\begin{aligned} (A \setminus B) \cap B &= \{x : x \in B, \ x \in A \setminus B\} \\ &= \{x : x \in B, \ x \in A, \ x \notin B\} = \emptyset \end{aligned}$$

   since there is no element x satisfying $x \in B$ and $x \notin B$.

11. Prove De Morgan's Law: $(A \cup B)^c = A^c \cap B^c$.

   Solution:
$$\begin{aligned} (A \cup B)^c &= \{x : x \notin A \cup B\} \\ &= \{x : x \notin A, \ x \notin B\} \\ &= \{x : x \in A^c, \ x \in B^c\} = A^c \cap B^c \end{aligned}$$

12. Prove: $B \setminus A = B \cap A^c$.

   Solution: $B \setminus A = \{x : x \in B, \ x \notin A\} = \{x : x \in B, \ x \in A^c\} = B \cap A^c$

13. Prove the Distributive Law: $A \cap (B \cup C) = (A \cap B) \cup (A \cap C)$.

    Solution:

$$
\begin{aligned}
A \cap (B \cup C) &= \{x : x \in A;\ x \in B \cup C\} \\
&= \{x : x \in A;\ x \in B \text{ or } x \in C\} \\
&= \{x : x \in A,\ x \in B;\text{ or } x \in A,\ x \in C\} \\
&= \{x : x \in A \cap B \text{ or } x \in A \cap C\} \\
&= (A \cap B) \cup (A \cap C)
\end{aligned}
$$

Observe that in the third step above we used the analogous logical law

$$p \wedge (q \vee r) = (p \wedge q) \vee (p \wedge r)$$

where $\wedge$ reads "and" and $\vee$ reads "or".

14. Prove: For any sets A and B, $A \cap B \subset A \subset A \cup B$.

    Solution:

    Let $x \in A \cap B$; then $x \in A$ and $x \in B$. In particular, $x \in A$. Accordingly, $A \cap B \subset A$. If $x \in A$, then $x \in A$ or $x \in B$, i.e. $x \in A \cup B$. Hence $A \subset A \cup B$. In other words, $A \cap B \subset A \subset A \cup B$.

15. Prove Theorem 1.3 (i): $A \subset B$ if and only if $A \cap B = A$.

    Solution:

    Suppose $A \subset B$. Let $x \in A$; then by hypothesis, $x \in B$. Hence $x \in A$ and $x \in B$, i.e. $x \in A \cap B$. Accordingly, $A \subset A \cap B$. But by the previous problem, $A \cap B \subset A$. Hence $A \cap B = A$.

    On the other hand, suppose $A \cap B = A$. Then in particular, $A \subset A \cap B$. But, by the previous problem, $A \cap B \subset B$. Hence, by Theorem 1.1, $A \subset B$.

PRODUCT SETS, RELATIONS, COMPOSITION OF RELATIONS

16. Let $A = \{a, b\}$, $B = \{2, 3\}$ and $C = \{3, 4\}$. Find: (i) $A \times (B \cup C)$, (ii) $(A \times B) \cup (A \times C)$.

    Solution:

    (i)   First compute $B \cup C = \{2, 3, 4\}$.   Then

$$A \times (B \cup C) = \{\langle a, 2\rangle, \langle a, 3\rangle, \langle a, 4\rangle, \langle b, 2\rangle, \langle b, 3\rangle, \langle b, 4\rangle\}$$

    (ii)   First find $A \times B$ and $A \times C$:

$$A \times B = \{\langle a, 2\rangle, \langle a, 3\rangle, \langle b, 2\rangle, \langle b, 3\rangle\}, \quad A \times C = \{\langle a, 3\rangle, \langle a, 4\rangle, \langle b, 3\rangle, \langle b, 4\rangle\}$$

    Then compute the union of the two sets:

$$(A \times B) \cup (A \times C) = \{\langle a, 2\rangle, \langle a, 3\rangle, \langle b, 2\rangle, \langle b, 3\rangle, \langle a, 4\rangle, \langle b, 4\rangle\}$$

    Observe, from (i) and (ii), that $A \times (B \cup C) = (A \times B) \cup (A \times C)$.

17. Prove: $A \times (B \cap C) = (A \times B) \cap (A \times C)$.

    Solution:

$$
\begin{aligned}
A \times (B \cap C) &= \{\langle x, y\rangle : x \in A,\ y \in B \cap C\} \\
&= \{\langle x, y\rangle : x \in A,\ y \in B,\ y \in C\} \\
&= \{\langle x, y\rangle : \langle x, y\rangle \in A \times B,\ \langle x, y\rangle \in A \times C\} \\
&= (A \times B) \cap (A \times C)
\end{aligned}
$$

18. Let R be the relation $<$ from $A = \{1, 2, 3, 4\}$ to $B = \{1, 3, 5\}$, i.e., $\langle a, b\rangle \in R$ iff $a < b$.

    (i)   Write R as a set of ordered pairs.

    (ii)   Plot R on a coordinate diagram of $A \times B$.

    (iii)   Find domain of R, range of R and R^{-1}.

    (iv)   Find $R \circ R^{-1}$.

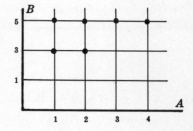

Solution:

(i) R consists of those ordered pairs $\langle a, b\rangle \in A \times B$ such that $a < b$; hence
$$R = \{\langle 1, 3\rangle, \langle 1, 5\rangle, \langle 2, 3\rangle, \langle 2, 5\rangle, \langle 3, 5\rangle, \langle 4, 5\rangle\}$$

(ii) R is displayed on the coordinate diagram of $A \times B$ as shown above.

(iii) The domain of R is the set of first coordinates of the pairs in R; hence domain of $R = \{1, 2, 3, 4\}$.
       The range of R is the set of second coordinates of the pairs in R; hence range of $R = \{3, 5\}$.
       R^{-1} can be obtained from R by reversing the pairs in R; hence
$$R^{-1} = \{\langle 3, 1\rangle, \langle 5, 1\rangle, \langle 3, 2\rangle, \langle 5, 2\rangle, \langle 5, 3\rangle, \langle 5, 4\rangle\}$$

(iv)   To find $R \circ R^{-1}$, construct diagrams of R^{-1} and R as shown below. Observe that R^{-1}, the
       second factor in the product $R \circ R^{-1}$, is constructed first. Then

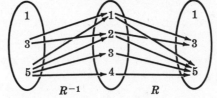

$$R \circ R^{-1} = \{\langle 3, 3\rangle, \langle 3, 5\rangle, \langle 5, 3\rangle, \langle 5, 5\rangle\}$$

19. Let T be the relation in the set of real numbers **R** defined by

       $x\,T\,y$ if both $x \in [n, n+1]$ and $y \in [n, n+1]$ for some integer n

Graph the relation T.

Solution:

     T consists of the shaded squares below.

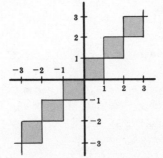

20. Let T be the relation in the set of real numbers **R** defined by $x\,T\,y$ iff $0 \le x - y \le 1$.

(i) Express T and T^{-1} as subsets of $\mathbf{R} \times \mathbf{R}$ and graph.

(ii) Show that $T \circ T^{-1} = \{\langle x, z\rangle : |x - z| \le 1\}$.

Solution:

(i) $T = \{\langle x, y\rangle : x, y \in \mathbf{R},\ 0 \le x - y \le 1\}$
$$T^{-1} = \{\langle x, y\rangle : \langle y, x\rangle \in T\} = \{\langle x, y\rangle : x, y \in \mathbf{R},\ 0 \le y - x \le 1\}$$

The relations T and T^{-1} are graphed below.

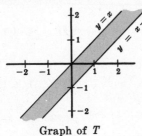

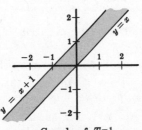

Graph of T Graph of T^{-1}

(ii)　By definition of composition of relations,

$$T \circ T^{-1} = \{\langle x, z \rangle : \exists y \in \mathbf{R} \text{ s.t. } \langle x, y \rangle \in T^{-1}, \langle y, z \rangle \in T \}$$
$$= \{\langle x, z \rangle : \exists y \in \mathbf{R} \text{ s.t. } \langle y, x \rangle, \langle y, z \rangle \in T \}$$
$$= \{\langle x, z \rangle : \exists y \in \mathbf{R} \text{ s.t. } 0 \le y - x \le 1, \ 0 \le y - z \le 1 \}$$

Let $S = \{\langle x, z \rangle : |x - z| \le 1 \}$. We want to show that $T \circ T^{-1} = S$.

Let $\langle x, z \rangle$ belong to $T \circ T^{-1}$. Then $\exists y$ s.t. $0 \le y - x, \ y - z \le 1$. But

$$0 \le y - x, \ y - z \le 1 \ \Rightarrow \ y - z \le 1$$
$$\Rightarrow \ y - z \le 1 + y - x$$
$$\Rightarrow \ x - z \le 1$$

Also,

$$0 \le y - x, \ y - z \le 1 \ \Rightarrow \ y - x \le 1$$
$$\Rightarrow \ y - x \le 1 + y - z$$
$$\Rightarrow \ -1 \le x - z$$

In other words, $0 \le y - x, \ y - z \le 1 \ \Rightarrow \ -1 \le x - z \le 1$ iff $|x - z| \le 1$.
Accordingly, $\langle x, z \rangle \in S$, i.e. $T \circ T^{-1} \subset S$.

Now let $\langle x, z \rangle$ belong to S; then $|x - z| \le 1$.

Let $y = \max(x, z)$; then $0 \le y - x \le 1$ and $0 \le y - z \le 1$.

Thus $\langle x, z \rangle$ also belongs to $T \circ T^{-1}$, i.e. $S \subset T \circ T^{-1}$. Hence $T \circ T^{-1} = S$.

21. Prove: For any two relations $R \subset X \times Y$ and $S \subset Y \times Z$, $(S \circ R)^{-1} = R^{-1} \circ S^{-1}$.

Solution:
$$(S \circ R)^{-1} = \{\langle z, x \rangle : \langle x, z \rangle \in S \circ R \}$$
$$= \{\langle z, x \rangle : \exists y \in Y \text{ s.t. } \langle x, y \rangle \in R, \langle y, z \rangle \in S \}$$
$$= \{\langle z, x \rangle : \exists y \in Y \text{ s.t. } \langle z, y \rangle \in S^{-1}, \langle y, x \rangle \in R^{-1} \}$$
$$= R^{-1} \circ S^{-1}$$

22. Prove: For any three relations $R \subset W \times X$, $S \subset X \times Y$ and $T \subset Y \times Z$, $(T \circ S) \circ R = T \circ (S \circ R)$.

Solution:
$$(T \circ S) \circ R = \{\langle w, z \rangle : \exists x \in X \text{ s.t. } \langle w, x \rangle \in R, \langle x, z \rangle \in T \circ S \}$$
$$= \{\langle w, z \rangle : \exists x \in X, \exists y \in Y \text{ s.t. } \langle w, x \rangle \in R, \langle x, y \rangle \in S, \langle y, z \rangle \in T \}$$
$$= \{\langle w, z \rangle : \exists y \in Y \text{ s.t. } \langle w, y \rangle \in S \circ R, \langle y, z \rangle \in T \}$$
$$= T \circ (S \circ R)$$

REFLEXIVE, SYMMETRIC, TRANSITIVE AND EQUIVALENCE RELATIONS

23. Prove: Let R be a relation in A, i.e. $R \subset A \times A$. Then:

(i)　R is reflexive iff $\Delta_A \subset R$;
(ii)　R is symmetric iff $R = R^{-1}$;
(iii)　R is transitive iff $R \circ R \subset R$;
(iv)　R reflexive implies $R \circ R \supset R$ and $R \circ R$ is reflexive;
(v)　R symmetric implies $R \circ R^{-1} = R^{-1} \circ R$;
(vi)　R transitive implies $R \circ R$ is transitive.

Solution:
(i)　Recall that the diagonal $\Delta_A = \{\langle a, a \rangle : a \in A\}$. Now R is reflexive iff, for every $a \in A$, $\langle a, a \rangle \in R$ iff $\Delta_A \subset R$.

(ii)　Follows directly from the definition of R^{-1} and symmetric.

(iii)　Let $\langle a, c \rangle \in R \circ R$; then $\exists b \in A$ such that $\langle a, b \rangle \in R$ and $\langle b, c \rangle \in R$. But, by transitivity, $\langle a, b \rangle, \langle b, c \rangle \in R$ implies $\langle a, c \rangle \in R$. Consequently, $R \circ R \subset R$.

On the other hand, suppose $R \circ R \subset R$. If $\langle a, b \rangle, \langle b, c \rangle \in R$, then $\langle a, c \rangle \in R \circ R \subset R$. In other words, R is transitive.

(iv)   Let $\langle a, b \rangle \in R$.   Now, $R \circ R \; = \; \{ \langle a, c \rangle : \; \exists b \in A \; \text{ s.t. } \; \langle a, b \rangle \in R, \; \langle b, c \rangle \in R \}$.

But $\langle a, b \rangle \in R$ and, since R is reflexive, $\langle b, b \rangle \in R$. Thus $\langle a, b \rangle \in R \circ R$, i.e. $R \subset R \circ R$. Furthermore, $\Delta_A \subset R \subset R \circ R$ implies $R \circ R$ is also reflexive.

(v) $\begin{aligned} R \circ R^{-1} \; &= \; \{ \langle a, c \rangle : \; \exists b \in A \; \text{ s.t. } \; \langle a, b \rangle \in R^{-1}, \; \langle b, c \rangle \in R \} \\ &= \; \{ \langle a, c \rangle : \; \exists b \in A \; \text{ s.t. } \; \langle a, b \rangle \in R, \; \langle b, c \rangle \in R^{-1} \} \\ &= \; R^{-1} \circ R \end{aligned}$

(vi)   Let $\langle a, b \rangle, \langle b, c \rangle \in R \circ R$. By (iii), $R \circ R \subset R$; hence $\langle a, b \rangle, \langle b, c \rangle \in R$. So $\langle a, c \rangle \in R \circ R$, i.e. $R \circ R$ is transitive.

24. Consider the relation $R = \{ \langle 1, 1 \rangle, \langle 2, 3 \rangle, \langle 3, 2 \rangle \}$ in $X = \{ 1, 2, 3 \}$. Determine whether or not R is (i) reflexive, (ii) symmetric, (iii) transitive.

Solution:

(i) R is not reflexive since $2 \in X$ but $\langle 2, 2 \rangle \notin R$.

(ii) R is symmetric since $R^{-1} = R$.

(iii) R is not transitive since $\langle 3, 2 \rangle \in R$ and $\langle 2, 3 \rangle \in R$ but $\langle 3, 3 \rangle \notin R$.

25. Consider the set $\mathbf{N} \times \mathbf{N}$, i.e. the set of ordered pairs of positive integers. Let R be the relation $\simeq$ in $\mathbf{N} \times \mathbf{N}$ which is defined by

$$\langle a, b \rangle \simeq \langle c, d \rangle \quad \text{iff} \quad ad = bc$$

Prove that R is an equivalence relation.

Solution:

Note that, for every $\langle a, b \rangle \in \mathbf{N} \times \mathbf{N}$, $\langle a, b \rangle \simeq \langle a, b \rangle$ since $ab = ba$; hence R is reflexive.

Suppose $\langle a, b \rangle \simeq \langle c, d \rangle$. Then $ad = bc$, which implies $cb = da$. Hence $\langle c, d \rangle \simeq \langle a, b \rangle$ and, therefore R is symmetric.

Now suppose $\langle a, b \rangle \simeq \langle c, d \rangle$ and $\langle c, d \rangle \simeq \langle e, f \rangle$. Then $ad = bc$ and $cf = de$. Thus

$$(ad)(cf) \; = \; (bc)(de)$$

and, by cancelling from both sides, $af = be$. Accordingly, $\langle a, b \rangle \simeq \langle e, f \rangle$ and R is transitive.

Since R is reflexive, symmetric and transitive, R is an equivalence relation.

Observe that if the ordered pair $\langle a, b \rangle$ is written as a fraction $\dfrac{a}{b}$, then the above relation R is, in fact, the usual definition of the equality of two fractions, i.e. $\dfrac{a}{b} = \dfrac{c}{d}$, iff $ad = bc$.

26. Prove Theorem 1.4: Let R be an equivalence relation in A and let $[a]$ be the equivalence class of $a \in A$. Then:

(i)   For every $a \in A$, $a \in [a]$.

(ii) $[a] = [b]$ if and only if $\langle a, b \rangle \in R$.

(iii) If $[a] \neq [b]$, then $[a]$ and $[b]$ are disjoint.

Solution:

Proof of (i).   Since R is reflexive, $\langle a, a \rangle \in R$ for every $a \in A$ and therefore $a \in [a]$.

Proof of (ii).   Suppose $\langle a, b \rangle \in R$. We want to show that $[a] = [b]$. Let $x \in [b]$; then $\langle b, x \rangle \in R$. But by hypothesis, $\langle a, b \rangle \in R$; hence by transitivity, $\langle a, x \rangle \in R$. Accordingly, $x \in [a]$, i.e. $[b] \subset [a]$. To prove that $[a] \subset [b]$, we observe that $\langle a, b \rangle \in R$ implies, by symmetry, that $\langle b, a \rangle \in R$. Then by a similar argument, we get $[a] \subset [b]$. So $[a] = [b]$.

On the other hand, if $[a] = [b]$, then by reflexivity, $b \in [b] = [a]$, i.e. $\langle a, b \rangle \in R$.

Proof of (iii).   We prove the equivalent contrapositive statement, i.e. if $[a] \cap [b] \neq \emptyset$, then $[a] = [b]$. If $[a] \cap [b] \neq \emptyset$, there exists an element $x \in A$ with $x \in [a] \cap [b]$. Hence $\langle a, x \rangle \in R$ and $\langle b, x \rangle \in R$. By symmetry, $\langle x, b \rangle \in R$ and, by transitivity, $\langle a, b \rangle \in R$. Consequently by (ii), $[a] = [b]$.

Supplementary Problems

SETS, ELEMENTS, SUBSETS

27. Determine which of the following sets is the empty set:

    (i) $\{x : 1 < x < 2,\ x \in \mathbf{R}\}$ (iii) $\{x : x \in \emptyset\}$

    (ii) $\{x : 1 < x < 2,\ x \in \mathbf{N}\}$ (iv) $\{x : x^2 < x,\ x \in \mathbf{R}\}$

28. Let $A = \{1, 2, \ldots, 8, 9\}$, $B = \{2, 4, 6, 8\}$, $C = \{1, 3, 5, 7, 9\}$, $D = \{3, 4, 5\}$ and $E = \{3, 5\}$. Which of these sets can equal X if we are given the following information?

    (i) X and B are disjoint,   (ii) $X \subset D$ and $X \not\subset B$,   (iii) $X \subset A$ and $X \not\subset C$,   (iv) $X \subset C$ and $X \not\subset A$.

29. State whether each of the following statements is true or false.

    (i) Every subset of a finite set is finite.   (ii) Every subset of an infinite set is infinite.

30. Discuss all inclusions and membership relations among the following three sets: $\emptyset$, $\{\emptyset\}$, $\{\emptyset, \{\emptyset\}\}$.

31. Prove that the closed interval $[a, b]$ is not a subset of the open interval (a, b).

32. Find the power set $\mathcal{P}(U)$ of $U = \{0, 1, 2\}$ and the power set $\mathcal{P}(V)$ of $V = \{0, \{1, 2\}\}$.

33. State whether each of the following is true or false. Here S is any non-empty set and 2^S is the power set of S.

    (i) $S \in 2^S$ (ii) $S \subset 2^S$ (iii) $\{S\} \in 2^S$ (iv) $\{S\} \subset 2^S$

SET OPERATIONS

34. Let $A = \{1, 2, 3, \{1, 2, 3\}\}$, $B = \{1, 2, \{1, 2\}\}$. Find: $A \cup B$, $A \cap B$, $A \setminus B$, $B \setminus A$.

35. In each of the Venn diagrams below shade: (i) $A \cap (B \cup C)$, (ii) $C \setminus (A \cap B)$.

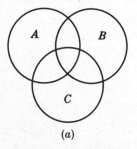
(a)

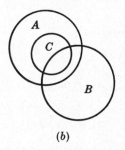
(b)

36. Prove and show by Venn diagrams: $A^c \setminus B^c = B \setminus A$.

37. (i) Prove $A \cap (B \setminus C) = (A \cap B) \setminus (A \cap C)$.

    (ii) Give an example to show that $A \cup (B \setminus C) \neq (A \cup B) \setminus (A \cup C)$.

38. Prove: $2^A \cap 2^B = 2^{A \cap B}$; $2^A \cup 2^B \subset 2^{A \cup B}$. Give an example to show that $2^A \cup 2^B \neq 2^{A \cup B}$.

39. Prove Theorem 1.3: Each of the following conditions is equivalent to $A \subset B$:

    (i) $A \cap B = A$, (ii) $A \cup B = B$, (iii) $B^c \subset A^c$, (iv) $A \cap B^c = \emptyset$, (v) $B \cup A^c = U$

    (*Note.* $A \cap B = A$ was already proven equivalent to $A \subset B$ in Problem 15.)

40. Prove that $A \subset B$ iff $(B \cap C) \cup A = B \cap (C \cup A)$ for any C.

PRODUCT SETS, RELATIONS, COMPOSITION OF RELATIONS

41. Prove: $A \times (B \cup C) = (A \times B) \cup (A \times C)$.

42. Using the definition of ordered pair, i.e. $\langle a, b \rangle = \{\{a\}, \{a, b\}\}$, prove that $\langle a, b \rangle = \langle c, d \rangle$ iff $a = c$ and $b = d$.

43. Determine the number of distinct relations from a set with m elements to a set with n elements, where m and n are positive integers.

44. Let R be the relation in the positive integers $\mathbf{N}$ defined by
$$R = \{\, \langle x, y \rangle : x, y \in \mathbf{N},\ x + 2y = 12 \,\}$$
(i) Write R as a set of ordered pairs. (ii) Find domain of R, range of R and R^{-1}. (iii) Find $R \circ R$.
(iv) Find $R^{-1} \circ R$.

45. Consider the relation $R = \{\, \langle 4, 5 \rangle,\ \langle 1, 4 \rangle,\ \langle 4, 6 \rangle,\ \langle 7, 6 \rangle,\ \langle 3, 7 \rangle \,\}$ in $\mathbf{N}$.
(i) Find domain of R, range of R and R^{-1}. (ii) Find $R \circ R$. (iii) Find $R^{-1} \circ R$.

46. Let U and V be the relations in $\mathbf{R}$ defined by $U = \{\langle x, y \rangle : x^2 + 2y = 5\}$ and $V = \{\langle x, y \rangle : 2x - y = 3\}$.
(i) Find $V \circ U$.   (ii) Find $U \circ V$.

47. Consider the relations $<$ and $\leq$ in $\mathbf{R}$. Show that $< \cup \Delta\ =\ \leq$ where Δ is the diagonal.

EQUIVALENCE RELATIONS

48. State whether each of the following statements is true or false. Assume R and S are (non-empty) relations in a set A.

   (1) If R is symmetric, then R^{-1} is symmetric.
   (2) If R is reflexive, then $R \cap R^{-1} \neq \emptyset$.
   (3) If R is symmetric, then $R \cap R^{-1} \neq \emptyset$.
   (4) If R and S are transitive, then $R \cup S$ is transitive.
   (5) If R and S are transitive, then $R \cap S$ is transitive.
   (6) If R and S are symmetric, then $R \cup S$ is symmetric.
   (7) If R and S are symmetric, then $R \cap S$ is symmetric.
   (8) If R and S are reflexive, then $R \cap S$ is reflexive.

49. Consider $\mathbf{N} \times \mathbf{N}$, the set of ordered pairs of positive integers. Let $\simeq$ be the relation in $\mathbf{N} \times \mathbf{N}$ defined by
$$\langle a, b \rangle \simeq \langle c, d \rangle \quad \text{iff} \quad a + d = b + c$$
(i) Prove $\simeq$ is an equivalence relation.   (ii) Find the equivalence class of $\langle 2, 5 \rangle$, i.e. $[\langle 2, 5 \rangle]$.

50. Let $\sim$ be the relation in $\mathbf{R}$ defined by $x \sim y$ iff $x - y$ is an integer. Prove that $\sim$ is an equivalence relation.

51. Let $\sim$ be the relation in the Cartesian plane $\mathbf{R}^2$ defined by $\langle x, y \rangle \sim \langle w, z \rangle$ iff $x = w$.
Prove that $\sim$ is an equivalence relation and graph several equivalence classes.

52. Let a and b be arbitrary real numbers. Furthermore, let $\sim$ be the relation in $\mathbf{R}^2$ defined by
$$\langle x, y \rangle \sim \langle w, z \rangle \quad \text{iff} \quad \exists k \in \mathbf{Z}\ \text{s.t.}\ x - w = ka,\ y - z = kb$$
Prove that $\sim$ is an equivalence relation and graph several equivalence classes.

Answers to Supplementary Problems

27. The sets in (ii) and (iii) are empty.

31. $a \in [a, b]$ but $a \notin (a, b)$.

32. $\mathcal{P}(V) = \{V, \{0\}, \{\{1, 2\}\}, \emptyset\}$

33. (i) T, (ii) F, (iii) F, (iv) T

34. $A \cup B = \{1, 2, 3, \{1, 2\}, \{1, 2, 3\}\}$, $A \cap B = \{1, 2\}$, $A \setminus B = \{3, \{1, 2, 3\}\}$, $B \setminus A = \{\{1, 2\}\}$.

35.

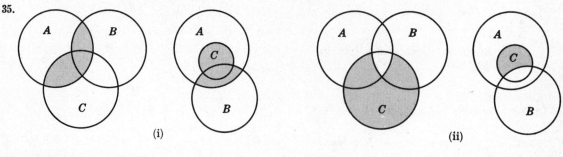

(i)   (ii)

37. (ii) $C = \varnothing,\ A = B \neq \varnothing$

38. Example: $A = \{1\},\ B = \{2\}$

43. 2^{mn}

44. (i) $R = \{\langle 10,1\rangle, \langle 8,2\rangle, \langle 6,3\rangle, \langle 4,4\rangle, \langle 2,5\rangle\}$

   (ii) domain of $R = \{10,8,6,4,2\}$, range of $R = \{1,2,3,4,5\}$,
   $R^{-1} = \{\langle 1,10\rangle, \langle 2,8\rangle, \langle 3,6\rangle, \langle 4,4\rangle, \langle 5,2\rangle\}$

   (iii) $R \circ R = \{\langle 8,5\rangle, \langle 4,4\rangle\}$

   (iv) $R^{-1} \circ R = \{\langle 10,10\rangle, \langle 8,8\rangle, \langle 6,6\rangle, \langle 4,4\rangle, \langle 2,2\rangle\}$

45. (i) domain of $R = \{4,1,7,3\}$, range of $R = \{5,4,6,7\}$, $R^{-1} = \{\langle 5,4\rangle, \langle 4,1\rangle, \langle 6,4\rangle, \langle 6,7\rangle, \langle 7,3\rangle\}$

   (ii) $R \circ R = \{\langle 1,5\rangle, \langle 1,6\rangle, \langle 3,6\rangle\}$

   (iii) $R^{-1} \circ R = \{\langle 4,4\rangle, \langle 1,1\rangle, \langle 4,7\rangle, \langle 7,4\rangle, \langle 7,7\rangle, \langle 3,3\rangle\}$

46. $V \circ U = \{\langle x,y\rangle : x^2 + y = 2\}$, $U \circ V = \{\langle x,y\rangle : 4x^2 - 12x + 2y + 4 = 0\}$

48. (1) T, (2) T, (3) T, (4) F, (5) T, (6) T, (7) T, (8) T

49. (ii) $[\langle 2,5\rangle] = \{\langle 1,4\rangle, \langle 2,5\rangle, \langle 3,6\rangle, \langle 4,7\rangle, \ldots, \langle n, n+3\rangle, \ldots\}$

51.

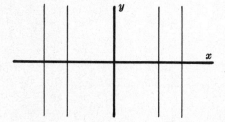

The equivalence classes are the vertical lines.

52.

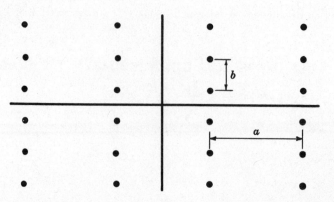

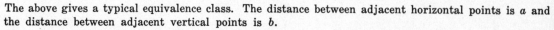

The above gives a typical equivalence class. The distance between adjacent horizontal points is a and the distance between adjacent vertical points is b.

Chapter 2

Functions

FUNCTIONS

Suppose that to each element of a set A there is assigned a unique element of a set B; the collection, f, of such assignments is called a *function* (or *mapping*) *from* (or *on*) A *into* B and is written

$$f : A \to B \quad \text{or} \quad A \overset{f}{\to} B$$

The unique element in B assigned to $a \in A$ by f is denoted by $f(a)$, and called the *value* of f at a or the *image* of a under f. The *domain* of f is A, the *co-domain* is B. To each function $f : A \to B$ there corresponds the relation in $A \times B$ given by

$$\{ \langle a, f(a) \rangle : a \in A \}$$

We call this set the *graph* of f. The *range* of f, denoted by $f[A]$, is the set of images, i.e. $f[A] = \{f(a) : a \in A\}$.

Two functions $f : A \to B$ and $g : A \to B$ are defined to be equal, written $f = g$, iff $f(a) = g(a)$ for every $a \in A$, i.e. iff they have the same graph. Accordingly, we do not distinguish between a function and its graph. A subset f of $A \times B$, i.e. a relation from A to B, is a function iff it possesses the following property:

[F] Each $a \in A$ appears as the first coordinate in exactly one ordered pair $\langle a, b \rangle$ in f.

The negation of $f = g$ is written $f \neq g$ and is the statement: $\exists a \in A$ for which $f(a) \neq g(a)$.

Example 1.1: Let $f : \mathbf{R} \to \mathbf{R}$ be the function which assigns to each real number its square, i.e. for each $x \in \mathbf{R}$, $f(x) = x^2$. Here f is a *real-valued function*. Its graph, $\{ \langle x, x^2 \rangle : x \in \mathbf{R} \}$, is displayed in Fig. 2-1 below. The range of f is the set of non-negative real numbers, i.e. $f[\mathbf{R}] = \{x : x \in \mathbf{R}, \ x \geq 0\}$.

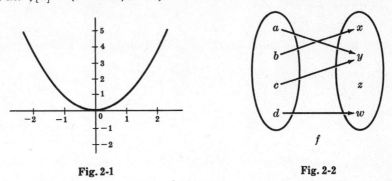

Fig. 2-1   Fig. 2-2

Example 1.2: Let $A = \{a, b, c, d\}$ and $B = \{x, y, z, w\}$. Then the diagram in Fig. 2-2 above defines a function f from A into B. Here $f[A] = \{x, y, w\}$. The graph of f is the relation

$$\{ \langle a, y \rangle, \langle b, x \rangle, \langle c, y \rangle, \langle d, w \rangle \}$$

Example 1.3: A function $f : A \to B$ is called a *constant function* if, for some $b_0 \in B$, $f(a) = b_0$ for all $a \in A$. Hence the range $f[A]$ of any constant function f is a singleton set, i.e. $f[A] = \{b_0\}$.

17

Consider now functions $f : A \to B$ and $g : B \to C$, illustrated below:

$$\boxed{A} \quad \xrightarrow{f} \quad \boxed{B} \quad \xrightarrow{g} \quad \boxed{C}$$

The function from A into C which maps the element $a \in A$ into the element $g(f(a))$ of C is called the *composition* or *product* of f and g and is denoted by $g \circ f$. Hence, by definition,

$$(g \circ f)(a) = g(f(a))$$

We remark that, if we view $f \subset A \times B$ and $g \subset B \times C$ as relations, we have already defined a product $g \cdot f$ (Chapter 1). However, these two products are the same in that if f and g are functions then $g \cdot f$ is a function and $g \cdot f = g \circ f$.

If $f : X \to Y$ and $A \subset X$, then the *restriction* of f to A, denoted by $f \,|\, A$, is the function from A into Y defined by

$$f \,|\, A(a) \equiv f(a) \quad \text{for all } a \in A$$

Equivalently, $f \,|\, A = f \cap (A \times Y)$. On the other hand, if $f : X \to Y$ is the restriction of some function $g : X^* \to Y$ where $X \subset X^*$, then g is called an *extension* of f.

ONE-ONE, ONTO, INVERSE AND IDENTITY FUNCTIONS

A function $f : A \to B$ is said to be *one-to-one* (or *one-one*, or 1-1) if distinct elements in A have distinct images, i.e. if

$$f(a) = f(a') \;\Rightarrow\; a = a'$$

A function $f : A \to B$ is said to be *onto* (or f is a function from A *onto* B, or f maps A *onto* B) if every $b \in B$ is the image of some $a \in A$, i.e. if

$$b \in B \;\Rightarrow\; \exists a \in A \text{ for which } f(a) = b$$

Hence if f is onto, $f[A] = B$.

In general, the inverse relation f^{-1} of a function $f \subset A \times B$ need not be a function. However, if f is both one-one and onto, then f^{-1} is a function from B onto A and is called the *inverse function*.

The diagonal $\Delta_A \subset A \times A$ is a function and called the *identity function* on A. It is also denoted by 1_A or 1. Here, $1_A(a) = a$ for every $a \in A$. Clearly, if $f : A \to B$, then

$$1_B \circ f = f = f \circ 1_A$$

Furthermore, if f is one-one and onto, and so has an inverse function f^{-1}, then

$$f^{-1} \circ f = 1_A \quad \text{and} \quad f \circ f^{-1} = 1_B$$

The converse is also true:

Proposition 2.1: Let $f : A \to B$ and $g : B \to A$ satisfy

$$g \circ f = 1_A \quad \text{and} \quad f \circ g = 1_B$$

Then $f^{-1} : B \to A$ exists and $g = f^{-1}$.

Example 2.1: Let $f : \mathbf{R} \to \mathbf{R}$, $g : \mathbf{R} \to \mathbf{R}$ and $h : \mathbf{R} \to \mathbf{R}$ be defined by

$$f(x) = e^x, \quad g(x) = x^3 - x \quad \text{and} \quad h(x) = x^2$$

The function f shown in Fig. 2-3(a) below is one-one; geometrically, this means that each horizontal line does not contain more than one point of f. The function g shown in Fig. 2-3(b) below is onto; geometrically this means that each horizontal line contains at least one point of g. The function h shown in Fig. 2-3(c) below is neither one-one nor onto, for $h(2) = h(-2) = 4$ and $h[\mathbf{R}]$ is a proper subset of $\mathbf{R}$, e.g. $-16 \notin h[\mathbf{R}]$.

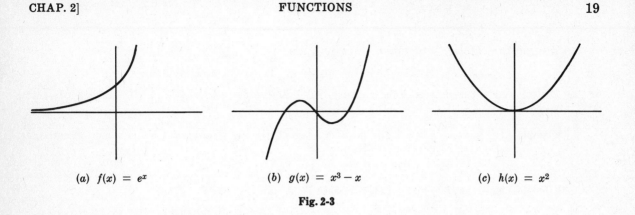

(a) $f(x) = e^x$ (b) $g(x) = x^3 - x$ (c) $h(x) = x^2$

Fig. 2-3

INDEXED SETS, CARTESIAN PRODUCTS

An *indexed class of sets*, denoted by

$$\{A_i : i \in I\}, \quad \{A_i\}_{i \in I} \quad \text{or simply} \quad \{A_i\}$$

assigns a set A_i to each $i \in I$, i.e. is a function from I into a class of sets. The set I is called the *index set*, the sets A_i are called *indexed sets*, and each $i \in I$ is called an *index*. When the index set I is the set of positive integers, the indexed class $\{A_1, A_2, \ldots\}$ is called a *sequence* (of sets).

> **Example 3.1:** For each $n \in \mathbf{N}$, the positive integers, let
> $$D_n = \{x : x \in \mathbf{N}, x \text{ is a multiple of } n\}$$
> Then $D_1 = \{1, 2, 3, \ldots\}, \quad D_2 = \{2, 4, 6, \ldots\}, \quad D_3 = \{3, 6, 9, \ldots\}, \quad \ldots$

The *Cartesian product* of an indexed class of sets, $\mathcal{A} = \{A_i : i \in I\}$, denoted by

$$\prod\{A_i : i \in I\} \quad \text{or} \quad \prod_{i \in I} A_i \quad \text{or simply} \quad \prod_i A_i$$

is the set of all functions $p : I \to \cup_i A_i$ such that $p(i) = a_i \in A_i$. We denote such an element of the Cartesian product by $p = \langle a_i : i \in I \rangle$. For each $i_0 \in I$ there exists a function π_{i_0}, called the i_0th *projection function*, from the product set $\prod_i A_i$ into the i_0th *coordinate set* A_{i_0} defined by

$$\pi_{i_0}(\langle a_i : i \in I \rangle) = a_{i_0}$$

> **Example 3.2:** Recall that $\mathbf{R}^3 = \mathbf{R} \times \mathbf{R} \times \mathbf{R}$ consists of all 3-tuples $p = \langle a_1, a_2, a_3 \rangle$ of real numbers. Now let R_1, R_2 and R_3 denote copies of $\mathbf{R}$. Then p can be viewed as a function on $I = \{1, 2, 3\}$ where $p(1) = a_1 \in R_1$, $p(2) = a_2 \in R_2$ and $p(3) = a_3 \in R_3$. In other words,
> $$\mathbf{R}^3 = \prod\{R_i : i \in I, R_i = \mathbf{R}\}$$

GENERALIZED OPERATIONS

The notion of union and intersection, originally defined for two sets, may be generalized to any class $\mathcal{A}$ of subsets of a universal set U. The union of the sets in $\mathcal{A}$, denoted by $\bigcup\{A : A \in \mathcal{A}\}$, is the set of elements which belong to at least one set in $\mathcal{A}$:

$$\bigcup\{A : A \in \mathcal{A}\} = \{x : x \in U, \exists A \in \mathcal{A} \text{ s.t. } x \in A\}$$

The intersection of the sets in $\mathcal{A}$, denoted by $\bigcap\{A : A \in \mathcal{A}\}$, is the set of elements which belong to every set in $\mathcal{A}$:

$$\bigcap\{A : A \in \mathcal{A}\} = \{x : x \in U, x \in A \text{ for every } A \in \mathcal{A}\}$$

For an indexed class of subsets of U, say $\mathcal{A} = \{A_i : i \in I\}$, we write

$$\bigcup\{A_i : i \in I\}, \quad \bigcup_{i \in I} A_i \quad \text{or} \quad \cup_i A_i$$

for the union of the sets in $\mathcal{A}$, and

$$\bigcap\{A_i : i \in I\}, \quad \bigcap_{i \in I} A_i \quad \text{or} \quad \cap_i A_i$$

for the intersection of the sets in $\mathcal{A}$. We will also write

$$\cup_{i=1}^{\infty} A_i = A_1 \cup A_2 \cup \cdots \quad \text{and} \quad \cap_{i=1}^{\infty} A_i = A_1 \cup A_2 \cup \cdots$$

for the union and intersection, respectively, of a sequence $\{A_1, A_2, \ldots\}$ of subsets of U.

> **Example 4.1:**   For each $n \in \mathbf{N}$, the positive integers, let $D_n = \{x : x \in \mathbf{N}, \ x \text{ is a multiple of } n\}$ (see Example 3.1). Then
> $$\cup\{D_i : i \geq 10\} = \{10, 11, 12, \ldots\} \quad \text{and} \quad \cap_{i=1}^{\infty} D_i = \varnothing$$

> **Example 4.2:**   Let $I = [0, 1]$ and, for each $i \in I$, let $A_i = [0, i]$. Then
> $$\cup_i A_i = [0, 1] \quad \text{and} \quad \cap_i A_i = \{0\}$$

The distributive laws and De Morgan's laws also hold for these generalized operations:

Theorem 2.2:   For any class of sets $\mathcal{A} = \{A_i\}$ and any set B,

$$\text{(i)} \ \ B \cup (\cap_i A_i) = \cap_i (B \cup A_i) \qquad \text{(ii)} \ \ B \cap (\cup_i A_i) = \cup_i (B \cap A_i)$$

Theorem 2.3:   Let $\mathcal{A} = \{A_i\}$ be any class of subsets of U. Then:

$$\text{(i)} \ \ (\cup_i A_i)^c = \cap_i A_i^c \qquad \text{(ii)} \ \ (\cap_i A_i)^c = \cup_i A_i^c$$

The following theorem will be used frequently.

Theorem 2.4:   Let A be any set and, for each $p \in A$, let G_p be a subset of A such that $p \in G_p \subset A$. Then $A = \cup\{G_p : p \in A\}$.

Remark: In the case of an empty class $\varnothing$ of subsets of a universal set U, it is convenient to define
$$\cup\{A : A \in \varnothing\} = \varnothing \quad \text{and} \quad \cap\{A : A \in \varnothing\} = U$$

Hence
$$\cup\{A_i : i \in \varnothing\} = \varnothing \quad \text{and} \quad \cap\{A_i : i \in \varnothing\} = U$$

ASSOCIATED SET FUNCTIONS

Let $f : X \to Y$. Then the *image* $f[A]$ of any subset A of X is the set of images of points in A, and the *inverse image* $f^{-1}[B]$ of any subset B of Y is the set of points in X whose images lie in B. That is,

$$f[A] = \{f(x) : x \in A\} \quad \text{and} \quad f^{-1}[B] = \{x : x \in X, f(x) \in B\}$$

> **Example 5.1:**   Let $f : \mathbf{R} \to \mathbf{R}$ be defined by $f(x) = x^2$. Then
> $$f[\{1, 3, 4, 7\}] = \{1, 9, 16, 49\}, \quad f[(1, 2)] = (1, 4)$$
> Also, $f^{-1}[\{4, 9\}] = \{-3, -2, 2, 3\}, \quad f^{-1}[(1, 4)] = (1, 2) \cup (-2, -1)$

Thus a function $f : X \to Y$ induces a function, also denoted by f, from the power set $\mathcal{P}(X)$ of X into the power set $\mathcal{P}(Y)$ of Y, and a function f^{-1} from $\mathcal{P}(Y)$ into $\mathcal{P}(X)$. The induced functions f and f^{-1} are called *set functions* since they are maps of classes (of sets) into classes.

We remark that the associated set function f^{-1} is not in general the inverse of the associated set function f. For example, if f is the function in Example 5.1, then

$$f^{-1} \circ f[(1, 2)] = f^{-1}[(1, 4)] = (1, 2) \cup (-2, -1)$$

Observe that different brackets are used to distinguish between a function and its associated set functions, i.e. $f(a)$ denotes a value of the original function, and $f[A]$ and $f^{-1}[B]$ denote values of the associated set functions.

The associated set functions possess various properties.   In particular we state:

Theorem 2.5:   Let $f: X \to Y$.   Then, for any subsets A and B of X,

$$\text{(i)} \quad f[A \cup B] \ = \ f[A] \cup f[B] \qquad\qquad \text{(iii)} \quad f[A \setminus B] \supset f[A] \setminus f[B]$$

$$\text{(ii)} \quad f[A \cap B] \subset f[A] \cap f[B] \qquad\qquad \text{(iv)} \quad A \subset B \text{ implies } f[A] \subset f[B]$$

and, more generally, for any indexed class $\{A_i\}$ of subsets of X,

$$\text{(i')} \quad f[\cup_i A_i] \ = \ \cup_i f[A_i] \qquad\qquad \text{(ii')} \quad f[\cap_i A_i] \ \subset \ \cap_i f[A_i]$$

The following example shows that the inclusions of (ii) and (iii) cannot in general be replaced by equality.

> **Example 5.2:**   Consider the subsets
> $$A \ = \ [1,2] \times [1,2] \quad \text{and} \quad B \ = \ [1,2] \times [3,4]$$
> of the plane $\mathbf{R}^2$ and the projection $\pi : \mathbf{R}^2 \to \mathbf{R}$, into the first coordinate set, i.e. the x-axis.   Observe that $\pi[A] = [1,2]$ and $\pi[B] = [1,2]$, and that $A \cap B = \emptyset$ implies $\pi[A \cap B] = \emptyset$.   Hence
> $$\pi[A] \cap \pi[B] \ = \ [1,2] \ \neq \ \pi[A \cap B] \ = \ \emptyset$$
> Furthermore, $A \setminus B = A$, so
> $$\pi[A \setminus B] \ = \ [1,2] \ \neq \ \emptyset \ = \ \pi[A] \setminus \pi[B]$$

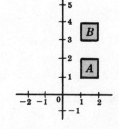

On the other hand, the inverse set function is much more "well-behaved" in the sense that equality holds in both cases.   Namely,

Theorem 2.6:   Let $f: X \to Y$.   Then for any subsets A and B of Y,

$$\text{(i)} \quad f^{-1}[A \cup B] \ = \ f^{-1}[A] \cup f^{-1}[B]$$

$$\text{(ii)} \quad f^{-1}[A \cap B] \ = \ f^{-1}[A] \cap f^{-1}[B]$$

$$\text{(iii)} \quad f^{-1}[A \setminus B] \ = \ f^{-1}[A] \setminus f^{-1}[B]$$

$$\text{(iv)} \quad A \subset B \text{ implies } f^{-1}[A] \subset f^{-1}[B]$$

and, more generally, for any indexed class $\{A_i\}$ of subsets of Y,

$$\text{(i')} \quad f^{-1}[\cup_i A_i] \ = \ \cup_i f^{-1}[A_i]$$

$$\text{(ii')} \quad f^{-1}[\cap_i A_i] \ = \ \cap_i f^{-1}[A_i]$$

Since $f^{-1}[Y] = X$, we have, as a special case of (iii),

Corollary 2.7:   Let $f: X \to Y$ and let $A \subset Y$.   Then $f^{-1}[A^c] = (f^{-1}[A])^c$.

Next follows an important relationship between the two set functions.

Theorem 2.8:   Let $f: X \to Y$ and let $A \subset X$ and $B \subset Y$.   Then:

$$\text{(i)} \quad A \subset f^{-1} \circ f[A] \qquad\qquad \text{(ii)} \quad B \supset f \circ f^{-1}[B]$$

As shown previously, the inclusion in (i) cannot in general be replaced by equality.

ALGEBRA OF REAL-VALUED FUNCTIONS

Let $\mathcal{F}(X, \mathbf{R})$ denote the collection of all real-valued functions defined on some set X. Many operations are inherited by $\mathcal{F}(X, \mathbf{R})$ from corresponding operations in $\mathbf{R}$.   Specifically, let $f: X \to \mathbf{R}$ and $g: X \to \mathbf{R}$ and let $k \in \mathbf{R}$:   then we define

$$(f + g): X \to \mathbf{R} \quad \text{by} \quad (f + g)(x) \ \equiv \ f(x) + g(x)$$

$$(k \cdot f): X \to \mathbf{R} \quad \text{by} \quad (k \cdot f)(x) \ \equiv \ k(f(x))$$

$$(|f|): X \to \mathbf{R} \quad \text{by} \quad (|f|)(x) \ \equiv \ |f(x)|$$

$$(fg): X \to \mathbf{R} \quad \text{by} \quad (fg)(x) \ \equiv \ f(x)\, g(x)$$

It is also convenient to identify the real number $k \in \mathbf{R}$ with the constant function $f(x) = k$ for every $x \in \mathbf{R}$. Then $(f + k) : X \to \mathbf{R}$ is the function

$$(f + k)(x) \equiv f(x) + k$$

Observe that $(fg) : X \to \mathbf{R}$ is not the composition of f and g discussed previously.

Example 6.1: Consider the functions

$$f = \{\langle a, 1 \rangle, \langle b, 3 \rangle\} \quad \text{and} \quad g = \{\langle a, 2 \rangle, \langle b, -1 \rangle\}$$

with domain $X = \{a, b\}$. Then

$$(3f - 2g)(a) \equiv 3f(a) - 2g(a) = 3(1) - 2(2) = -1$$
$$(3f - 2g)(b) \equiv 3f(b) - 2g(b) = 3(3) - 2(-1) = 11$$

that is, $3f - 2g = \{\langle a, -1 \rangle, \langle b, 11 \rangle\}$

Also, since $|g|(x) \equiv |g(x)|$ and $(g + 3)(x) \equiv g(x) + 3$,

$$|g| = \{\langle a, 2 \rangle, \langle b, 1 \rangle\} \quad \text{and} \quad g + 3 = \{\langle a, 5 \rangle, \langle b, 2 \rangle\}$$

The collection $\mathcal{F}(X, \mathbf{R})$ with the above operations possesses various properties of which some are included in the next theorem.

Theorem 2.9: The collection $\mathcal{F}(X, \mathbf{R})$ of all real-valued functions defined on a non-empty set X together with the above operations satisfies the following axioms of a *real linear vector space*:

[V_1] The operation of addition of functions f and g satisfies:

(1) $(f + g) + h = f + (g + h)$

(2) $f + g = g + f$

(3) $\exists 0 \in \mathcal{F}(X, \mathbf{R})$, i.e. $0 : X \to \mathbf{R}$, such that $f + 0 = f$.

(4) For each $f \in \mathcal{F}(X, \mathbf{R})$, $\exists -f \in \mathcal{F}(X, \mathbf{R})$, i.e. $-f : X \to \mathbf{R}$, such that $f + (-f) = 0$.

[V_2] The operation of scalar multiplication $k \cdot f$ of a function f by a real number k satisfies:

(1) $k \cdot (k' \cdot f) = (kk') \cdot f$

(2) $1 \cdot f = f$

[V_3] The operations of addition and scalar multiplication satisfy:

(1) $k \cdot (f + g) = k \cdot f + k \cdot g$

(2) $(k + k') \cdot f = k \cdot f + k' \cdot f$

Example 6.2: Let $X = \{1, 2, \ldots, m\}$. Then each function $f \in \mathcal{F}(X, \mathbf{R})$ may be written as an ordered m-tuple $\langle f(1), \ldots, f(m) \rangle$. Furthermore, if

$$f = \langle a_1, \ldots, a_m \rangle \quad \text{and} \quad g = \langle b_1, \ldots, b_m \rangle$$

then $f + g = \langle a_1 + b_1, a_2 + b_2, \ldots, a_m + b_m \rangle$

and, for any $k \in \mathbf{R}$, $k \cdot f = \langle ka_1, \ldots, ka_m \rangle$

In this case, the real linear (vector) space $\mathcal{F}(X, \mathbf{R})$ is called *m-dimensional Euclidean space*.

Example 6.3: A function $f \in \mathcal{F}(X, \mathbf{R})$ is said to be *bounded* iff

$$\exists M \in \mathbf{R} \quad \text{such that} \quad |f(x)| \leq M \text{ for every } x \in X$$

Let $\beta(X, \mathbf{R})$ denote the collection of all bounded functions in $\mathcal{F}(X, \mathbf{R})$. Then $\beta(X, \mathbf{R})$ possesses the following properties:

(i)   If $f, g \in \beta(X, \mathbf{R})$, then $f + g \in \beta(X, \mathbf{R})$.

(ii) If $f \in \beta(X, \mathbf{R})$ and $k \in \mathbf{R}$, then $k \cdot f \in \beta(X, \mathbf{R})$.

Any subset of $\mathcal{F}(X, \mathbf{R})$ satisfying (i) and (ii) is called a (linear) *subspace* of $\mathcal{F}(X, \mathbf{R})$.

Solved Problems

FUNCTIONS

1. State whether or not each of the diagrams defines a function from $A = \{a, b, c\}$ into $B = \{x, y, z\}$.

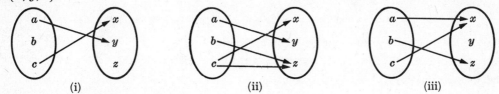

(i)   (ii)   (iii)

Solution:

(i)   No.   There is nothing assigned to the element $b \in A$.

(ii)   No.   Two elements, x and z, are assigned to $c \in A$.

(iii) Yes.

2. Let $X = \{1, 2, 3, 4\}$. State whether or not each of the following relations is a function from X into X.

(i) $f = \{\langle 2, 3 \rangle, \langle 1, 4 \rangle, \langle 2, 1 \rangle, \langle 3, 2 \rangle, \langle 4, 4 \rangle\}$

(ii) $g = \{\langle 3, 1 \rangle, \langle 4, 2 \rangle, \langle 1, 1 \rangle\}$

(iii) $h = \{\langle 2, 1 \rangle, \langle 3, 4 \rangle, \langle 1, 4 \rangle, \langle 2, 1 \rangle, \langle 4, 4 \rangle\}$

Solution:

Recall that a subset f of $X \times X$ is a function $f : X \to X$ iff each $x \in X$ appears as the first coordinate in exactly one ordered pair in f.

(i)   No.   Two different ordered pairs $\langle 2, 3 \rangle$ and $\langle 2, 1 \rangle$ in f have the same first coordinate.

(ii)   No.   The element $2 \in X$ does not appear as the first coordinate in any ordered pair in g.

(iii) Yes.   Although $2 \in X$ appears as the first coordinate in two ordered pairs in h, these two ordered pairs are equal.

3. Consider the functions
$$f = \{\langle 1, 3 \rangle, \langle 2, 5 \rangle, \langle 3, 3 \rangle, \langle 4, 1 \rangle, \langle 5, 2 \rangle\}$$
$$g = \{\langle 1, 4 \rangle, \langle 2, 1 \rangle, \langle 3, 1 \rangle, \langle 4, 2 \rangle, \langle 5, 3 \rangle\}$$

from $X = \{1, 2, 3, 4, 5\}$ into X.

(i)   Determine the range of f and of g.

(ii)   Find the composition functions $g \circ f$ and $f \circ g$.

Solution:

(i)   Recall that the range of a function is the set of image values, i.e. the set of second coordinates. Hence

$$\text{range of } f = \{3, 5, 1, 2\} \quad \text{and} \quad \text{range of } g = \{4, 1, 2, 3\}$$

(ii)   Use the definition of the composition function and compute:

$$(g \circ f)(1) \equiv g(f(1)) = g(3) = 1 \qquad (f \circ g)(1) \equiv f(g(1)) = f(4) = 1$$
$$(g \circ f)(2) \equiv g(f(2)) = g(5) = 3 \qquad (f \circ g)(2) \equiv f(g(2)) = f(1) = 3$$
$$(g \circ f)(3) \equiv g(f(3)) = g(3) = 1 \qquad (f \circ g)(3) \equiv f(g(3)) = f(1) = 3$$
$$(g \circ f)(4) \equiv g(f(4)) = g(1) = 4 \qquad (f \circ g)(4) \equiv f(g(4)) = f(2) = 5$$
$$(g \circ f)(5) \equiv g(f(5)) = g(2) = 1 \qquad (f \circ g)(5) \equiv f(g(5)) = f(3) = 3$$

In other words,
$$g \circ f = \{\langle 1, 1 \rangle, \langle 2, 3 \rangle, \langle 3, 1 \rangle, \langle 4, 4 \rangle, \langle 5, 1 \rangle\}$$
$$f \circ g = \{\langle 1, 1 \rangle, \langle 2, 3 \rangle, \langle 3, 3 \rangle, \langle 4, 5 \rangle, \langle 5, 3 \rangle\}$$

Observe that $f \circ g \neq g \circ f$.

4. Let the functions $f : \mathbf{R} \to \mathbf{R}$ and $g : \mathbf{R} \to \mathbf{R}$ be defined by
$$f(x) = 2x + 1, \quad g(x) = x^2 - 2$$
Find formulas defining the product functions $g \circ f$ and $f \circ g$.

Solution:

Compute $g \circ f : \mathbf{R} \to \mathbf{R}$ as follows:
$$(g \circ f)(x) \equiv g(f(x)) = g(2x + 1) = (2x + 1)^2 - 2 = 4x^2 + 4x - 1$$

Observe that the same answer can be found by writing
$$y = f(x) = 2x + 1, \quad z = g(y) = y^2 - 2$$
and then eliminating y from the two equations:
$$z = y^2 - 2 = (2x + 1)^2 - 2 = 4x^2 + 4x - 1$$

Now compute $f \circ g : \mathbf{R} \to \mathbf{R}$:
$$(f \circ g)(x) \equiv f(g(x)) = f(x^2 - 2) = 2(x^2 - 2) + 1 = 2x^2 - 3$$

5. Prove the associative law for composition of functions, i.e. if $f : A \to B$, $g : B \to C$ and $h : C \to D$, then $(h \circ g) \circ f = h \circ (g \circ f)$.

Solution:

Since the associative law was proven for composition of relations in general, this result follows. We also give a direct proof:
$$((h \circ g) \circ f)(a) = (h \circ g)(f(a)) = h(g(f(a))), \quad \forall a \in A$$
$$(h \circ (g \circ f))(a) = h((g \circ f)(a)) = h(g(f(a))), \quad \forall a \in A$$

Hence $(h \circ g) \circ f = h \circ (g \circ f)$.

ONE-ONE AND ONTO FUNCTIONS

6. Let $f : A \to B$, $g : B \to C$. Prove:

(i)   If f and g are onto, then $g \circ f : A \to C$ is onto.

(ii) If f and g are one-one, then $g \circ f : A \to C$ is one-one.

Solution:

(i)   Let $c \in C$. Since g is onto, $\exists b \in B$ s.t. $g(b) = c$. Since f is onto, $\exists a \in A$ s.t. $f(a) = b$. But then $(g \circ f)(a) = g(f(a)) = c$, i.e. $g \circ f$ is also onto.

(ii) Suppose $(g \circ f)(a) = (g \circ f)(a')$; i.e. $g(f(a)) = g(f(a'))$. So $f(a) = f(a')$ since g is one-one; hence $a = a'$ since f is one-one. Accordingly, $g \circ f$ is also one-one.

7. Let $A = [-1, 1]$ and let $f : A \to A$, $g : A \to A$ and $h : A \to A$ be defined by
$$f(x) = \sin x, \quad g(x) = \sin \pi x, \quad h(x) = \sin \frac{\pi}{2} x$$

State whether or not each of the functions is (i) one-one, (ii) onto, (iii) bijective (i.e. one-one and onto).

Solution:

The graphs of the functions are as follows:

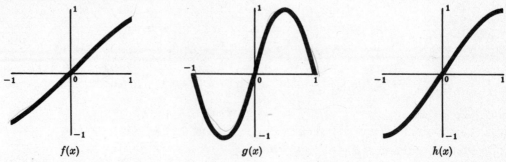

$f(x)$ $g(x)$ $h(x)$

The function f is one-one; each horizontal line does not contain more than one point of f. It is not onto since, for example, $\sin x \neq 1$ for any $x \in A$. On the other hand, g is onto; each horizontal line contains at least one point of f. But g is not one-one since, for example, $g(-1) = g(0) = 0$. The function h is both one-one and onto; each horizontal line contains exactly one point of h.

8. **Prove:** Let $f : A \to B$ and $g : B \to C$ be one-one and onto; then $(g \circ f)^{-1} : C \to A$ exists and equals $f^{-1} \circ g^{-1} : C \to A$.

   Solution:

   Utilizing Proposition 2.1, we show that:

   $$(f^{-1} \circ g^{-1}) \circ (g \circ f) = 1_A \quad \text{and} \quad (g \circ f) \circ (f^{-1} \circ g^{-1}) = 1_B$$

   Using the associative law for composition of functions,

   $$\begin{aligned}
   (f^{-1} \circ g^{-1}) \circ (g \circ f) &= f^{-1} \circ (g^{-1} \circ (g \circ f)) \\
   &= f^{-1} \circ ((g^{-1} \circ g) \circ f) \\
   &= f^{-1} \circ (1 \circ f) \\
   &= f^{-1} \circ f \\
   &= 1_A
   \end{aligned}$$

   since $g^{-1} \circ g = 1$ and $1 \circ f = f = f \circ 1$. Similarly,

   $$\begin{aligned}
   (g \circ f) \circ (f^{-1} \circ g^{-1}) &= g \circ (f \circ (f^{-1} \circ g^{-1})) \\
   &= g \circ ((f \circ f^{-1}) \circ g^{-1}) \\
   &= g \circ (1 \circ g^{-1}) \\
   &= g \circ g^{-1} \\
   &= 1_B
   \end{aligned}$$

9. When will a projection function $\pi_{i_0} : \prod \{A_i : i \in I\} \to A_{i_0}$, $A_{i_0} \neq \emptyset$, be an onto function?

   Solution:

   A projection function is always onto, providing the Cartesian product $\prod \{A_i : i \in I\}$ is non-empty, i.e. provided no A_i is the empty set.

INDEXED SETS, GENERALIZED OPERATIONS

10. Let $A_n = \{x : x \text{ is a multiple of } n\}$, where $n \in \mathbf{N}$, the positive integers, and let $B_i = [i, i+1]$, where $i \in \mathbf{Z}$, the integers. Find: (i) $A_3 \cap A_5$; (ii) $\bigcup \{A_i : i \in P\}$, where P is the set of prime numbers; (iii) $B_3 \cap B_4$; (iv) $\bigcup \{B_i : i \in \mathbf{Z}\}$; (v) $(\bigcup \{B_i : i \geq 7\}) \cap A_5$.

    Solution:

    (i)   Those numbers which are multiples of both 3 and 5 are the multiples of 15; hence $A_3 \cap A_5 = A_{15}$.

    (ii) Every positive integer except 1 is a multiple of at least one prime number; hence $\bigcup \{A_i : i \in P\} = \{2, 3, 4, \ldots\} = \mathbf{N} \setminus \{1\}$.

    (iii) $B_3 \cap B_4 = \{x : 3 \leq x \leq 4, \ 4 \leq x \leq 5\} = \{4\}$

    (iv) Since every real number belongs to at least one interval $[i, i+1]$, $\bigcup \{B_i : i \in \mathbf{Z}\} = \mathbf{R}$, the set of real numbers.

    (v) $(\bigcup \{B_i : i \geq 7\}) \cap A_5 = \{x : x \text{ is a multiple of } 5, \ x \geq 7\} = A_5 \setminus \{5\} = \{10, 15, 20, \ldots\}$.

11. Let $D_n = (0, 1/n)$, where $n \in \mathbf{N}$, the positive integers. Find:

    (i) $D_3 \cup D_7$ (iii) $D_s \cup D_t$ (v) $\bigcup \{D_i : i \in A \subset \mathbf{N}\}$

    (ii) $D_3 \cap D_{20}$ (iv) $D_s \cap D_t$ (vi) $\bigcap \{D_i : i \in \mathbf{N}\}$

    Solution:

    (i) Since $(0, 1/7) \subset (0, 1/3)$, $D_3 \cup D_7 = D_3$.

    (ii) Since $(0, 1/20) \subset (0, 1/3)$, $D_3 \cap D_{20} = D_{20}$.

(iii) Let $m = \min \{s, t\}$, i.e. the smaller of the two numbers s and t; then D_m equals D_s or D_t and contains the other. So $D_s \cup D_t = D_m$.

(iv) Let $M = \max \{s, t\}$, i.e. the larger of the two numbers. Then $D_s \cap D_t = D_M$.

(v) Let $a \in A$ be the smallest number in A. Then $\bigcup \{D_i : i \in A \subset \mathbf{N}\} = D_a$.

(vi) If $x \in \mathbf{R}$, then $\exists i \in \mathbf{N}$ s.t. $x \notin (0, 1/i)$. Hence $\bigcap \{D_i : i \in \mathbf{N}\} = \emptyset$.

12. Prove (Distributive Law) Theorem 2.2 (ii): $B \cap (\cup_{i \in I} A_i) = \cup_{i \in I} (B \cap A_i)$.

Solution:
$$
\begin{aligned}
B \cap (\cup_{i \in I} A_i) &= \{x : x \in B, \ x \in \cup_{i \in I} A_i\} \\
&= \{x : x \in B, \ \exists i_0 \in I \ \text{s.t.} \ x \in A_{i_0}\} \\
&= \{x : \exists i_0 \in I \ \text{s.t.} \ x \in B \cap A_{i_0}\} \\
&= \cup_{i \in I} (B \cap A_i)
\end{aligned}
$$

13. Prove: Let $\{A_i : i \in I\}$ be an indexed class of sets and let $i_0 \in I$. Then
$$\cap_{i \in I} A_i \subset A_{i_0} \subset \cup_{i \in I} A_i$$

Solution:

Let $x \in \cap_{i \in I} A_i$; then $x \in A_i$ for every $i \in I$. In particular, $x \in A_{i_0}$. Hence $\cap_{i \in I} A_i \subset A_{i_0}$. Now let $y \in A_{i_0}$. Since $i_0 \in I$, $y \in \cup_{i \in I} A_i$. Hence $A_{i_0} \subset \cup_{i \in I} A_i$.

14. Prove Theorem 2.4: Let A be any set and, for each $p \in A$, let G_p be a subset of A such that $p \in G_p \subset A$. Then $A = \bigcup \{G_p : p \in A\}$.

Solution:

Let $x \in \bigcup \{G_p : p \in A\}$. Then $\exists p_0 \in A$ s.t. $x \in G_{p_0} \subset A$; hence $x \in A$, so $\bigcup \{G_p : p \in A\} \subset A$. (In other words, if each G_p is a subset of A, then the union of the G_p is also a subset of A.)

Now let $y \in A$. Then $y \in G_y$, so $y \in \bigcup \{G_p : p \in A\}$. Thus $A \subset \bigcup \{G_p : p \in A\}$ and the two sets are equal.

ASSOCIATED SET FUNCTIONS

15. Let $A = \{1, 2, 3, 4, 5\}$ and let $f : A \to A$ be defined by the diagram:

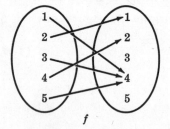

f

Find (i) $f[\{1, 3, 5\}]$, (ii) $f^{-1}[\{2, 3, 4\}]$, (iii) $f^{-1}[\{3, 5\}]$.

Solution:

(i) $f[\{1, 3, 5\}] = \{f(1), f(3), f(5)\} = \{4\}$

(ii) $f^{-1}[\{2, 3, 4\}] = \{4, 1, 3, 5\}$

(iii) $f^{-1}[\{3, 5\}] = \emptyset$ since no element of A has 3 or 5 as an image.

16. Consider the function $f : \mathbf{R} \to \mathbf{R}$ defined by $f(x) = x^2$. Find:
(i) $f^{-1}[\{25\}]$, (ii) $f^{-1}[\{-9\}]$, (iii) $f^{-1}[\{x : x \leq 0\}]$, (iv) $f^{-1}[\{x : 4 \leq x \leq 25\}]$.

Solution:

(i) $f^{-1}[\{25\}] = \{5, -5\}$ since $f(5) = 25$, $f(-5) = 25$ and since the square of no other number is 25.

(ii) $f^{-1}[\{-9\}] = \emptyset$ since the square of no real number is -9.

(iii) $f^{-1}[\{x : x \leqq 0\}] = \{0\}$ since $f(0) = 0 \leqq 0$ and since the square of every other real number is greater than 0.

(iv) $f^{-1}[\{x : 4 \leqq x \leqq 25\}]$ consists of those numbers x such that $4 \leqq x^2 \leqq 25$. Accordingly,

$$f^{-1}[\{x : 4 \leqq x \leqq 25\}] = [2,5] \cup [-5,-2]$$

17. Prove: Let $f : X \to Y$ be one-one. Then the associated set function $f : \mathcal{P}(X) \to \mathcal{P}(Y)$ is also one-one.

Solution:

If $X = \emptyset$, then $\mathcal{P}(X) = \{\emptyset\}$; hence $f : \mathcal{P}(X) \to \mathcal{P}(Y)$ is one-one, for no two different members of $\mathcal{P}(X)$ can have the same image, as there are no two different members in $\mathcal{P}(X)$.

If $X \neq \emptyset$, $\mathcal{P}(X)$ has at least two members. Let $A, B \in \mathcal{P}(X)$, but $A \neq B$. Then $\exists p \in X$ s.t. $p \in A$, $p \notin B$ (or $p \in B$, $p \notin A$). Thus $f(p) \in f[A]$ and, since f is one-one, $f(p) \notin f[B]$ (or $f(p) \in f[B]$ and $f(p) \notin f[A]$). Hence $f[A] \neq f[B]$, and so the induced set function is also one-one.

18. Prove (Theorem 2.5, (i) and (iii)):

(a) $f[A \cup B] = f[A] \cup f[B]$, (b) $f[A] \setminus f[B] \subset f[A \setminus B]$.

Solution:

(a) We first show $f[A \cup B] \subset f[A] \cup f[B]$. Let $y \in f[A \cup B]$, i.e. $\exists x \in A \cup B$ s.t. $f(x) = y$. Then either $x \in A$ or $x \in B$, but

$$x \in A \quad \text{implies} \quad f(x) = y \in f[A]$$

or

$$x \in B \quad \text{implies} \quad f(x) = y \in f[B]$$

In either case, $y \in f[A] \cup f[B]$.

We now prove the reverse inclusion, i.e. $f[A] \cup f[B] \subset f[A \cup B]$. Let $y \in f[A] \cup f[B]$. Then $y \in f[A]$ or $y \in f[B]$, but

$$y \in f[A] \quad \text{implies} \quad \exists x \in A \text{ s.t. } f(x) = y$$

$$y \in f[B] \quad \text{implies} \quad \exists x \in B \text{ s.t. } f(x) = y$$

In either case, $y = f(x)$ with $x \in A \cup B$, i.e. $y \in f[A \cup B]$.

(b) Let $y \in f[A] \setminus f[B]$. Then $\exists x \in A$ s.t. $f(x) = y$, but $y \notin \{f(x) : x \in B\}$. Hence $x \notin B$, or $x \in B \setminus A$. Accordingly, $y \in f[A \setminus B]$.

19. Prove (Theorem 2.6, (ii) and (iii)):

(a) $f^{-1}[A \cap B] = f^{-1}[A] \cap f^{-1}[B]$, (b) $f^{-1}[A \setminus B] = f^{-1}[A] \setminus f^{-1}[B]$.

Solution:

(a) We first show $f^{-1}[A \cap B] \subset f^{-1}[A] \cap f^{-1}[B]$. Let $x \in f^{-1}[A \cap B]$. Then $f(x) \in A \cap B$ so $f(x) \in A$ and $f(x) \in B$, or $x \in f^{-1}[A]$ and $x \in f^{-1}[B]$. Hence $x \in f^{-1}[A] \cap f^{-1}[B]$.

For the reverse inclusion, let $x \in f^{-1}[A] \cap f^{-1}[B]$. Then $f(x) \in A$ and $f(x) \in B$, i.e. $f(x) \in A \cap B$. Hence $x \in f^{-1}[A \cap B]$.

(b) To show $f^{-1}[A \setminus B] \subset f^{-1}[A] \setminus f^{-1}[B]$, assume $x \in f^{-1}[A \setminus B]$. Then $f(x) \in A \setminus B$, i.e. $f(x) \in A$ and $f(x) \notin B$. Thus $x \in f^{-1}[A]$ but $x \notin f^{-1}[B]$, i.e. $x \in f^{-1}[A] \setminus f^{-1}[B]$.

For the reverse inclusion, let $x \in f^{-1}[A] \setminus f^{-1}[B]$. Then $f(x) \in A$ but $f(x) \notin B$, i.e. $f(x) \in A \setminus B$. Hence $x \in f^{-1}[A \setminus B]$.

ALGEBRA OF REAL-VALUED FUNCTIONS

20. Let $X = \{a, b, c\}$ and let $f, g \in \mathcal{F}(X, \mathbf{R})$ be as follows:

$$f = \{ \langle a, 1 \rangle, \langle b, -2 \rangle, \langle c, 3 \rangle \}, \qquad g = \{ \langle a, -2 \rangle, \langle b, 0 \rangle, \langle c, 1 \rangle \}$$

Find: (i) $f + 2g$, (ii) $fg - 2f$, (iii) $f + 4$, (iv) $|f|$, (v) f^2.

Solution:

(i) Compute as follows: $(f + 2g)(a) \equiv f(a) + 2g(a) = 1 - 4 = -3$

$$(f + 2g)(b) \equiv f(b) + 2g(b) = -2 + 0 = -2$$

$$(f + 2g)(c) \equiv f(c) + 2g(c) = 3 + 2 = 5$$

In other words, $f + 2g = \{\langle a, -3 \rangle, \langle b, -2 \rangle, \langle c, 5 \rangle\}$.

(ii)   Similarly, $(fg - 2f)(a) \equiv f(a)\, g(a) - 2f(a) = (1)(-2) - 2(1) = -4$

$(fg - 2f)(b) \equiv f(b)\, g(b) - 2f(b) = (-2)(0) - 2(-2) = 4$

$(fg - 2f)(c) \equiv f(c)\, g(c) - 2f(c) = (3)(1) - 2(3) = -3$

That is, $fg - 2f = \{\langle a, -4 \rangle, \langle b, 4 \rangle, \langle c, -3 \rangle\}$

(iii)   Since, by definition, $(f + 4)(x) \equiv f(x) + 4$, add 4 to each image value, i.e. to the second coordinate in each pair in f. Thus

$$f + 4 = \{\langle a, 5 \rangle, \langle b, 2 \rangle, \langle c, 7 \rangle\}$$

(iv)   Since $|f|(x) \equiv |f(x)|$, replace the second coordinate of each pair in f by its absolute value. Thus

$$|f| = \{\langle a, 1 \rangle, \langle b, 2 \rangle, \langle c, 3 \rangle\}$$

(v)   Since $f^2(x) = (ff)(x) \equiv f(x)\, f(x) = (f(x))^2$, replace the second coordinate of each pair in f by its square. Thus

$$f^2 = \{\langle a, 1 \rangle, \langle b, 4 \rangle, \langle c, 9 \rangle\}$$

21. Let $\hat{0} \in \mathcal{F}(X, \mathbf{R})$ be defined by $\hat{0}(x) = 0$ for all $x \in X$.

Prove: For any $f \in \mathcal{F}(X, \mathbf{R})$, (i) $f + \hat{0} = f$ and (ii) $f\hat{0} = \hat{0}$.

Solution:

(i) $(f + \hat{0})x \equiv f(x) + \hat{0}(x) = f(x) + 0 = f(x)$ for every $x \in X$; hence $f + \hat{0} = f$. Observe that $\hat{0}$ satisfies the conditions of the 0 in the axiom $[\mathbf{V_1}]$ of Theorem 2.9.

(ii) $(f\hat{0})(x) \equiv f(x)\,\hat{0}(x) = f(x) \cdot (0) = 0 = \hat{0}(x)$ for all $x \in X$; hence $f\hat{0} = \hat{0}$.

22. Prove: $\mathcal{F}(X, \mathbf{R})$ satisfies the axiom $[\mathbf{V_3}]$ of Theorem 2.9, i.e. if $f, g \in \mathcal{F}(X, \mathbf{R})$ and $k, k' \in \mathbf{R}$, then:

(i) $k \cdot (f + g) = k \cdot f + k \cdot g$,   (ii) $(k + k') \cdot f = k \cdot f + k' \cdot f$.

Solution:

(i) $[k \cdot (f + g)](x) = k[(f + g)(x)] = k[f(x) + g(x)] = k(f(x)) + k(g(x))$

$(k \cdot f + k \cdot g)(x) = (k \cdot f)(x) + (k \cdot g)(x) = k(f(x)) + k(g(x))$

for all $x \in X$; hence $k \cdot (f + g) = k \cdot f + k \cdot g$. Observe that we use the fact that k, $f(x)$ and $g(x)$ are real numbers and satisfy the distributive law.

(ii) $((k + k') \cdot f)(x) = (k + k')\, f(x) = k(f(x)) + k'(f(x))$

$(k \cdot f + k' \cdot f)(x) = (k \cdot f)(x) + (k' \cdot f)(x) = k(f(x)) + k'(f(x))$

for all $x \in X$; so $(k + k') \cdot f = k \cdot f + k' \cdot f$.

Supplementary Problems

FUNCTIONS

23. Let $f : \mathbf{R} \to \mathbf{R}$ and $g : \mathbf{R} \to \mathbf{R}$ be defined by $\quad f(x) = \begin{cases} 2x - 5 & \text{if } x > 2 \\ x^2 - 2|x| & \text{if } x \leq 2 \end{cases}, \quad g(x) = 3x + 1$.

Find   (i) $f(-2)$, (ii) $g(-3)$, (iii) $f(4)$, (iv) $(g \circ f)(1)$, (v) $(f \circ g)(2)$, (vi) $(f \circ f)(3)$.

24. Let $f : \mathbf{R} \to \mathbf{R}$ and $g : \mathbf{R} \to \mathbf{R}$ be defined by $f(x) = x^2 + 3x + 1$, $g(x) = 2x - 3$.

Find formulas which define the composition functions (i) $f \circ g$, (ii) $g \circ f$, (iii) $f \circ f$.

25. Let $k : X \to X$ be a constant function. Prove that for any function $f : X \to X$, $k \circ f = k$. What can be said about $f \circ k$?

26. Consider the function $f(x) = x$ where $x \in \mathbf{R}$, $x \geq 0$. State whether or not each of the following functions is an extension of f.

   (i) $g_1(x) = |x|$ for all $x \in \mathbf{R}$ (iii) $g_3(x) = (x + |x|)/2$ for all $x \in \mathbf{R}$

   (ii) $g_2(x) = x$ where $x \in [-1, 1]$ (iv) $1_{\mathbf{R}} : \mathbf{R} \to \mathbf{R}$

27. Let $A \subset X$ and let $f : X \to Y$. The *inclusion function* j from A into X, denoted by $j : A \subset X$, is defined by $j(a) = a$ for all $a \in A$. Show that $f \,|\, A$, the restriction of f to A, equals the composition $f \circ j$, i.e. $f \,|\, A = f \circ j$.

ONE-ONE, ONTO, INVERSE AND IDENTITY FUNCTIONS

28. Prove: For any function $f : A \to B$, $f \circ 1_A = f = 1_B \circ f$.

29. Prove: If $f : A \to B$ is both one-one and onto, then $f^{-1} \circ f = 1_A$ and $f \circ f^{-1} = 1_B$.

30. Prove: If $f : A \to B$ and $g : B \to A$ satisfy $g \circ f = 1_A$, then f is one-one and g is onto.

31. Prove Proposition 2.1: Let $f : A \to B$ and $g : B \to A$ satisfy $g \circ f = 1_A$ and $f \circ g = 1_B$. Then $f^{-1} : B \to A$ exists and $g = f^{-1}$.

32. Under what conditions will the projection $\pi_{i_0} : \prod \{A_i : i \in I\} \to A_{i_0}$ be one-to-one?

33. Let $f : (-1, 1) \to \mathbf{R}$ be defined by $f(x) = x/(1 - |x|)$. Prove that f is both one-one and onto.

34. Let R be an equivalence relation in a non-empty set A. The natural function η from A into the quotient set A/R is defined by $\eta(a) = [a]$, the equivalence class of a. Prove that η is an onto function.

35. Let $f : A \to B$. The relation R in A defined by $a \, R \, a'$ iff $f(a) = f(a')$ is an equivalence relation. Let $\hat{f}$ denote the correspondence from the quotient set A/R into the range $f[A]$ of f by $\hat{f} : [a] \to f(a)$.

   (i)   Prove that $\hat{f} : A/R \to f[A]$ is a function which is both one-one and onto.

   (ii) Prove that $f = j \circ \hat{f} \circ \eta$, where $\eta : A \to A/R$ is the natural function and $j : f[A] \subset B$ is the inclusion function.

$$ A \xrightarrow{\eta} A/R \xrightarrow{\hat{f}} f[A] \xrightarrow{j} B $$

INDEXED SETS AND GENERALIZED OPERATIONS

36. Let $A_n = \{x : x \text{ is a multiple of } n\} = \{n, 2n, 3n, \ldots\}$, where $n \in \mathbf{N}$, the positive integers. Find:

   (i) $A_2 \cap A_7$; (ii) $A_6 \cap A_8$; (iii) $A_3 \cup A_{12}$; (iv) $A_3 \cap A_{12}$; (v) $A_s \cup A_{st}$, where $s, t \in \mathbf{N}$; (vi) $A_s \cap A_{st}$, where $s, t \in \mathbf{N}$. (vii) Prove: If $J \subset \mathbf{N}$ is infinite, then $\cap \{A_i : i \in J\} = \varnothing$.

37. Let $B_i = (i, i + 1]$, an open-closed interval, where $i \in \mathbf{Z}$, the integers. Find:

   (i) $B_4 \cup B_5$ (iii) $\cup_{i=4}^{20} B_i$ (v) $\cup_{i=0}^{15} B_{s+i}$

   (ii) $B_6 \cap B_7$ (iv) $B_s \cup B_{s+1} \cup B_{s+2}$, $s \in \mathbf{Z}$ (vi) $\cup_{i \in \mathbf{Z}} B_{s+i}$

38. Let $D_n = [0, 1/n]$, $S_n = (0, 1/n]$ and $T_n = [0, 1/n)$ where $n \in \mathbf{N}$, the positive integers. Find:

   (i) $\cap \{D_n : n \in \mathbf{N}\}$,   (ii) $\cap \{S_n : n \in \mathbf{N}\}$,   (iii) $\cap \{T_n : n \in \mathbf{N}\}$.

39. Prove DeMorgan's Laws: (i) $(\cup_i A_i)^c = \cap_i A_i^c$,   (ii) $(\cap_i A_i)^c = \cup_i A_i^c$.

40. Let $\mathcal{A} = \{A_i : i \in I\}$ be an indexed class of sets and let $J \subset K \subset I$. Prove:

   (i) $\bigcup\{A_i : i \in J\} \subset \bigcup\{A_i : i \in K\}$,   (ii) $\bigcap\{A_i : i \in J\} \supset \bigcap\{A_i : i \in K\}$

ASSOCIATED SET FUNCTIONS

41. Let $f : \mathbf{R} \to \mathbf{R}$ be defined by $f(x) = x^2 + 1$. Find: (i) $f[\{-1, 0, 1\}]$, (ii) $f^{-1}[\{10, 17\}]$, (iii) $f[(-2, 2)]$, (iv) $f^{-1}[(5, 10)]$, (v) $f[\mathbf{R}]$, (vi) $f^{-1}[\mathbf{R}]$.

42. Prove: A function $f : X \to Y$ is one-one if and only if $f[A \cap B] = f[A] \cap f[B]$, for all subsets A and B of X.

43. Prove: Let $f : X \to Y$. Then, for any subsets A and B of X,

   (a) $f[A \cap B] \subset f[A] \cap f[B]$,   (b) $A \subset B$ implies $f[A] \subset f[B]$

44. Prove: Let $f : X \to Y$. Then, for any subsets A and B of Y,

   (a) $f^{-1}[A \cup B] = f^{-1}[A] \cup f^{-1}[B]$,   (b) $A \subset B$ implies $f^{-1}[A] \subset f^{-1}[B]$

45. Prove Theorem 2.8: Let $f : X \to Y$ and let $A \subset X$ and $B \subset Y$. Then

   (i) $A \subset f^{-1} \circ f[A]$,   (ii) $B \supset f \circ f^{-1}[B]$

46. Prove: Let $f : X \to Y$ be onto. Then the associated set function $f : \mathcal{P}(X) \to \mathcal{P}(Y)$ is also onto.

47. Prove: A function $f : X \to Y$ is both one-one and onto if and only if $f[A^c] = (f[A])^c$ for every subset A of X.

48. Prove: A function $f : X \to Y$ is one-one if and only if $A = f^{-1} \circ f[A]$ for every subset A of X.

ALGEBRA OF REAL-VALUED FUNCTIONS

49. Let $X = \{a, b, c\}$ and let f and g be the following real valued functions on X:

$$f = \{\langle a, 2\rangle, \langle b, -3\rangle, \langle c, -1\rangle\}, \qquad g = \{\langle a, -2\rangle, \langle b, 0\rangle, \langle c, 1\rangle\}$$

   Find   (i) $3f$, (ii) $2f - 5g$, (iii) fg, (iv) $|f|$, (v) f^3, (vi) $|3f - fg|$.

50. Let A be any subset of a universal set U. Then the real-valued function $\chi_A : U \to \mathbf{R}$ defined by

$$\chi_A(x) = \begin{cases} 1 & \text{if } x \in A \\ 0 & \text{if } x \notin A \end{cases}$$

   is called the *characteristic function* of A. Prove:

   (i) $\chi_{A \cap B} = \chi_A \chi_B$,   (ii) $\chi_{A \cup B} = \chi_A + \chi_B - \chi_{A \cap B}$,   (iii) $\chi_{A \setminus B} = \chi_A - \chi_{A \cap B}$.

51. Prove: $\mathcal{F}(X, \mathbf{R})$ satisfies the axiom $[\mathbf{V_2}]$ of Theorem 2.9; i.e. if $f \in \mathcal{F}(X, \mathbf{R})$ and $k, k' \in \mathbf{R}$, then
   (i) $k \cdot (k' \cdot f) = (kk') \cdot f$,   (ii) $1 \cdot f = f$

52. For each $k \in \mathbf{R}$, let $\hat{k} \in \mathcal{F}(X, \mathbf{R})$ denote the constant function $\hat{k}(x) = k$ for all $x \in X$.

   (i)   Show that the collection C of constant functions, i.e. $C = \{\hat{k} : k \in \mathbf{R}\}$, is a linear subspace of $\mathcal{F}(X, \mathbf{R})$.

   (ii) Let $\alpha : C \to \mathbf{R}$ be defined by $\alpha(\hat{k}) = k$. Show that α is both one-one and onto and that, for any $k, k' \in \mathbf{R}$,
   $$\alpha(\hat{k} + \hat{k}') = \alpha(\hat{k}) + \alpha(\hat{k}')$$

Answers to Supplementary Problems

23. (i) 0, (ii) −8, (iii) 3, (iv) −2, (v) 9, (vi) −1

24. (i) $(f \circ g)(x) = 4x^2 - 6x + 1$, (ii) $(g \circ f)(x) = 2x^2 + 6x - 1$, (iii) $(f \circ f)(x) = x^4 + 6x^3 + 14x^2 + 15x + 5$

25. The function $f \circ k$ is a constant function.

26. (i) yes, (ii) no, (iii) yes, (iv) yes

32. A_i is a singleton set, say $A_i = \{a_i\}$, for $i \neq i_0$.

36. (i) A_{14}, (ii) A_{24}, (iii) A_3, (iv) A_{12}, (v) A_s, (vi) A_{st}

37. (i) $(4, 6]$, (ii) $\emptyset$, (iii) $(4, 21]$, (iv) $(s, s+3]$, (v) $(s, s+16]$, (vi) R

38. (i) $\{0\}$, (ii) $\emptyset$, (iii) $\{0\}$

41. (i) $\{1, 2\}$, (ii) $\{3, -3, 4, -4\}$, (iii) $(1, 5)$, (iv) $(-3, -2), (2, 3)$, (v) $\{x : x \geq 1\}$, (vi) $\mathbf{R}$

49. (i) $3f = \{\langle a, 6 \rangle, \langle b, -9 \rangle, \langle c, -3 \rangle\}$

    (ii) $2f - 5g = \{\langle a, 14 \rangle, \langle b, -6 \rangle, \langle c, -7 \rangle\}$

    (iii) $fg = \{\langle a, -4 \rangle, \langle b, 0 \rangle, \langle c, -1 \rangle\}$

    (iv) $|f| = \{\langle a, 2 \rangle, \langle b, 3 \rangle, \langle c, 1 \rangle\}$

    (v) $f^3 = \{\langle a, 8 \rangle, \langle b, -27 \rangle, \langle c, -1 \rangle\}$

    (vi) $|3f - fg| = \{\langle a, 10 \rangle, \langle b, 9 \rangle, \langle c, 2 \rangle\}$

Chapter 3

Cardinality, Order

EQUIVALENT SETS

A set A is called *equivalent* to a set B, written $A \sim B$, if there exists a function $f : A \to B$ which is one-one and onto. The function f is then said to define a *one-to-one correspondence* between the sets A and B.

A set is *finite* iff it is empty or equivalent to $\{1, 2, \ldots, n\}$ for some $n \in \mathbf{N}$; otherwise it is said to be *infinite*. Clearly two finite sets are equivalent iff they contain the same number of elements. Hence, for finite sets, equivalence corresponds to the usual meaning of two sets containing the same number of elements.

> **Example 1.1:** Let $\mathbf{N} = \{1, 2, 3, \ldots\}$ and $E = \{2, 4, 6, \ldots\}$. The function $f : \mathbf{N} \to E$ defined by $f(x) = 2x$ is both one-one and onto; hence $\mathbf{N}$ is equivalent to E.

> **Example 1.2:** The function $f : (-1, 1) \to \mathbf{R}$ defined by $f(x) = x/(1 - |x|)$ is both one-one and onto. Hence the open interval $(-1, 1)$ is equivalent to $\mathbf{R}$, the set of real numbers.

Observe that an infinite set can be equivalent to a proper subset of itself. This property is true of infinite sets generally.

Proposition 3.1: The relation in any collection of sets defined by $A \sim B$ is an equivalence relation.

DENUMERABLE AND COUNTABLE SETS

Let $\mathbf{N}$ be the set of positive integers $\{1, 2, 3, \ldots\}$. A set X is called *denumerable* and is said to have cardinality $\aleph_0$ (read: *aleph-null*) iff it is equivalent to $\mathbf{N}$. A set is called *countable* iff it is finite or denumerable.

> **Example 2.1:** The set of terms in any infinite sequence
>
> $$a_1, a_2, a_3, \ldots$$
>
> of distinct terms is denumerable, for a sequence is essentially a function $f(n) = a_n$ whose domain is $\mathbf{N}$. So if the a_n are distinct, the function is one-one and onto. Accordingly, each of the following sets is denumerable:
>
> $$\{1, \tfrac{1}{2}, \tfrac{1}{3}, \ldots\}, \quad \{1, -2, 3, -4, \ldots\}, \quad \{\langle 1, 1 \rangle, \langle 4, 8 \rangle, \langle 9, 27 \rangle, \ldots, \langle n^2, n^3 \rangle, \ldots\}$$

> **Example 2.2:** Consider the product set $\mathbf{N} \times \mathbf{N}$ as exhibited below.

The set $\mathbf{N} \times \mathbf{N}$ can be written in an infinite sequence of distinct elements as follows:

$$\langle 1, 1 \rangle, \; \langle 2, 1 \rangle, \; \langle 1, 2 \rangle, \; \langle 1, 3 \rangle, \; \langle 2, 2 \rangle, \; \ldots$$

(Note that the sequence is determined by "following the arrows" in the above diagram.) Thus we see that $\mathbf{N} \times \mathbf{N}$ is denumerable.

Example 2.3:   Let $M = \{0, 1, 2, 3, \ldots\} = \mathbf{N} \cup \{0\}$. Now each positive integer $a \in \mathbf{N}$ can be written uniquely in the form $a = 2^r(2s+1)$ where $r, s \in M$. The function $f : \mathbf{N} \to M \times M$ defined by $f(a) = \langle r, s \rangle$

where r and s are as above, is one-one and onto. Hence $M \times M$ is denumerable. Note that $\mathbf{N} \times \mathbf{N}$ is a subset of $M \times M$.

The following theorems concern denumerable and countable sets.

Theorem 3.2:   Every infinite set contains a denumerable subset.

Theorem 3.3:   Every subset of a countable set is countable.

Lemma 3.4:   Let $\{A_1 \, A_2, \ldots\}$ be a denumerable disjoint class of denumerable sets. Then $\cup_{i=1}^{\infty} A_i$ is also denumerable.

Theorem 3.5:   Let $\{A_i : i \in I\}$ be a countable class of countable sets, i.e. I is countable and A_i is countable for each $i \in I$. Then $\bigcup \{A_i : i \in I\}$ is countable.

A set which is neither finite nor denumerable is said to be *non-denumerable* or *non-countable*.

THE CONTINUUM

Not every infinite set is denumerable; in fact, the next theorem gives a specific and extremely important example.

Theorem 3.6:   The unit interval $[0, 1]$ is non-denumerable.

A set X is said to have the *power of the continuum* or is said to have *cardinality* c iff it is equivalent to the unit interval $[0, 1]$.

We show, in a solved problem, that every interval, open or closed, has cardinality c. By Example 1.2, the open interval $(-1, 1)$ is equivalent to $\mathbf{R}$.   Hence,

Proposition 3.7: $\mathbf{R}$, the set of real numbers, has cardinality c.

SCHROEDER-BERNSTEIN THEOREM

We write $A \precsim B$ if A is equivalent to a subset of B, i.e.,

$$A \precsim B \quad \text{iff} \quad \exists \; B^* \subset B \quad \text{such that} \quad A \sim B^*$$

We also write $A \prec B$ if $A \precsim B$ but $A \not\sim B$, i.e. A is not equivalent to B.

Example 3.1:   Since $\mathbf{N}$ is a subset of $\mathbf{R}$, we may write $\mathbf{N} \precsim \mathbf{R}$. On the other hand, by Proposition 3.7, $\mathbf{R}$ is not denumerable, i.e. $\mathbf{R} \not\sim \mathbf{N}$. Accordingly, $\mathbf{N} \prec \mathbf{R}$.

Given any pair of sets A and B, then at least one of the following must be true:
(i) $A \sim B$,   (ii) $A \prec B$ or $B \prec A$,   (iii) $A \precsim B$ and $B \precsim A$,   (iv) $A \not\prec B$, $A \not\sim B$ and $B \not\prec A$

The celebrated Schroeder-Bernstein Theorem states that, in Case (iii) above, A is equivalent to B.   Namely,

Theorem (Schroeder-Bernstein) 3.8:   If $A \precsim B$ and $B \precsim A$, then $A \sim B$.

The Schroeder-Bernstein Theorem can be restated as follows:

Theorem 3.8:   Let $X \supset Y \supset X_1$ and let $X \sim X_1$. Then $X \sim Y$.

We remark that Case (iv) above is impossible. That is,

Theorem (Law of Trichotomy) 3.9: Given any pair of sets A and B, either $A \prec B$, $A \sim B$ or $B \prec A$.

CONCEPT OF CARDINALITY

If A is equivalent to B, i.e. $A \sim B$, then we say that A and B have the same *cardinal number* or *cardinality*. We write $\#(A)$ for "the cardinal number (or cardinality) of A". So

$$\#(A) = \#(B) \quad \text{iff} \quad A \sim B$$

On the other hand, if $A \prec B$ then we say that A has *cardinality less than* B or B has *cardinality greater than* A. That is,

$$\#(A) < \#(B) \quad \text{iff} \quad A \prec B$$

So $\#(A) \leqq \#(B)$ iff $A \precsim B$. Accordingly, the Schroeder-Bernstein Theorem can be restated as follows:

Theorem 3.8: If $\#(A) \leqq \#(B)$ and $\#(B) \leqq \#(A)$, then $\#(A) = \#(B)$.

The cardinal number of each of the sets

$$\emptyset, \quad \{\emptyset\}, \quad \{\emptyset, \{\emptyset\}\}, \quad \{\emptyset, \{\emptyset\}, \{\emptyset, \{\emptyset\}\}\}, \quad \ldots$$

is denoted by $0, 1, 2, 3, \ldots$, respectively, and is called a *finite* cardinal. The cardinal numbers of $\mathbf{N}$ and $[0, 1]$ are denoted by

$$\aleph_0 = \#(N), \quad \mathbf{c} = \#([0, 1])$$

Accordingly, we may write $\quad 0 < 1 < 2 < 3 < \cdots < \aleph_0 < \mathbf{c}$

CANTOR'S THEOREM AND THE CONTINUUM HYPOTHESIS

It is natural to ask if there are infinite cardinal numbers other than $\aleph_0$ and $\mathbf{c}$. The answer is yes. In fact, Cantor's Theorem determines a set with cardinality greater than any given set. Namely,

Theorem (Cantor) 3.10: The power set $\mathcal{P}(A)$ of any set A has cardinality greater than A.

It is also natural to ask if there exists a set whose cardinality lies between $\aleph_0$ and $\mathbf{c}$. The conjecture that the answer to this question is negative is known as the Continuum Hypothesis. That is,

Continuum Hypothesis: There does not exist a set A with the property that $\aleph_0 < \#(A) < \mathbf{c}$.

In 1963 it was shown that the Continuum Hypothesis is independent of our axioms of set theory in somewhat the same sense that Euclid's Fifth Postulate on parallel lines is independent of the other axioms of geometry.

PARTIALLY ORDERED SETS

A relation $\precsim$ in a set A is called a *partial order* (or *order*) on A iff, for every $a, b, c \in A$: (i) $a \precsim a$; (ii) $a \precsim b$ and $b \precsim a$ implies $a = b$; and (iii) $a \precsim b$ and $b \precsim c$ implies $a \precsim c$. The set A together with the partial order, i.e. the pair $(A, \precsim)$, is called a *partially ordered set*.

Recall that a relation is reflexive iff it satisfies (i), and transitive iff it satisfies (iii). A relation is said to be *anti-symmetric* iff it satisfies (ii). In other words, a partial order is a reflexive, anti-symmetric, transitive relation.

Example 4.1: Set inclusion is a partial order in any class of sets since: (i) $A \subset A$ for any set A; (ii) $A \subset B$ and $B \subset A$ implies $A = B$; and (iii) $A \subset B$ and $B \subset C$ implies $A \subset C$.

Example 4.2: Let A be any set of real numbers. Then the relation in A defined by $x \le y$ is a partial order and is called the *natural order* in A.

Example 4.3: Let $X = \{a, b, c, d, e\}$. Then the diagram

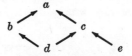

defines a partial order in X as follows: $x \lesssim y$ iff $x = y$ or if one can go from x to y in the diagram, always moving in the indicated direction, i.e. upward.

If $a \lesssim b$ in an ordered set, then we say that a *precedes* or is *smaller* than b and that b *follows* or *dominates* or is *larger* than a. Furthermore, we write $a \prec b$ if $a \lesssim b$ but $a \ne b$.

A partially ordered set A is said to be *totally* (or *linearly*) *ordered* if, for every $a, b \in A$, either $a \lesssim b$ or $b \lesssim a$. **R**, the set of real numbers, with the natural order defined by $x \le y$ is an example of a totally ordered set.

Example 4.4: Let A and B be totally ordered. Then the product set $A \times B$ can be totally ordered as follows:

$$\langle a, b \rangle \prec \langle a', b' \rangle \quad \text{if} \quad a \prec a' \quad \text{or if} \quad a = a' \text{ and } b \prec b'$$

This order is called the *lexicographical order* of $A \times B$ since it is similar to the way words are arranged in a dictionary.

Remark: If a relation R in a set A defines a partial order, i.e. is reflexive, anti-symmetric and transitive, then the inverse relation R^{-1} is also a partial order; it is called the *inverse order*.

SUBSETS OF ORDERED SETS

Let A be a subset of a partially ordered set X. Then the order in X induces an order in A in the following natural way: If $a, b \in A$, then $a \lesssim b$ as elements in A iff $a \lesssim b$ as elements in X. More precisely, if R is a partial order in X, then the relation $R_A = R \cap (A \times A)$, called the *restriction* of R to A, is a partial order in A. The ordered set (A, R_A) is called a *(partially ordered) subset* of the ordered set (X, R).

Some subsets of a partially ordered set X may, in fact, be totally ordered. Clearly if X itself is totally ordered, every subset of X will also be totally ordered.

Example 5.1: Consider the partial order in $W = \{a, b, c, d, e\}$ defined by the diagram

The sets $\{a, c, d\}$ and $\{b, e\}$ are totally ordered subsets; the sets $\{a, b, c\}$ and $\{d, e\}$ are not totally ordered subsets.

FIRST AND LAST ELEMENTS

Let X be an ordered set. An element $a_0 \in X$ is a *first* or *smallest* element of X iff $a_0 \lesssim x$ for all $x \in X$. Analogously, an element $b_0 \in X$ is a *last* or *largest* element of X iff $x \lesssim b_0$ for all $x \in X$.

Example 6.1:   Let $X = \{a, b, c, d, e\}$ be ordered by the diagram

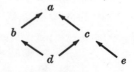

Then a is a last element since a follows every element. Note that X has no first element. The element d is not a first element since d does not precede e.

Example 6.2:   The positive integers $\mathbf{N}$ with the natural order has 1 as a first element. The integers $\mathbf{Z}$ with the natural order has no first element and no last element.

MAXIMAL AND MINIMAL ELEMENTS

Let X be an ordered set. An element $a_0 \in X$ is *maximal* iff $a_0 \precsim x$ implies $x = a$, i.e. if no element follows a_0 except itself. Similarly, an element $b_0 \in X$ is *minimal* iff $x \precsim b_0$ implies $x = b_0$, i.e. if no element precedes b_0 except itself.

Example 7.1:   Let $X = \{a, b, c, d, e\}$ be ordered by the diagram in Example 6.1. Then both d and e are minimal elements. The element a is a maximal element.

Example 7.2:   Although $\mathbf{R}$ with the natural order is totally ordered it has no minimal and no maximal elements.

Example 7.3:   Let $A = \{a_1, a_2, \ldots, a_m\}$ be a finite totally ordered set. Then A contains precisely one minimal element and precisely one maximal element, denoted respectively by

$$\min \{a_1, \ldots, a_m\} \quad \text{and} \quad \max \{a_1, \ldots, a_m\}$$

UPPER AND LOWER BOUNDS

Let A be a subset of a partially ordered set X. An element $m \in X$ is a *lower bound* of A iff $m \precsim x$ for all $x \in A$, i.e. if m precedes every element in A. If some lower bound of A follows every other lower bound of A, then it is called the *greatest lower bound* (g.l.b.) or *infimum* of A and is denoted by $\inf (A)$.

Similarly, an element $M \in X$ is an *upper bound* of A iff $x \precsim M$ for all $x \in A$, i.e. if M follows every element in A. If some upper bound of A precedes every other upper bound of A, then it is called the *least upper bound* (l.u.b.) or *supremum* of A and is denoted by $\sup (A)$.

A is said to be *bounded above* if it has an upper bound, and *bounded below* if it has a lower bound. If A has both an upper and lower bound, then it is said to be *bounded*.

Example 8.1:   Let $X = \{a, b, c, d, e, f, g\}$ be ordered by the following diagram:

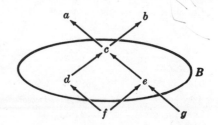

Let $B = \{c, d, e\}$. Then a, b and c are upper bounds of B, and f is the only lower bound of B. Note that g is not a lower bound of B since g does not precede d. Furthermore, $c = \sup (B)$ belongs to B, while $f = \inf (B)$ does not belong to B.

Example 8.2:   Let A be a bounded set of real numbers. Then a fundamental theorem about real numbers states that, under the natural order, $\inf (A)$ and $\sup (A)$ exist.

Example 8.3:   Let **Q** be the set of rational numbers. Let

$$B \;=\; \{x : x \in \mathbf{Q},\ x > 0,\ 2 < x^2 < 3\}$$

that is, B consists of those rational points which lie between $\sqrt{2}$ and $\sqrt{3}$ on the real line. Then B has an infinite number of upper and lower bounds, but $\inf(B)$ and $\sup(B)$ do not exist. Note that the real numbers $\sqrt{2}$ and $\sqrt{3}$ do not belong to **Q** and cannot be considered as upper or lower bounds of B.

ZORN'S LEMMA

Zorn's Lemma is one of the most important tools in mathematics; it asserts the existence of certain types of elements although no constructive process is given to find these elements.

Zorn's Lemma 3.11:   Let X be a non-empty partially ordered set in which every totally ordered subset has an upper bound. Then X contains at least one maximal element.

Remark.   Zorn's Lemma is equivalent to the classical Axiom of Choice and the Well-ordering Principle. The proof of this fact, which uses the concept of ordinal numbers, is beyond the scope of this text.

Solved Problems

EQUIVALENT SETS, DENUMERABLE SETS

1. Consider the concentric circles

$$C_1 \;=\; \{\langle x, y\rangle : x^2 + y^2 = a^2\}, \quad C_2 \;=\; \{\langle x, y\rangle : x^2 + y^2 = b^2\}$$

where, say $0 < a < b$. Establish, geometrically, a one-to-one correspondence between C_1 and C_2.

Solution:

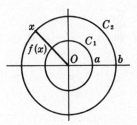

Let $x \in C_2$. Consider the function $f : C_2 \to C_1$, where $f(x)$ is the point of intersection of the radius from the center of C_2 (and C_1) to x, and C_1, as shown in the adjacent diagram.

Note that f is both one-one and onto. Thus f defines a one-to-one correspondence between C_1 and C_2.

2. Prove: The set of rational numbers is denumerable.

Let $\mathbf{Q}^+$ be the set of positive rational numbers and let $\mathbf{Q}^-$ be the set of negative rational numbers. Then $\mathbf{Q} = \mathbf{Q}^- \cup \{0\} \cup \mathbf{Q}^+$ is the set of rational numbers.

Let the function $f : \mathbf{Q}^+ \to \mathbf{N} \times \mathbf{N}$ be defined by

$$f(p/q) \;=\; \langle p, q\rangle$$

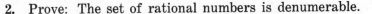

where p/q is any positive rational number expressed as the ratio of two positive integers. Note f is one-one; hence $\mathbf{Q}^+$ is equivalent to a subset of $\mathbf{N} \times \mathbf{N}$. But $\mathbf{N} \times \mathbf{N}$ is denumerable (see Example 2.2); hence $\mathbf{Q}^+$ is also denumerable. Similarly $\mathbf{Q}^-$ is denumerable. Accordingly, by Theorem 3.5, the union of $\mathbf{Q}^-$, $\{0\}$ and $\mathbf{Q}^+$, i.e. the set of rational numbers, is also denumerable.

3. Prove Proposition 3.1: The relation in any collection of sets defined by $A \sim B$ is an equivalence relation. That is, (i) $A \sim A$ for any set A; (ii) if $A \sim B$ then $B \sim A$; and (iii) if $A \sim B$ and $B \sim C$ then $A \sim C$.

Solution:

(i)   The identity function $1_A : A \to A$ is one-one and onto; hence $A \sim A$.

(ii)   If $A \sim B$, then there exists $f : A \to B$ which is one-one and onto. But then f has an inverse $f^{-1} : B \to A$ which is also one-one and onto. Hence

$$A \sim B \quad \text{implies} \quad B \sim A$$

(iii) If $A \sim B$ and $B \sim C$, then there exist functions $f : A \to B$ and $g : B \to C$ which are one-one and onto. Thus the composition function $g \circ f : A \to C$ is also one-one and onto. Hence

$$A \sim B \quad \text{and} \quad B \sim C \quad \text{implies} \quad A \sim C$$

4. Prove: The collection P of all polynomials

$$p(x) \;=\; a_0 + a_1 x + \cdots + a_m x^m$$

with integral coefficients, i.e. where $a_0, a_1, \ldots, a_m$ are integers, is denumerable.

Solution:

For each pair of positive integers $\langle n, m \rangle \in \mathbf{N} \times \mathbf{N}$, let P_{nm} denote the set of polynomials $p(x)$ of degree m in which

$$|a_0| + |a_1| + \cdots + |a_m| \;=\; n$$

Observe that P_{nm} is finite. Accordingly,

$$P \;=\; \mathbf{U}\{P_{nm} : \langle n, m \rangle \in \mathbf{N} \times \mathbf{N}\}$$

is countable since it is a countable union of countable sets. In particular, since P is not finite, P is denumerable.

5. A real number r is called an *algebraic number* if r is a solution to a polynomial equation

$$p(x) \;=\; a_0 + a_1 x + \cdots + a_m x^m$$

with integral coefficients. Prove that the set A of algebraic numbers is denumerable.

Solution:

Note, by the preceding problem, that the set E of polynomial equations is denumerable:

$$E \;=\; \{p_1(x) = 0, \; p_2(x) = 0, \; p_3(x) = 0, \; \ldots\}$$

Let

$$A_i \;=\; \{x : x \text{ is a solution of } p_i(x) = 0\}$$

Since a polynomial of degree n can have at most n roots, each A_i is finite. Hence $A = \mathbf{U}\{A_i : i \in \mathbf{N}\}$ is denumerable.

6. Prove Theorem 3.2: Every infinite set X contains a subset D which is denumerable.

Solution:

Let $f : \mathcal{P}(X) \to X$ be a choice function, i.e. for each non-empty subset A of X, $f(A) \in A$. (Such a function exists by virtue of the Axiom of Choice.) Consider the following sequence:

$$a_1 \;=\; f(X)$$
$$a_2 \;=\; f(X \setminus \{a_1\})$$
$$a_3 \;=\; f(X \setminus \{a_1, a_2\})$$
$$\cdots\cdots\cdots\cdots\cdots\cdots\cdots\cdots$$
$$a_n \;=\; f(X \setminus \{a_1, \ldots, a_{n-1}\})$$
$$\cdots\cdots\cdots\cdots\cdots\cdots\cdots\cdots$$

Since X is infinite, $X \setminus \{a_1, \ldots, a_{n-1}\}$ is not empty for every $n \in N$. Furthermore, since f is a choice function,

$$a_n \neq a_i \quad \text{for} \quad i < n$$

Thus the a_n are distinct and $D = \{a_1, a_2, \ldots\}$ is a denumerable subset of X.

Essentially, the choice function f "chooses" an element $a_1 \in X$, then chooses an element a_2 from those elements which "remain" in X, etc. Since X is infinite, the set of elements which "remain" in X is non-empty.

7. Prove: Let X be any set and let $C(X)$ be the collection of characteristic functions on X, i.e. the collection of functions $f: X \to \{1, 0\}$. Then the power set of X is equivalent to $C(X)$, i.e. $\mathcal{P}(X) \sim C(X)$.

Solution:

Let A be any subset of X, i.e. $A \in \mathcal{P}(X)$. Let $f: \mathcal{P}(X) \to C(X)$ be defined by

$$f(A) = \chi_A = \begin{cases} 0 & \text{if } x \notin A \\ 1 & \text{if } x \in A \end{cases}$$

Then f is one-one and onto. Hence $\mathcal{P}(X) \sim C(X)$.

8. Prove: A subset of a denumerable set is either finite or denumerable, i.e. is countable.

Solution:

Let $X = \{a_1, a_2, \ldots\}$ be any denumerable set and let A be a subset of X. If $A = \emptyset$, then A is finite. If $A \neq \emptyset$, then let n_1 be the least positive integer such that $a_{n_1} \in A$; let n_2 be the least positive integer such that $n_2 > n_1$ and $a_{n_2} \in A$; etc. Then $A = \{a_{n_1}, a_{n_2}, \ldots\}$. If the set of integers $\{n_1, n_2, \ldots\}$ is bounded, then A is finite. Otherwise A is denumerable.

9. Prove Theorem 3.3: Every subset of a countable set is countable.

Solution:

If X is countable, then X is either finite or denumerable. In either case, its subsets are countable.

10. Prove Lemma 3.4: Let $\{A_1, A_2, \ldots\}$ be a denumerable disjoint class of denumerable sets. Then $\cup_{i=1}^{\infty} A_i$ is denumerable.

Solution:

Since the sets A_i are denumerable, we can write

$$A_1 = \{a_{11}, a_{12}, a_{13}, \ldots\}$$
$$A_2 = \{a_{21}, a_{22}, a_{23}, \ldots\}$$
$$\cdots\cdots\cdots\cdots\cdots\cdots$$
$$A_n = \{a_{n1}, a_{n2}, a_{n3}, \ldots\}$$
$$\cdots\cdots\cdots\cdots\cdots\cdots$$

Then $\cup_{i=1}^{\infty} A_i = \{a_{ij} : \langle i, j \rangle \in \mathbf{N} \times \mathbf{N}\}$. The function $f: \cup_{i=1}^{\infty} A_i \to \mathbf{N} \times \mathbf{N}$ defined by $f(a_{ij}) = \langle i, j \rangle$ is clearly one-one and onto. Hence $\cup_{i=1}^{\infty} A_i$ is denumerable since $\mathbf{N} \times \mathbf{N}$ is denumerable.

11. Prove: Let A be an infinite set, let $B = \{b_1, b_2, \ldots\}$ be denumerable, and let A and B be disjoint. Then $A \cup B \sim A$.

Solution:

Since A is infinite, A contains a denumerable subset $D = \{d_1, d_2, \ldots\}$. Let $f: A \cup B \to A$ be defined by the following diagram:

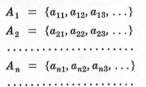

$$A \cup B = (A \setminus D) \cup (D \cup B) = (A \setminus D) \cup \{d_1, d_2, d_3, \ldots, b_1, b_2, b_3, \ldots\}$$

$$A = (A \setminus D) \cup D = (A \setminus D) \cup \{d_1, d_2, d_3, d_4, d_5, d_6, \ldots\}$$

In other words,

$$f(x) = \begin{cases} x & \text{if } x \in A \setminus D \\ d_{2n-1} & \text{if } x = d_n \\ d_{2n} & \text{if } x = b_n \end{cases}$$

Observe that f is one-one and onto; hence $A \cup B \sim A$.

CONTINUUM, CARDINALITY

12. Prove that the intervals $(0, 1)$, $[0, 1)$ and $(0, 1]$ have cardinality c, i.e. is equivalent to $[0, 1]$.

Solution:

(i)   Note that $[0, 1] = \{0, 1, 1/2, 1/3, \ldots\} \cup A$, $(0, 1) = \{1/2, 1/3, 1/4, \ldots\} \cup A$

where $A = [0, 1] \setminus \{0, 1, 1/2, 1/3, \ldots\} = (0, 1) \setminus \{1/2, 1/3, 1/4, \ldots\}$

Consider the function $f : [0, 1] \to (0, 1)$ defined by the following diagram

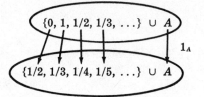

In other words,

$$f(x) = \begin{cases} 1/2 & \text{if } x = 0 \\ 1/(n + 2) & \text{if } x = 1/n,\ n \in \mathbf{N} \\ x & \text{if } x \neq 0,\ 1/n,\ n \in \mathbf{N}, \text{ i.e. if } x \in A \end{cases}$$

The function f is one-one and onto. Accordingly, $[0, 1] \sim (0, 1)$.

(ii) The function $f : [0, 1] \to [0, 1)$ defined by

$$f(x) = \begin{cases} 1/(n + 1) & \text{if } x = 1/n,\ n \in \mathbf{N} \\ x & \text{if } x \neq 1/n,\ n \in \mathbf{N} \end{cases}$$

is one-one and onto. (It is similar to the function in Part (i)). Hence $[0, 1] \sim [0, 1)$.

(iii) Let $f : [0, 1) \to (0, 1]$ be defined by $f(x) = 1 - x$. Then f is one-one and onto. Hence $[0, 1) \sim (0, 1]$ and, by transitivity, $[0, 1] \sim (0, 1]$.

In other words, $(0, 1)$, $[0, 1)$ and $(0, 1]$ have cardinality c.

13. Prove: Each of the following intervals has the power of the continuum, i.e. has cardinality c: $[a, b]$, (a, b), $[a, b)$ and $(a, b]$. Here $a < b$.

Solution:

Let each of the following functions be defined by $f(x) = a + (b - a)x$:

$$[0, 1] \xrightarrow{f} [a, b] \qquad [0, 1) \xrightarrow{f} [a, b) \qquad (0, 1) \xrightarrow{f} (a, b) \qquad (0, 1] \xrightarrow{f} (a, b]$$

Each function is one-one and onto. Hence by the preceding problem and Proposition 3.1, each interval is equivalent to $[0, 1]$, i.e. has cardinality c.

14. Prove Theorem 3.6: The unit interval $A = [0, 1]$ is non-denumerable.

Solution:

Method 1.   Assume the contrary; then

$$A = \{x_1, x_2, x_3, \ldots\}$$

i.e. the elements of A can be written in a sequence.

Each element in A can be written in the form of an infinite decimal as follows:

$$x_1 = 0.\, a_{11}\, a_{12}\, a_{13} \ldots a_{1n} \ldots$$
$$x_2 = 0.\, a_{21}\, a_{22}\, a_{23} \ldots a_{2n} \ldots$$
$$x_3 = 0.\, a_{31}\, a_{32}\, a_{33} \ldots a_{3n} \ldots$$
$$\cdots\cdots\cdots\cdots\cdots\cdots\cdots\cdots\cdots$$
$$x_n = 0.\, a_{n1}\, a_{n2}\, a_{n3} \ldots a_{nn} \ldots$$
$$\cdots\cdots\cdots\cdots\cdots\cdots\cdots\cdots\cdots$$

where $a_{ij} \in \{0, 1, \ldots, 9\}$ and where each decimal contains an infinite number of non-zero elements, i.e. for those numbers which can be written in the form of a decimal in two ways, e.g.,

$$1/2 = .5000\ldots = .4999\ldots$$

we write the infinite decimal in which all except a finite set of digits are nines.

Now construct the real number

$$y = 0. \, b_1 \, b_2 \, b_3 \ldots b_n \ldots$$

which will belong to A, in the following way: choose b_1 so $b_1 \neq a_{11}$ and $b_1 \neq 0$, choose b_2 so $b_2 \neq a_{22}$ and $b_2 \neq 0$, etc.

Observe that $y \neq x_1$ since $b_1 \neq a_{11}$, $y_2 \neq x_2$ since $b_2 \neq a_{22}$, etc., that is, $y \neq x_n$, for $n \in \mathbf{N}$. Hence $y \notin A$, which is impossible. Thus the assumption that A is denumerable has led to a contradiction. Consequently, A is non-denumerable.

Method 2. Assume the contrary. Then, as above,

$$A = \{x_1, x_2, x_3, \ldots\}$$

Now construct a sequence of closed intervals as follows: Consider the following three closed sub-intervals of $A = [0, 1]$,

$$[0, \tfrac{1}{3}], \quad [\tfrac{1}{3}, \tfrac{2}{3}], \quad [\tfrac{2}{3}, 1] \tag{1}$$

each having length $\tfrac{1}{3}$. Now x_1 cannot belong to all three intervals. Let $I_1 = [a_1, b_1]$ be one of the intervals in (1) such that $x_1 \notin I_1$.

Now consider the following three closed sub-intervals of $I_1 = [a_1, b_1]$,

$$[a_1, a_1 + \tfrac{1}{9}], \quad [a_1 + \tfrac{1}{9}, a_1 + \tfrac{2}{9}], \quad [a_1 + \tfrac{2}{9}, b_1] \tag{2}$$

each having length $\tfrac{1}{9}$. Similarly, let I_2 be one of the intervals in (2) such that $x_2 \notin I_2$.

By continuing in this manner, we obtain a sequence of closed intervals

$$I_1 \supset I_2 \supset I_3 \supset \cdots \tag{3}$$

such that $x_n \notin I_n$ for all $n \in N$. By the *Nested Interval Property* (see Appendix A) of the real numbers, there exists a real number $y \in A = [0, 1]$ such that y belongs to every interval in (3). But

$$y \in A = \{x_1, x_2, \ldots\} \quad \text{implies} \quad y = x_{m_0} \quad \text{for some} \quad m_0 \in \mathbf{N}$$

Then by our construction $y = x_{m_0} \notin I_{m_0}$, which contradicts the fact that y belongs to every interval in (3). Thus our assumption that A is denumerable has led to a contradiction. In other words, A is non-denumerable.

15. Prove Theorem (Schroeder-Bernstein) 3.8: Let $X \supset Y \supset X_1$ and let $X \sim X_1$; then $X \sim Y$.

Solution:

Since $X \sim X_1$, there exists a function $f : X \to X_1$ which is one-one and onto. But $X \supset Y$; hence the restriction of f to Y, which we shall also denote by f, is also one-one. So Y is equivalent to a subset of X_1, i.e. $Y \sim Y_1$ where

$$X \supset Y \supset X_1 \supset Y_1$$

and $f : Y \to Y_1$ is one-one and onto. But now $Y \supset X_1$; hence, for similar reasons, $X \sim X_2$ where

$$X \supset Y \supset X_1 \supset Y_1 \supset X_2$$

and $f : X_1 \to X_2$ is one-one and onto. Consequently, there exist equivalent sets $X_1, X_2, X_3, \ldots$ and equivalent sets $Y_1, Y_2, Y_3, \ldots$ such that

$$X \supset Y \supset X_1 \supset Y_1 \supset X_2 \supset Y_2 \supset \cdots$$

Let

$$B = X \cap Y \cap X_1 \cap Y_1 \cap X_2 \cap Y_2 \cap \cdots$$

Then

$$X = (X \setminus Y) \cup (Y \setminus X_1) \cup (X_1 \setminus Y_1) \cup \cdots \cup B$$

$$Y = (Y \setminus X_1) \cup (X_1 \setminus Y_1) \cup (Y_1 \setminus X_2) \cup \cdots \cup B$$

Note further that

$$(X \setminus Y) \sim (X_1 \setminus Y_1) \sim (X_2 \setminus Y_2) \sim \cdots$$

Specifically, the function $f : (X_n \setminus Y_n) \to (X_{n+1} \setminus Y_{n+1})$ is one-one and onto.

Consider the function $g : X \to Y$ defined by the following diagram:

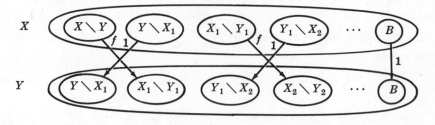

In other words,

$$g(x) = \begin{cases} f(x) & \text{if } x \in X_i \setminus Y_i \text{ or } x \in X \setminus Y \\ x & \text{if } x \in Y_i \setminus X_i \text{ or } x \in B \end{cases}$$

Then g is one-one and onto. Therefore $X \sim Y$.

16. Prove Theorem (Cantor) 3.10: The power set $\mathcal{P}(A)$ of any arbitrary set A has cardinality greater than A, i.e. $A \prec \mathcal{P}(A)$ and hence $\#(A) < \#(\mathcal{P}(A))$.

Solution:

The function $g : A \to \mathcal{P}(A)$ which sends each element $a \in A$ into the singleton set $\{a\}$, i.e. $g(a) = \{a\}$, is one-one; hence $A \precsim \mathcal{P}(A)$.

If we show that A is not equivalent to $\mathcal{P}(A)$, then the theorem will follow. Suppose the contrary, i.e. let there exist a function $f : A \to \mathcal{P}(A)$ which is one-one and onto. Call $a \in A$ a "bad" element if a is not a member of the set which is its image, i.e. if $a \notin f(a)$. Let B be the set of "bad" elements, i.e.,

$$B = \{x : x \in A, x \notin f(x)\}$$

Observe that B is a subset of A, that is, $B \in \mathcal{P}(A)$. Since $f : A \to \mathcal{P}(A)$ is onto, there exists an element $b \in A$ with the property that $f(b) = B$. Question: Is b "bad" or "good"? If $b \in B$ then, by definition of B, $b \notin f(b) = B$ which is a contradiction. Likewise, if $b \notin B$, then $b \in f(b) = B$ which is also a contradiction. Thus the original assumption, that $A \sim \mathcal{P}(A)$, has led to a contradiction. Accordingly $A \sim \mathcal{P}(A)$ is false, and so the theorem is true.

ORDERED SETS AND SUBSETS

17. Let **N**, the positive integers, be ordered as follows: each pair of elements $a, a' \in \mathbf{N}$ can be written uniquely in the form

$$a = 2^r(2s+1), \quad a' = a^r(as'+1)$$

where $r, r', s, s' \in \{0, 1, 2, 3, \ldots\}$. Let

$$a \prec a' \text{ if } r < r' \text{ or if } r = r' \text{ but } s < s'$$

Insert the correct symbol, $<$ or $>$, between each of the following pairs of numbers. (Here $x \succ y$ iff $y \prec x$.)

(i) 5___14,   (ii) 6___9,   (iii) 3___20,   (iv) 14___21

Solution:

The elements in **N** can be written as follows:

r \ s	0	1	2	3	4	5	6	7	
0	1	3	5	7	9	11	13	15	...
1	2	6	10	14	18	22	26	30	...
2	4	12	20	28	36	44	52	60	...
.	.	.	.	.	.	.	.	.	
.	.	.	.	.	.	.	.	.	
.	.	.	.	.	.	.	.	.	

Then a number in a higher row precedes a number in a lower row and, if two numbers are in the same row, the number to the left precedes the number to the right. Accordingly,

(i) $5 \prec 14$,   (ii) $6 \succ 9$,   (iii) $3 \prec 20$,   (iv) $14 \succ 21$

18. Let $A = \{a, b, c\}$ be ordered as in the diagram on the right. Let $\mathcal{A}$ be the collection of all non-empty totally ordered subsets of A, and let $\mathcal{A}$ be partially ordered by set inclusion. Construct a diagram of the order of $\mathcal{A}$.

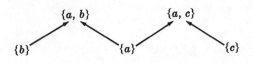

Solution:

　　The totally ordered subsets of A are: $\{a\}, \{b\}, \{c\}, \{a, b\}, \{a, c\}$. Since $\mathcal{A}$ is ordered by set inclusion, the order of $\mathcal{A}$ is the following:

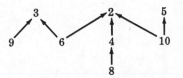

19. Let $A = \{2, 3, 4, \ldots\} = \mathbf{N} \setminus \{1\}$, and let A be ordered by "x divides y". (i) Determine the minimal elements of A. (ii) Determine the maximal elements of A.

Solution:

(i)　If $p \in A$ is a prime number, then only p divides p (since $1 \notin A$); hence all prime numbers are minimal elements. Furthermore, if $a \in A$ is not prime, then there is a number $b \in A$ such that b divides a, i.e. $b \prec a$; hence a is not minimal. In other words, the minimal elements are precisely the prime numbers.

(ii)　There are no maximal elements since, for every $a \in A$, a divides $2a$, for example.

20. Let $B = \{2, 3, 4, 5, 6, 8, 9, 10\}$ be ordered by "x is a multiple of y". (i) Find all maximal elements of B. (ii) Find all minimal elements of B.

Solution:

　　Construct a diagram of the order of B as follows:

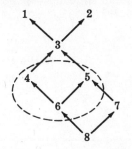

(i) The maximal elements are 2, 3 and 5. (ii) The minimal elements are 6, 8, 9 and 10.

21. Let $W = \{1, 2, \ldots, 7, 8\}$ be ordered as follows:

Consider the subset $V = \{4, 5, 6\}$ of W. (i) Find the set of upper bounds of V. (ii) Find the set of lower bounds of V. (iii) Does $\sup(V)$ exist? (iv) Does $\inf(V)$ exist?

Solution:

(i)　Each of the elements in $\{1, 2, 3\}$, and only these elements, follows every element in V and hence is an upper bound.

(ii)　Only 6 and 8 precede every element in V; hence $\{6, 8\}$ is the set of lower bounds.

(iii)　Since 3 is a first element in the set of upper bounds of V, $\sup(V) = 3$. Note that $3 \notin V$.

(iv)　Since 6 is a last element in the set of lower bounds of V, $\inf(V) = 6$. Note that $6 \in V$.

22. Let $\mathscr{A}$ be a collection of sets partially ordered by set inclusion, and let $\mathscr{B}$ be a sub-collection of $\mathscr{A}$. (i) Prove that if $A \in \mathscr{A}$ is an upper bound of $\mathscr{B}$, then $\bigcup \{B : B \in \mathscr{B}\} \subset A$. (ii) Is $\bigcup \{B : B \in \mathscr{B}\}$ an upper bound of $\mathscr{B}$?

Solution:

(i)   Let $x \in \bigcup \{B : B \in \mathscr{B}\}$; then $\exists B_0 \in \mathscr{B}$ s.t. $x \in B_0$. But A is an upper bound of $\mathscr{B}$; so $B_0 \subset A$ and hence $x \in A$. Accordingly, $\bigcup \{B : B \in \mathscr{B}\} \subset A$.

(ii)   Even though $\mathscr{B}$ is a subcollection of $\mathscr{A}$, it need not be true that the union of members of $\mathscr{B}$, i.e. $\bigcup \{B : B \in \mathscr{B}\}$, is a member of $\mathscr{A}$. In other words, $\bigcup \{B : B \in \mathscr{B}\}$ is an upper bound of $\mathscr{B}$ if and only if it belongs to $\mathscr{A}$.

APPLICATIONS OF ZORN'S LEMMA

23. Prove: Let X be a partially ordered set. Then there exists a totally ordered subset of X which is not a proper subset of any other totally ordered subset of X.

Solution:

Let $\mathscr{A}$ be the class of all totally ordered subsets of X. Let $\mathscr{A}$ be partially ordered by set inclusion. We want to show, by Zorn's Lemma, that $\mathscr{A}$ possesses a maximal element. So suppose $\mathscr{B} = \{B_i : i \in I\}$ is a totally ordered subclass of $\mathscr{A}$. Let $A = \bigcup \{B_i : i \in I\}$.

Observe that $\qquad\qquad B_i \subset X$ for all $B_i \in \mathscr{B} \quad$ implies $\quad A \subset X$

We next show that A is totally ordered. Let $a, b \in A$; then

$$\exists B_j, \ B_k \in \mathscr{B} \quad \text{such that} \quad a \in B_j, \ b \in B_k$$

But $\mathscr{B}$ is totally ordered by set inclusion; hence one of them, say B_j, is a subset of the other. Consequently, $a, b \in B_k$. Recall that $B_k \in \mathscr{B}$ is a totally ordered subset of X; so either $a \precsim b$ or $b \precsim a$. Then A is a totally ordered subset of X, and so $A \in \mathscr{A}$.

But $B_i \subset A$ for all $B_i \in \mathscr{B}$; hence A is an upper bound of $\mathscr{B}$. Since every totally ordered subset of $\mathscr{A}$ has an upper bound in $\mathscr{A}$, by Zorn's Lemma, $\mathscr{A}$ has a maximal element, i.e. a totally ordered subset of X which is not a proper subset of any other totally ordered subset of X.

24. Prove: Let R be a relation from A to B, i.e. $R \subset A \times B$, and suppose the domain of R is A. Then there exists a subset f^* of R such that f^* is a function from A into B.

Solution:

Let $\mathscr{A}$ be the class of subsets of R such that each $f \in \mathscr{A}$ is a function from a subset of A into B. Partially order $\mathscr{A}$ by set inclusion. Observe that if $f : A_1 \to B$ is a subset of $g : A_2 \to B$ then $A_1 \subset A_2$.

Now suppose $\mathscr{B} = \{f_i : A_i \to B\}_{i \in I}$ is a totally ordered subset of $\mathscr{A}$. Then (see Problem 44) $f = \cup_i f_i$ is a function from $\cup_i A_i$ into B. Furthermore, $f \subset R$. Hence f is an upper bound of $\mathscr{B}$. By Zorn's Lemma, $\mathscr{A}$ possesses a maximal element $f^* : A^* \to B$. If we show that $A^* = A$, then the theorem is proven.

Suppose $A^* \neq A$. Then $\exists a \in A$ s.t. $a \notin A^*$. By hypothesis, the domain of R is A; hence there exists an ordered pair $\langle a, b \rangle \in R$. Then $f^* \cup \{\langle a, b \rangle\}$ is a function from $A^* \cup \{a\}$ into B. But this contradicts the fact that f^* is a maximal element in $\mathscr{A}$. So $A^* = A$, and the theorem is proven.

Supplementary Problems

EQUIVALENT SETS, CARDINALITY

25. Prove: Every infinite set is equivalent to a proper subset of itself.

26. Prove: If A and B are denumerable, then $A \times B$ is denumerable.

27. Prove: The set of points in the plane R^2 with rational coordinates is denumerable.

28. A real number x is called *transcendental* if x is not algebraic, i.e. if x is not a solution to a polynomial equation
$$p(x) = a_0 + a_1 x + \cdots + a_m x^m = 0$$
with integral coefficients (see Problem 5). For example, π and e are transcendental numbers.
- (i)   Prove that the set T of transcendental numbers is non-denumerable.
- (ii) Prove that T has the power of the continuum, i.e. has cardinality $\mathbf{c}$.

29. An operation of multiplication is defined for cardinal numbers as follows:
$$\#(A)\,\#(B) = \#(A \times B)$$

- (i)   Show that the operation is well-defined, i.e.,
$$\#(A) = \#(A') \text{ and } \#(B) = \#(B') \quad \text{implies} \quad \#(A)\,\#(B) = \#(A')\,\#(B')$$
or, equivalently, $A \sim A'$ and $B \sim B'$ implies $(A \times B) \sim (A' \times B')$

- (ii) Prove: (a) $\aleph_0 \aleph_0 = \aleph_0$,   (b) $\aleph_0 \mathbf{c} = \mathbf{c}$,   (c) $\mathbf{c}\,\mathbf{c} = \mathbf{c}$.

30. An operation of addition is defined for cardinal numbers as follows:
$$\#(A) + \#(B) = \#(A \times \{1\} \cup B \times \{2\})$$
- (i)   Show that if $A \cap B = \emptyset$, then $\#(A) + \#(B) = \#(A \cup B)$.
- (ii) Show that the operation is well-defined, i.e.,
$$\#(A) = \#(A') \text{ and } \#(B) = \#(B') \quad \text{implies} \quad \#(A) + \#(B) = \#(A') + \#(B')$$

31. An operation of powers is defined for cardinal numbers as follows:
$$\#(A)^{\#(B)} = \#(\{f \,:\, f : B \to A\})$$
- (i)   Show that if $\#(A) = m$ and $\#(B) = n$ are finite cardinals, then
$$\#(A)^{\#(B)} = m^n$$
i.e. the operation of powers for cardinals corresponds, in the case of finite cardinals, to the usual operation of powers of positive integers.
- (ii) Show that the operation is well-defined, i.e.,
$$\#(A) = \#(A') \text{ and } \#(B) = \#(B') \quad \text{implies} \quad \#(A)^{\#(B)} = \#(A')^{\#(B')}$$
- (iii) Prove: For any set A, $\#(\mathcal{P}(A)) = 2^{\#(A)}$.

32. Let $\sim$ be the equivalence relation in $\mathbf{R}$ defined by $x \sim y$ iff $x - y$ is rational. Determine the cardinality of the quotient set $\mathbf{R}/\!\sim$.

33. Prove: The cardinal number of the class of all functions from $[0, 1]$ into $\mathbf{R}$ is $2^{\mathbf{c}}$.

34. Prove that the following two statements of the Schroeder-Bernstein Theorem 3.8 are equivalent:
- (i)   If $A \precsim B$ and $B \precsim A$, then $A \sim B$.
- (ii) If $X \supset Y \supset X_1$ and $X \sim X_1$, then $X \sim Y$.

35. Prove Theorem 3.9: Given any pair of sets A and B, either $A \prec B$, $A \sim B$ or $B \prec A$.
(*Hint.* Use Zorn's Lemma.)

ORDERED SETS AND SUBSETS

36. Let $A = (\mathbf{N}, \leqq)$, the positive integers with the natural order; and let $B = (\mathbf{N}, \geqq)$, the positive integers with the inverse order. Furthermore, let $A \times B$ denote the lexicographical ordering of $\mathbf{N} \times \mathbf{N}$ according to the order of A and then B. Insert the correct symbol, $<$ or $>$, between each pair of elements of $\mathbf{N} \times \mathbf{N}$.

- (i) $\langle 3, 8 \rangle$____$\langle 1, 1 \rangle$,   (ii) $\langle 2, 1 \rangle$____$\langle 2, 8 \rangle$,   (iii) $\langle 3, 3 \rangle$____$\langle 3, 1 \rangle$,   (iv) $\langle 4, 9 \rangle$____$\langle 7, 15 \rangle$.

37. Let $X = \{1, 2, 3, 4, 5, 6\}$ be ordered as in the adjacent diagram. Consider the subset $A = \{2, 3, 4\}$ of X. (i) Find the maximal elements of X. (ii) Find the minimal elements of X. (iii) Does X have a first element? (iv) Does X have a last element? (v) Find the set of upper bounds of A. (vi) Find the set of lower bounds of A. (vii) Does sup (A) exist? (viii) Does inf (A) exist?

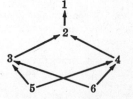

38. Consider Q, the set of rational numbers, with the natural order, and its subset $A = \{x : x \in Q, x^3 < 3\}$. (i) Is A bounded above? (ii) Is A bounded below? (iii) Does sup (A) exist? (iv) Does inf (A) exist?

39. Let $\mathbf{N}$, the positive integers, be ordered by "x divides y", and let $A \subset \mathbf{N}$. (i) Does inf (A) exist? (ii) Does sup (A) exist?

40. Prove: Every finite partially ordered set has a maximal element.

41. Give an example of an ordered set which has exactly one maximal element but does not have a last element.

42. Prove: If R is a partial order on A, then R^{-1} is also a partial order on A.

ZORN'S LEMMA

43. Consider the proof of the following statement: There exists a finite set of positive integers which is not a proper subset of any other finite set of positive integers.

    Proof. Let $\mathcal{A}$ be the class of all finite sets of positive integers. Partially order $\mathcal{A}$ by set inclusion. Now let $\mathcal{B} = \{B_i : i \in I\}$ be a totally ordered subclass of $\mathcal{A}$. Consider the set $A = \cup_i B_i$. Observe that $B_i \subset A$ for every $B_i \in \mathcal{B}$; hence A is an upper bound of $\mathcal{B}$.

    Since every totally ordered subset of $\mathcal{A}$ has an upper bound, by Zorn's Lemma, $\mathcal{A}$ has a maximal element, a finite set which is not a proper subset of another finite set.

    Question: Since the statement is clearly false, which step in the proof is incorrect?

44. Prove the following fact which was assumed in the proof in Problem 24: Let $\{f_i : A_i \to B\}$ be a class of functions which is totally ordered by set inclusion. Then $\cup_i f_i$ is a function from $\cup_i A_i$ into B.

45. Prove that the following two statements are equivalent:

    (i)   (Axiom of Choice.) The product $\prod \{A_i : i \in I\}$ of a non-empty class of non-empty sets is non-empty.

    (ii) If $\mathcal{A}$ is a non-empty class of non-empty disjoint sets, then there exists a subset $B \subset \mathbf{U}\{A : A \in \mathcal{A}\}$ such that the intersection of B and each set $A \in \mathcal{A}$ consists of exactly one element.

46. Prove: If every totally ordered subset of an ordered set X has a lower bound in X, then X has a minimal element.

Answers to Supplementary Problems

32. *c*

36. (i) $\succ$, (ii) $\succ$, (iii) $\prec$, (iv) $\prec$

37. (i) $\{1\}$; (ii) $\{5, 6\}$; (iii) No; (iv) Yes, 1; (v) $\{1, 2\}$; (vi) $\{5, 6\}$; (vii) Yes, 2; (viii) No

38. (i) Yes, (ii) No, (iii) No, (iv) No

39. (i) inf (A) exists iff $A \neq \emptyset$. (ii) sup (A) exists iff A is finite.

41. *a*

$$\uparrow$$
$$1 \longrightarrow 2 \longrightarrow 3 \longrightarrow 4 \longrightarrow \cdots$$

Here a is maximal but a is not a last element.

Chapter 4

Topology of the Line and Plane

REAL LINE

The set of *real numbers*, denoted by **R**, plays a dominant role in mathematics and, in particular, in analysis. In fact, many concepts in topology are abstractions of properties of sets of real numbers. The set **R** can be characterized by the statement that **R** is a *complete, Archimedean ordered field*. These notions are explained in the Appendix. Here we use the order relation in **R** to define the "usual topology" for **R**.

We assume the reader is familiar with the geometric representation of **R** by means of the points on a straight line. As in Fig. 4-1, a point, called the origin, is chosen to represent 0 and another point, usually to the right of 0, to represent 1. Then there is a natural way to pair off the points on the line and the real numbers, i.e. each point will represent a unique real number and each real number will be represented by a unique point. For this reason we refer to the line as the *real line* or *real axis*. Furthermore, we will use the words point and number interchangeably.

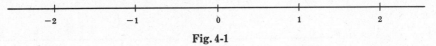

Fig. 4-1

OPEN SETS IN R

Let A be a set of real numbers. A point $p \in A$ is an *interior point* of A iff p belongs to some open interval S_p which is contained in A:

$$p \in S_p \subset A$$

The set A is *open* (or $\mathcal{U}$-open) iff each of its points is an interior point. (The significance of $\mathcal{U}$ in $\mathcal{U}$-open will appear in the next chapter.)

> **Example 1.1:** An open interval $A = (a, b)$ is an open set, for we may choose $S_p = A$ for each $p \in A$.
>
> **Example 1.2:** The real line **R**, itself, is open since any open interval S_p must be a subset of **R**, i.e. $p \in S_p \subset \mathbf{R}$.

Observe that a set is not open iff there exists a point in the set that is not an interior point.

> **Example 1.3:** The closed interval $B = [a, b]$ is not an open set, for any open interval containing a or b must contain points outside of B. Hence the end points a and b are not interior points of B.
>
> **Example 1.4:** The empty set $\emptyset$ is open since there is no point in $\emptyset$ which is not an interior point.
>
> **Example 1.5:** The infinite open intervals, i.e. the subsets of **R** defined and denoted by
> $$\{x : x \in \mathbf{R}, \ x > a\} = (a, \infty), \quad \{x : x \in \mathbf{R}, \ x < a\} = (-\infty, a),$$
> $$\{x : x \in \mathbf{R}\} = \mathbf{R} = (-\infty, \infty)$$
> are open sets. On the other hand, the infinite closed intervals, i.e. the subsets of **R** defined and denoted by
> $$\{x : x \in \mathbf{R}, \ x \geqq a\} = [a, \infty), \quad \{x : x \in \mathbf{R}, \ x \leqq a\} = (-\infty, a]$$
> are not open sets, since $a \in \mathbf{R}$ is not an interior point of either $[a, \infty)$ or $(-\infty, a]$.

We state two fundamental theorems about open sets.

Theorem 4.1: The union of any number of open sets in **R** is open.

Theorem 4.2: The intersection of any finite number of open sets in **R** is open.

The next example shows that the finiteness condition in the preceding theorem cannot be removed.

> **Example 1.6:** Consider the class of open intervals and, hence, open sets
> $$\{A_n = (-1/n, 1/n) : n \in \mathbf{N}\}, \quad \text{i.e.} \quad \{(-1,1), (-\tfrac{1}{2}, \tfrac{1}{2}), (-\tfrac{1}{3}, \tfrac{1}{3}), \ldots\}$$
> Observe that the intersection
> $$\cap_{n=1}^{\infty} A_n = \{0\}$$
> of the open intervals consists of the single point 0 which is not an open set. In other words, an arbitrary intersection of open sets need not be open.

ACCUMULATION POINTS

Let A be a subset of **R**, i.e. a set of real numbers. A point $p \in \mathbf{R}$ is an *accumulation point* or *limit point* of A iff every open set G containing p contains a point of A different from p; i.e.,
$$G \text{ open}, \ p \in G \quad \text{implies} \quad A \cap (G \setminus \{p\}) \neq \emptyset$$
The set of accumulation points of A, denoted by A', is called the *derived set* of A.

> **Example 2.1:** Let $A = \{1, \tfrac{1}{2}, \tfrac{1}{3}, \tfrac{1}{4}, \ldots\}$. The point 0 is an accumulation point of A since any open set G with $0 \in G$ contains an open interval $(-a_1, a_2) \subset G$ with $-a_1 < 0 < a_2$ which contains points of A.
>
>
> Observe that the limit point 0 of A does not belong to A. Observe also that A does not contain any other limit points; hence the derived set of A is the singleton set $\{0\}$, i.e. $A' = \{0\}$.

> **Example 2.2:** Consider the set **Q** of rational numbers. Every real number $p \in \mathbf{R}$ is a limit point of **Q** since every open set contains rational numbers, i.e. points of **Q**.

> **Example 2.3:** The set of integers $\mathbf{Z} = \{\ldots, -2, -1, 0, 1, 2, \ldots\}$ does not have any points of accumulation. In other words, the derived set of **Z** is the empty set $\emptyset$.

Remark: The reader should not confuse the concept "limit point of a set" with the different, though related, concept "limit of a sequence". Some of the solved and supplementary problems will show the relationship between these two concepts.

BOLZANO-WEIERSTRASS THEOREM

The existence or non-existence of accumulation points for various sets is an important question in topology. Not every set, even if it is infinite as in Example 2.3, has a limit point. There does exist, however, an important general case which gives a positive answer.

Theorem (Bolzano-Weierstrass) 4.3: Let A be a bounded, infinite set of real numbers. Then A has at least one accumulation point.

CLOSED SETS

A subset A of **R**, i.e. a set of real numbers, is a *closed set* iff its complement A^c is an open set. A closed set can also be described in terms of its accumulation points.

Theorem 4.4: A subset A of **R** is closed if and only if A contains each of its points of accumulation.

Example 3.1: The closed interval $[a, b]$ is a closed set since its complement $(-\infty, a) \cup (b, \infty)$, the union of two open infinite intervals, is open.

Example 3.2: The set $A = \{1, \frac{1}{2}, \frac{1}{3}, \frac{1}{4}, \ldots\}$ is not closed since, as seen in Example 2.1, 0 is a limit point of A but does not belong to A.

Example 3.3: The empty set $\emptyset$ and the entire line **R** are closed sets since their complements **R** and $\emptyset$, respectively, are open sets.

Sets may be neither open nor closed as seen in the next example.

Example 3.4: Consider the open-closed interval $A = (a, b]$. Note that A is not open since $b \in A$ is not an interior point of A, and is not closed since $a \notin A$ but is a limit point of A.

HEINE-BOREL THEOREM

One of the most important properties of a closed and bounded interval is given in the next theorem. Here a class of sets, $\mathcal{A} = \{A_i\}$, is said to *cover* a set A if A is contained in the union of the members of $\mathcal{A}$, i.e. $A \subset \cup_i A_i$.

Theorem (Heine-Borel) 4.5: Let $A = [c, d]$ be a closed and bounded interval, and let $\mathcal{G} = \{G_i : i \in I\}$ be a class of open intervals which covers A, i.e. $A \subset \cup_i G_i$. Then $\mathcal{G}$ contains a finite subclass, say $\{G_{i_1}, \cdots, G_{i_m}\}$, which also covers A, i.e.,

$$A \subset G_{i_1} \cup G_{i_2} \cup \ldots \cup G_{i_m}$$

Both conditions, closed and bounded, must be satisfied by A or else the theorem is not true. We show this by the next two examples.

Example 4.1: Consider the open, bounded unit interval $A = (0, 1)$. Observe that the class

$$\mathcal{G} = \left\{ G_n = \left(\frac{1}{n+2}, \frac{1}{n} \right) : n \in \mathbf{N} \right\}$$

of open intervals covers A, i.e.,

$$A \subset (\tfrac{1}{3}, 1) \cup (\tfrac{1}{4}, \tfrac{1}{2}) \cup (\tfrac{1}{5}, \tfrac{1}{3}) \cup \cdots$$

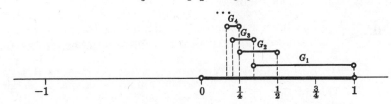

But the union of no finite subclass of $\mathcal{G}$ contains A.

Example 4.2: Consider the closed infinite interval $A = [1, \infty)$. The class

$$\mathcal{G} = \{(0, 2), (1, 3), (2, 4), \ldots\}$$

of open intervals covers A, but no finite subclass does.

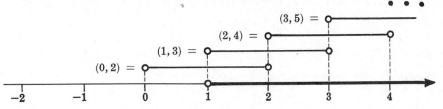

SEQUENCES

A *sequence*, denoted by

$$\langle s_1, s_2, \ldots \rangle, \quad \langle s_n : n \in \mathbf{N} \rangle \quad \text{or} \quad \langle s_n \rangle$$

is a function whose domain is $\mathbf{N} = \{1, 2, 3, \ldots\}$, i.e. a sequence assigns a point s_n to each positive integer $n \in \mathbf{N}$. The image s_n or $s(n)$ of $n \in \mathbf{N}$ is called the nth *term* of the sequence.

Example 5.1: The sequences

$$\langle s_n \rangle = \langle 1, 3, 5, \ldots \rangle, \quad \langle t_n \rangle = \langle -\tfrac{1}{2}, \tfrac{1}{4}, -\tfrac{1}{8}, \tfrac{1}{16}, \ldots \rangle, \quad \langle u_n \rangle = \langle 1, 0, 1, 0, \ldots \rangle$$

can be defined, respectively, by the formulas

$$s(n) = 2n - 1, \quad t(n) = (-1)^n/2^n, \quad u(n) = \tfrac{1}{2}(1 + (-1)^{n+1}) = \begin{cases} 1 & \text{if } n \text{ is odd} \\ 0 & \text{if } n \text{ is even} \end{cases}$$

A sequence $\langle s_n : n \in \mathbf{N} \rangle$ is said to be bounded if its range $\{s_n : n \in \mathbf{N}\}$ is a bounded set.

Example 5.2: Consider the three sequences in Example 5.1. The range of $\langle s_n \rangle$ is $\{1, 3, 5, \ldots\}$; so $\langle s_n \rangle$ is not a bounded sequence. The range of $\langle t_n \rangle$ is $\{-\tfrac{1}{2}, \tfrac{1}{4}, -\tfrac{1}{8}, \ldots\}$ which is bounded; hence $\langle t_n \rangle$ is a bounded sequence. The range of $\langle u_n \rangle$ is the finite set $\{0, 1\}$; so $\langle u_n \rangle$ is also a bounded sequence.

Observe that $\langle s_n : n \in \mathbf{N} \rangle$ denotes a sequence and is a function. On the other hand, $\{s_n : n \in \mathbf{N}\}$ denotes the range of the sequence and is a set.

CONVERGENT SEQUENCES

The usual definition of a convergent sequence is stated as follows:

Definition: The sequence $\langle a_1, a_2, \ldots \rangle$ of real numbers converges to $b \in R$ or, equivalently, the real number b is the limit of the sequence $\langle a_n : n \in \mathbf{N} \rangle$, denoted by

$$\lim_{n \to \infty} a_n = b, \quad \lim a_n = b \quad \text{or} \quad a_n \to b$$

if for every $\epsilon > 0$ there exists a positive integer n_0 such that

$$n > n_0 \quad \text{implies} \quad |a_n - b| < \epsilon$$

Observe that $|a_n - b| < \epsilon$ means that $b - \epsilon < a_n < b + \epsilon$, or, equivalently, that a_n belongs to the open interval $(b - \epsilon, \ b + \epsilon)$ containing b. Furthermore, since each term after the n_0th lies inside the interval $(b - \epsilon, \ b + \epsilon)$, only the terms before a_{n_0}, and there are only a finite number of them, can lie outside the interval $(b - \epsilon, \ b + \epsilon)$. Hence we can restate the preceding definition as follows.

Definition: The sequence $\langle a_n : n \in \mathbf{N} \rangle$ converges to b if every open set containing b contains *almost all*, i.e. all but a finite number, of the terms of the sequence.

Example 6.1: A constant sequence $\langle a_0, a_0, a_0, \ldots \rangle$, such as $\langle 1, 1, 1, \ldots \rangle$ or $\langle -3, -3, -3, \ldots \rangle$, converges to a_0 since each open set containing a_0 contains every term of the sequence.

Example 6.2: Each of the sequences

$$\langle 1, \tfrac{1}{2}, \tfrac{1}{3}, \tfrac{1}{4}, \ldots \rangle, \quad \langle 1, 0, \tfrac{1}{2}, 0, \tfrac{1}{3}, 0, \tfrac{1}{4}, 0, \ldots \rangle, \quad \langle 1, -\tfrac{1}{2}, \tfrac{1}{3}, -\tfrac{1}{4}, \ldots \rangle$$

converges to 0 since any open interval containing 0 contains almost all of the terms of each of the sequences.

Example 6.3: Consider the sequence $\langle \tfrac{1}{2}, \tfrac{1}{4}, \tfrac{3}{4}, \tfrac{1}{8}, \tfrac{7}{8}, \tfrac{1}{16}, \tfrac{15}{16}, \ldots \rangle$, i.e. the sequence

$$a_n = \begin{cases} \dfrac{1}{2^{(n+2)/2}} & \text{if } n \text{ is even} \\ 1 - \dfrac{1}{2^{(n+1)/2}} & \text{if } n \text{ is odd} \end{cases}$$

The points are displayed below:

Observe that any open interval containing either 0 or 1 contains an infinite number of the terms of the sequence. Neither 0 nor 1, however, is a limit of the sequence. Observe, though, that 0 and 1 are accumulation points of the *range* of the sequence, that is, of the *set* $\{\tfrac{1}{2}, \tfrac{1}{4}, \tfrac{3}{4}, \tfrac{1}{8}, \tfrac{7}{8}, \ldots\}$.

SUBSEQUENCES

Consider a sequence $\langle a_1, a_2, a_3, \ldots \rangle$. If $\langle i_n \rangle$ is a sequence of positive integers such that $i_1 < i_2 < \cdots$, then

$$\langle a_{i_1}, a_{i_2}, a_{i_3}, \ldots \rangle$$

is called a *subsequence* of $\langle a_n : n \in \mathbf{N} \rangle$.

> **Example 7.1:** Consider the sequence $\langle a_n \rangle = \langle 1, \frac{1}{2}, \frac{1}{3}, \frac{1}{4}, \ldots \rangle$. Observe that $\langle 1, \frac{1}{2}, \frac{1}{4}, \frac{1}{8}, \ldots \rangle$ is a subsequence of $\langle a_n \rangle$, but that $\langle \frac{1}{2}, 1, \frac{1}{4}, \frac{1}{3}, \frac{1}{6}, \frac{1}{5}, \ldots \rangle$ is not a subsequence of $\langle a_n \rangle$ since 1 appears before $\frac{1}{2}$ in the original sequence.

> **Example 7.2:** Although the sequence $\langle \frac{1}{2}, \frac{1}{4}, \frac{3}{4}, \frac{1}{8}, \frac{7}{8}, \ldots \rangle$ of Example 6.3 does not converge, it does have convergent subsequences such as $\langle \frac{1}{2}, \frac{1}{4}, \frac{1}{8}, \frac{1}{16}, \ldots \rangle$ and $\langle \frac{1}{2}, \frac{3}{4}, \frac{7}{8}, \frac{15}{16}, \ldots \rangle$. On the other hand, the sequence $\langle 1, 3, 5, \ldots \rangle$ does not have any convergent subsequences.

As seen in the preceding example, sequences may or may not have convergent subsequences. There does exist a very important general case which gives a positive answer.

Theorem 4.6: Every bounded sequence of real numbers contains a convergent subsequence.

CAUCHY SEQUENCES

A sequence $\langle a_n : n \in \mathbf{N} \rangle$ of real numbers is a *Cauchy sequence* iff for every $\epsilon > 0$ there exists a positive integer n_0 such that

$$n, m > n_0 \quad \text{implies} \quad |a_n - a_m| < \epsilon$$

In other words, a sequence is a Cauchy sequence iff the terms of the sequence become arbitrarily close to each other as n gets large.

> **Example 8.1:** Let $\langle a_n : n \in \mathbf{N} \rangle$ be a Cauchy sequence of integers, i.e. each term of the sequence belongs to $\mathbf{Z} = \{\ldots, -1, 0, 1, \ldots\}$. Then the sequence must be of the form
>
> $$\langle a_1, a_2, \ldots, a_{n_0}, b, b, b, \ldots \rangle$$
>
> i.e. the sequence is constant after some n_0th term. For if we choose $\epsilon = \frac{1}{2}$, then
>
> $$a_n, a_m \in \mathbf{Z} \text{ and } |a_n - a_m| < \tfrac{1}{2} \quad \text{implies} \quad a_n = a_m$$

> **Example 8.2:** We show that every convergent sequence is a Cauchy sequence. Let $a_n \to b$ and let $\epsilon > 0$. Then there exists $n_0 \in \mathbf{N}$ sufficiently large such that
>
> $$n > n_0 \text{ implies } |a_n - b| < \tfrac{1}{2}\epsilon \quad \text{and} \quad m > n_0 \text{ implies } |a_m - b| < \tfrac{1}{2}\epsilon$$
>
> Consequently, $n, m > n_0$ implies
>
> $$|a_n - a_m| = |a_n - b + b - a_m| \leq |a_n - b| + |b - a_m| < \tfrac{1}{2}\epsilon + \tfrac{1}{2}\epsilon = \epsilon$$
>
> Hence $\langle a_n \rangle$ is a Cauchy sequence.

COMPLETENESS

A set A of real numbers is said to be *complete* if every Cauchy sequence $\langle a_n \in A : n \in \mathbf{N} \rangle$ of points in A converges to a point in A.

> **Example 9.1:** The set $\mathbf{Z} = \{\ldots, -2, -1, 0, 1, 2, \ldots\}$ of integers is complete. For, as seen in Example 8.1, a Cauchy sequence $\langle a_n : n \in \mathbf{N} \rangle$ of points in $\mathbf{Z}$ is of the form
>
> $$\langle a_1, a_2, \ldots, a_{n_0}, b, b, b, \ldots \rangle$$
>
> which converges to the point $b \in \mathbf{Z}$.

> **Example 9.2:** The set $\mathbf{Q}$ of rational numbers is not complete. For we can choose a sequence of rational numbers, such as $\langle 1, 1.4, 1.41, 1.412, \ldots \rangle$ which converges to the real number $\sqrt{2}$, which is not rational, i.e. which does not belong to $\mathbf{Q}$.

A fundamental property of the entire set **R** of real numbers is that **R** is complete. Namely,

Theorem (Cauchy) 4.7: Every Cauchy sequence of real numbers converges to a real number.

CONTINUOUS FUNCTIONS

The usual $\epsilon - \delta$ definition of a continuous function is stated as follows:

| **Definition:** | A function $f : \mathbf{R} \to \mathbf{R}$ is *continuous* at a point x_0 if for every $\epsilon > 0$ there exists a $\delta > 0$ such that |

$$|x - x_0| < \delta \quad \text{implies} \quad |f(x) - f(x_0)| < \epsilon$$

The function f is a *continuous function* if it is continuous at every point.

Observe that $|x - x_0| < \delta$ means that $x_0 - \delta < x < x_0 + \delta$, or equivalently that x belongs to the open interval $(x_0 - \delta, x_0 + \delta)$. Similarly, $|f(x) - f(x_0)| < \epsilon$ means that $f(x)$ belongs to the open interval $(f(x_0) - \epsilon, f(x_0) + \epsilon)$. Accordingly, the statement

$$|x - x_0| < \delta \quad \text{implies} \quad |f(x) - f(x_0)| < \epsilon$$

is equivalent to the statement

$$x \in (x_0 - \delta, x_0 + \delta) \quad \text{implies} \quad f(x) \in (f(x_0) - \epsilon, f(x_0) + \epsilon)$$

which is equivalent to the statement

$$f[(x_0 - \delta, x_0 + \delta)] \quad \text{is contained in} \quad (f(x_0) - \epsilon, f(x_0) + \epsilon)$$

Hence we can restate the previous definition as follows.

| **Definition:** | A function $f : \mathbf{R} \to \mathbf{R}$ is continuous at a point $p \in \mathbf{R}$ if for any open set $V_{f(p)}$ containing $f(p)$ there exists an open set U_p containing p such that $f[U_p] \subset V_{f(p)}$. The function f is a continuous function if it is continuous at every point. |

The Venn diagram below may be helpful in visualizing this definition.

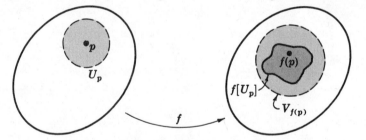

A continuous function can be completely characterized in terms of open sets as follows:

Theorem 4.8: A function is continuous if and only if the inverse image of every open set is open.

Observe that Theorem 4.8 also states that a function is not continuous iff there exists an open set whose inverse image is not open.

Example 10.1: Consider the function $f : \mathbf{R} \to \mathbf{R}$ defined by

$$f(x) = \begin{cases} x - 1 & \text{if } x \le 3 \\ \tfrac{1}{2}(x + 5) & \text{if } x > 3 \end{cases}$$

and illustrated in the adjacent diagram. Note that the inverse of the open interval $(1, 3)$ is the open-closed interval $(2, 3]$ which is not an open set. Hence the function f is not continuous.

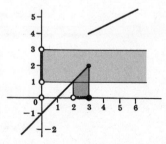

We now state one important property of continuous functions which we will refer to later in the text.

Theorem 4.9:   Let $f : \mathbf{R} \to \mathbf{R}$ be continuous on a closed interval $[a, b]$. Then the function assumes every value between $f(a)$ and $f(b)$.

In other words, if y_0 is a real number for which $f(a) \leqq y_0 \leqq f(b)$ or $f(b) \leqq y_0 \leqq f(a)$, then

$$\exists x_0 \in \mathbf{R} \quad \text{such that} \quad a \leqq x_0 \leqq b \text{ and } f(x_0) = y_0$$

This theorem is known as the *Weierstrass Intermediate Value Theorem*.

Remark: A function $f : \mathbf{R} \to \mathbf{R}$ is said to be continuous on a subset D of $\mathbf{R}$ if it is continuous at each point in D.

TOPOLOGY OF THE PLANE

An *open disc* D in the plane $\mathbf{R}^2$ is the set of points inside a circle, say, with center $p = \langle a_1, a_2 \rangle$ and radius $\delta > 0$, i.e.,

$$D \;=\; \{\langle x, y \rangle : (x - a_1)^2 + (y - a_2)^2 < \delta^2\} \;=\; \{q \in \mathbf{R}^2 : d(p, q) < \delta\}$$

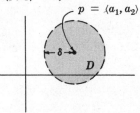

Here $d(p, q)$ denotes the usual distance between two points $p = \langle a_1, a_2 \rangle$ and $q = \langle b_1, b_2 \rangle$ in $\mathbf{R}^2$:

$$d(p, q) \;=\; \sqrt{(a_1 - b_1)^2 + (a_2 - b_2)^2}$$

The open disc plays a role in the topology of the plane $\mathbf{R}^2$ that is analogous to the role of the open interval in the topology of the line $\mathbf{R}$.

Let A be a subset of $\mathbf{R}^2$. A point $p \in A$ is an *interior point* of A iff p belongs to some open disc D_p which is contained in A:

$$p \in D_p \subset A$$

The set A is *open* (or *$\mathcal{U}$-open*) iff each of its points is an interior point.

Example 11.1:   Clearly an open disc, the entire plane $\mathbf{R}^2$ and the empty set $\emptyset$ are open subsets of $\mathbf{R}^2$. We now show that the intersection of any two open discs, say

$$D_1 \;=\; \{q \in \mathbf{R}^2 : d(p_1, q) < \delta_1\} \quad \text{and} \quad D_2 \;=\; \{q \in \mathbf{R}^2 : d(p_2, q) < \delta_2\}$$

is also an open set. For let $p_0 \in D_1 \cap D_2$ so

$$d(p_1, p_0) < \delta_1 \quad \text{and} \quad d(p_2, p_0) < \delta_2$$

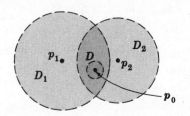

Set $r \;=\; \min\{\delta_1 - d(p_1, p_0),\; \delta_2 - d(p_2, p_0)\} > 0$

and let $D \;=\; \{q \in \mathbf{R}^2 : d(p_0, q) < \tfrac{1}{2}r\}$

Then $p_0 \in D \subset D_1 \cap D_2$ or, p_0 is an interior point of $D_1 \cap D_2$.

A point $p \in \mathbf{R}^2$ is an *accumulation point* or *limit point* of a subset A of $\mathbf{R}^2$ iff every open set G containing p contains a point of A different from p, i.e.,

$$G \subset \mathbf{R}^2 \text{ open}, \quad p \in G \quad \text{implies} \quad A \cap (G \setminus \{p\}) \neq \emptyset$$

Example 11.2:   Consider the following subset of $\mathbf{R}^2$:

$$A \;=\; \left\{ \langle x,y \rangle \,:\, y = \sin\frac{1}{x}, \; x > 0 \right\}$$

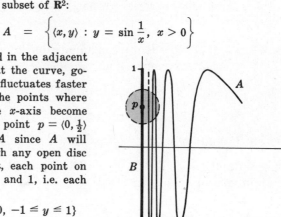

The set A is illustrated in the adjacent diagram. Observe that the curve, going from right to left, fluctuates faster and faster, i.e. that the points where the curve crosses the x-axis become closer and closer. The point $p = \langle 0, \frac{1}{2} \rangle$ is a limit point of A since A will eventually pass through any open disc containing p. In fact, each point on the y-axis between -1 and 1, i.e. each point in the set

$$B \;=\; \{ \langle x,y \rangle \,:\, x = 0, \; -1 \le y \le 1 \}$$

is a limit point of A.

A subset A of $\mathbf{R}^2$ is *closed* iff its complement A^c is an open subset of $\mathbf{R}^2$.

A sequence $\langle p_1, p_2, \ldots \rangle$ of points in $\mathbf{R}^2$ *converges* to the point $q \in \mathbf{R}^2$ iff every open set containing q contains almost all of the terms of the sequence. Convergence in the plane $\mathbf{R}^2$ can be characterized in terms of convergence in $\mathbf{R}$ as follows.

Proposition 4.10:   Consider the sequence $\langle p_1 = \langle a_1, b_1 \rangle,\; p_2 = \langle a_2, b_2 \rangle,\; \ldots \rangle$ of point in $\mathbf{R}^2$ and the point $q = \langle a, b \rangle \in \mathbf{R}^2$. Then

$$p_n \to q \quad \text{if and only if} \quad a_n \to a \text{ and } b_n \to b$$

A function $f : \mathbf{R}^2 \to \mathbf{R}^2$ is *continuous* at a point $p \in \mathbf{R}^2$ iff for any open set $V_{f(p)}$ containing $f(p)$ there exists an open set U_p containing p such that $f[U_p] \subset V_{f(p)}$.

We list theorems for the plane $\mathbf{R}^2$ which are analagous to theorems for the line $\mathbf{R}$ stated earlier in this chapter.

Theorem 4.1*:   The union of any number of open subsets of $\mathbf{R}^2$ is open.

Theorem 4.2*:   The intersection of any finite number of open subsets of $\mathbf{R}^2$ is open.

Theorem 4.4*:   A subset A of $\mathbf{R}^2$ is closed if and only if A contains each of its accumulation points.

Theorem 4.8*:   A function $f : \mathbf{R}^2 \to \mathbf{R}^2$ is continuous if and only if the inverse image of every open set is open.

Solved Problems

OPEN SETS, ACCUMULATION POINTS

1.   Determine the accumulation points of each set of real numbers:

(i) $\mathbf{N}$;   (ii) $(a, b]$;   (iii) $\mathbf{Q}^c$, the set of irrational points.

Solution:

(i) $\mathbf{N}$, the set of positive integers, does not have any limit points. For if a is any real number, we can find a $\delta > 0$ so small that the open set $(a - \delta, \, a + \delta)$ contains no point of $\mathbf{N}$ other than a.

(ii)   Every point p in the closed interval $[a, b]$ is a limit point of the open-closed interval $(a, b]$, since every open interval containing $p \in [a, b]$ will contain points of $(a, b]$ other than p.

(iii)   Every real number $p \in \mathbf{R}$ is a limit point of $\mathbf{Q}^c$ since every open interval containing $p \in \mathbf{R}$ will contain points of $\mathbf{Q}^c$, i.e. irrational numbers, other than p.

2. Recall that A' denotes the derived set, i.e. set of limit points, of a set A. Find sets A such that (i) A and A' are disjoint, (ii) A is a proper subset of A', (iii) A' is a proper subset of A, (iv) $A = A'$.

Solution:

(i)   The set $A = \{1, \frac{1}{2}, \frac{1}{3}, \ldots\}$ has 0 as its only point of accumulation. Hence $A' = \{0\}$ and A and A' are disjoint.

(ii)   Let $A = (a, b]$, an open-closed interval. As seen in the preceding problem $A' = [a, b]$, the closed interval, and so $A \subset A'$.

(iii) Let $A = \{0, 1, \frac{1}{2}, \frac{1}{3}, \ldots\}$. Then 0, which belongs to A, is the only limit point of A. Hence $A' = \{0\}$ and $A' \subset A$.

(iv)   Let $A = [a, b]$, a closed interval. Then each point in A is a limit point of A and they are the only limit points. So $A = A' = [a, b]$.

3. Prove Theorem 4.1*: The union of any number of open subsets of $\mathbf{R}^2$ is open.

Solution:

Let $\mathcal{A}$ be a class of open subsets of $\mathbf{R}^2$, let $H = \bigcup\{G : G \in \mathcal{A}\}$, and let $p \in H$. The theorem is proved if we show that p is an interior point of H, i.e. there exists an open disc D_p containing p such that D_p is contained in H.

Since $\quad p \in H = \bigcup\{G : G \in \mathcal{A}\}$,
$$\exists G_0 \in \mathcal{A} \quad \text{such that} \quad p \in G_0$$
But G_0 is an open set; hence there exists an open disc D_p containing p such that
$$p \in D_p \subset G_0$$
Since G_0 is a subset of $H = \bigcup\{G : G \in \mathcal{A}\}$, D_p is also a subset of H. Thus H is open.

4. Prove: Every open subset G of the plane $\mathbf{R}^2$ is the union of open discs.

Solution:

Since G is open, for each point $p \in G$ there is an open disc D_p such that $p \in D_p \subset G$. Then $G = \bigcup\{D_p : p \in G\}$.

5. Prove Theorem 4.2*: The intersection of any finite number of open subsets of $\mathbf{R}^2$ is open.

Solution:

We prove the theorem in the case of two open subsets of $\mathbf{R}^2$. The theorem will then follow by induction.

Let G and H be open subsets of $\mathbf{R}^2$ and let $p \in G \cap H$; so $p \in G$ and $p \in H$. Hence there exist open discs D_1 and D_2 such that
$$p \in D_1 \subset G \quad \text{and} \quad p \in D_2 \subset H$$
Then $p \in D_1 \cap D_2 \subset G \cap H$. By Example 11.1, the intersection of any two open discs is open; so there exists an open disc D such that
$$p \in D \subset D_1 \cap D_2 \subset G \cap H$$
Hence p is an interior point of $G \cap H$ and, so, $G \cap H$ is open.

6. Prove: Let $p \in G$, an open subset of $\mathbf{R}^2$. Then there exists an open disc D with center p such that $p \in D \subset G$.

Solution:

By definition of an interior point, there exists an open disc $D_1 = \{q \in \mathbf{R}^2 : d(p_1, q) < \delta\}$, with center p_1 and radius δ, such that $p \in D_1 \subset G$. So $d(p_1, p) < \delta$. Set
$$r = \delta - d(p_1, p) > 0$$

and let $\quad\quad D = \{q \in \mathbf{R}^2 : d(p, q) < \frac{1}{2}r\}$

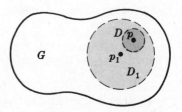

Then, as indicated in the diagram, $p \in D \subset D_1 \subset G$ where D is an open disc with center p.

7. **Prove:** Let p be an accumulation point of a subset A of the plane $\mathbf{R}^2$. Then every open set containing p contains an infinite number of points of A.

   Solution:

   Suppose G is an open set containing p and containing only a finite number of points, say $a_1, \ldots, a_m$, of A different from p. By the preceding problem, there exists an open disc D_p with center p and, say, radius δ such that $p \in D_p \subset G$. Choose $r > 0$ to be less than δ and less than the distance from p to any of the points $a_1, \ldots, a_m$; and let

   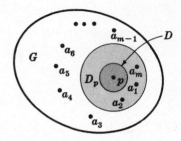

   $$D = \{q \in \mathbf{R}^2 : d(p, q) < \tfrac{1}{2}r\}$$

   Then the open disc D containing p does not contain $a_1, \ldots, a_m$ and, since $D \subset D_p \subset G$, does not contain any other points of A different from p.

   The last statement contradicts the fact that p is a limit point of A. Hence every open set containing p contains an infinite number of points of A.

   Remark: A similar statement is true for the real line $\mathbf{R}$, i.e. if $a \in \mathbf{R}$ is a limit point of $A \subset \mathbf{R}$, then every open subset of $\mathbf{R}$ containing a contains an infinite number of points of A.

8. **Prove:** Consider any open disc D_p with center $p \in \mathbf{R}^2$ and radius δ. Then there exists an open disc D such that (i) the center of D has rational coordinates, (ii) the radius of D is rational, and (iii) $p \in D \subset D_p$.

   Solution:

   Suppose $p = \langle a, b \rangle$. Then there exist rational numbers c and d such that

   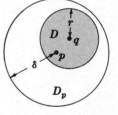

   $$a < c < a + \tfrac{1}{6}\delta \quad \text{and} \quad b < d < b + \tfrac{1}{6}\delta$$

   Let $q = \langle c, d \rangle$. Note that $d(p, q) < \tfrac{1}{3}\delta$. Now choose a rational number r such that $\tfrac{1}{3}\delta < r < \tfrac{2}{3}\delta$; and let D be the open disc with center q, which has rational coordinates, and radius r which is rational. Then, as indicated in the diagram, $p \in D \subset D_p$.

9. **Prove:** Every open subset G of the plane $\mathbf{R}^2$ is the union of a countable number of open discs.

   Solution:

   Since G is open, for each point $p \in G$ there exists an open disc D_p with center p such that $p \in D_p \subset G$. But, by the preceding problem, for each disc D_p there exists an open disc E_p such that (i) the center of E_p has rational coordinates, (ii) the radius of E_p is rational, and (iii) $p \in E_p \subset D_p$. So

   $$p \in E_p \subset D_p \subset G$$

   Accordingly, $$G = \bigcup\{E_p : p \in G\}$$

   The theorem now follows from the fact that there are only a countable number of open discs whose center has rational coordinates and whose radius is rational.

10. **Prove Theorem (Bolzano-Weierstrass) 4.3:** Let A be a bounded infinite set of real numbers. Then A contains at least one accumulation point.

    Solution:

    Since A is bounded, A is a subset of a closed interval $I_1 = [a_1, b_1]$. Bisect I_1 at $\tfrac{1}{2}(a_1 + b_1)$. Note that both of the closed subintervals of I_1,

    $$[a_1, \tfrac{1}{2}(a_1 + b_1)] \quad \text{and} \quad [\tfrac{1}{2}(a_1 + b_1), b_1] \tag{1}$$

    cannot contain a finite number of points of A since A is infinite. Let $I_2 = [a_2, b_2]$ be one of the intervals in (1) which contains an infinite number of points of A.

Now bisect I_2. As before, one of the two closed intervals

$$[a_2, \tfrac{1}{2}(a_2 + b_2)] \quad \text{and} \quad [\tfrac{1}{2}(a_2 + b_2), b_2]$$

must contain an infinite number of points of A. Call that interval I_3.

Continuing this procedure we obtain a sequence of nested closed intervals

$$I_1 \supset I_2 \supset I_3 \supset \cdots$$

such that each interval I_n contains an infinite number of points of A and

$$\lim |I_n| = 0$$

where $|I_n|$ denotes the length of the interval I_n.

By the Nested Interval Property of the real numbers (see Appendix A), there exists a point p in each interval I_n. We show that p is a limit point of A and then the theorem will follow.

Let $S_p = (a, b)$ be an open interval containing p. Since $\lim |I_n| = 0$,

$$\exists n_0 \in \mathbf{N} \quad \text{such that} \quad |I_{n_0}| < \min (p - a, b - p)$$

Then the interval I_{n_0} is a subset of the open interval $S_p = (a, b)$ as indicated in the diagram below.

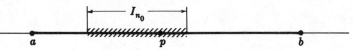

Since I_{n_0} contains an infinite number of points of A, so does the open interval S_p. Thus each open interval containing p contains points of A other than p, i.e. p is a limit point of A.

CLOSED SETS

11. Prove: A set F is closed if and only if its complement F^c is open.

Solution:

Note that $(F^c)^c = F$; so F is the complement of F^c. Thus, by definition, F is closed iff F^c is open.

12. Prove: The union of a finite number of closed sets is closed.

Solution:

Let $F_1, \ldots, F_m$ be closed sets and let $F = F_1 \cup \cdots \cup F_m$. By DeMorgan's Law,

$$F^c = (F_1 \cup \cdots \cup F_m)^c = F_1^c \cap F_2^c \cap \cdots \cap F_m^c$$

So F^c is the intersection of a finite number of open sets F_i^c, and thus F^c is also open. Hence its complement $F^{cc} = F$ is closed.

13. Prove: The intersection of any number of closed sets is closed.

Solution:

Let $\{F_i\}$ be a class of closed sets and let $F = \cap_i F_i$. By DeMorgan's Law,

$$F^c = (\cap_i F_i)^c = \cup_i F_i^c$$

So F^c is the union of open sets and, hence, is open itself. Consequently, $F^{cc} = F$ is closed.

14. Prove Theorem 4.4*: A subset of $\mathbf{R}^2$ is closed if and only if it contains each of its accumulation points.

Solution:

Suppose p is a limit point of a closed set F. Then every open disc containing p contains points of F other than p. Hence there cannot be an open disc D_p containing p which is completely contained in the complement of F. In other words, p is not an interior point of F^c. But F^c is open since F is closed; so p does not belong to F^c, i.e. $p \in F$.

On the other hand, suppose a set A contains each of its limit points. We claim that A is closed or, equivalently, that its complement A^c is open. Let $p \in A^c$. Since A contains each of its limit points, p is not a limit point of A. Hence there exists at least one open disc D_p containing p such that D_p does not contain any points of A. So $D_p \subset A^c$, and hence p is an interior point of A^c. Since each point $p \in A^c$ is an interior point, A^c is open and so A is closed.

15. Prove: The derived set A', i.e. set of accumulation points, of an arbitrary subset A of $\mathbf{R}^2$ is closed.

Solution:

Let p be a limit point of A'. By Theorem 4.4*, the theorem is proved if we show that $p \in A'$, that is, that p is also a limit point of A.

Let G_p be an open set containing p. Since p is a limit point of A', G_p contains at least one point $q \in A'$ different from p. But G_p is an open set containing $q \in A'$; hence G_p contains (infinitely many) points of A. So,

$$\exists a \in A \quad \text{such that} \quad a \ne p, \ a \ne q, \ \text{and} \ a \in G_p$$

That is, each open set containing p contains points of A other than p; so $p \in A'$.

16. Prove: Let A be a closed and bounded set of real numbers and let $\sup(A) = p$. Then $p \in A$.

Solution:

Suppose $p \notin A$. Let G be an open set containing p. Then G contains an open interval (b, c) containing p, i.e. such that $b < p < c$. Since $\sup(A) = p$ and $p \notin A$,

$$\exists a \in A \quad \text{such that} \quad b < a < p < c$$

for otherwise b would be an upper bound for A. So $a \in (b, c) \subset G$. Thus each open set containing p contains a point of A different from p; hence p is a limit point of A. But A is closed; hence, by Theorem 4.4*, $p \in A$.

17. Prove Theorem (Heine-Borel) 4.5:

Let $I_1 = [c_1, d_1]$ be covered by a class $G = \{(a_i, b_i) : i \in I\}$ of open intervals. Then G contains a finite subclass which also covers I_1.

Solution:

Assume that no finite subclass of G covers I_1. We bisect $I_1 = [c_1, d_1]$ at $\frac{1}{2}(c_1 + d_1)$ and consider the two closed intervals

$$[c_1, \tfrac{1}{2}(c_1 + d_1)] \quad \text{and} \quad [\tfrac{1}{2}(c_1 + d_1), d_1] \tag{1}$$

At least one of these two intervals cannot be covered by a finite subclass of G or else the whole interval I_1 will be covered by a finite subclass of G. Let $I_2 = [c_2, d_2]$ be one of the two intervals in *(1)* which cannot be covered by a finite subclass of G. We now bisect I_2. As before, one of the two closed intervals

$$[c_2, \tfrac{1}{2}(c_2 + d_2)] \quad \text{and} \quad [\tfrac{1}{2}(c_2 + d_2), d_2]$$

cannot be covered by a finite subclass of G. Call that interval I_3.

We continue this procedure and obtain a sequence of nested closed intervals $I_1 \supset I_2 \supset I_3 \supset \cdots$ such that each interval I_n cannot be covered by a finite subclass of G and $\lim |I_n| = 0$ where $|I_n|$ denotes the length of the interval I_n.

By the Nested Interval Property of the real numbers (see Appendix), there exists a point p in each interval I_n. In particular, $p \in I_1$. Since G is a cover of I_1, there exists an open interval (a_{i_0}, b_{i_0}) in G which contains p. Hence $a_{i_0} < p < b_{i_0}$. Since $\lim |I_n| = 0$,

$$\exists n_0 \in N \quad \text{such that} \quad |I_{n_0}| < \min(p - a_{i_0}, b_{i_0} - p)$$

Then, as indicated in the diagram below, the interval I_{n_0} is a subset of the one interval (a_{i_0}, b_{i_0}) in G.

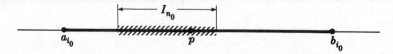

But this contradicts our choice of I_{n_0}. Thus the original assumption that no finite subclass of G covers I_1 is false and the theorem is true.

SEQUENCES

18. Write the first six terms of each of the following sequences:

(i) $s(n) = \begin{cases} n-1 & \text{if } n \text{ is odd} \\ n^2 & \text{if } n \text{ is even} \end{cases}$ (ii) $t(n) = \begin{cases} 1 & \text{if } n = 1 \\ 2 & \text{if } n = 2 \\ t(n-1) + t(n-2) & \text{if } n > 2 \end{cases}$

Solution:

(i)   Two formulas are used to define this function. Substitute 1, 3 and 5 into $s(n) = n-1$ to get $s_1 = 0$, $s_3 = 2$ and $s_5 = 4$. Then substitute 2, 4 and 6 into $s(n) = n^2$ to get $s_2 = 4$, $s_4 = 16$ and $s_6 = 36$. Thus we have $\langle 0, 4, 2, 16, 4, 36, \ldots \rangle$.

(ii)   Here the function is defined recursively. Each term after the second is found by adding the two previous terms. Thus:

$$t_1 = 1 \qquad\qquad t_4 = t_3 + t_2 = 3 + 2 = 5$$
$$t_2 = 2 \qquad\qquad t_5 = t_4 + t_3 = 5 + 3 = 8$$
$$t_3 = t_2 + t_1 = 2 + 1 = 3 \qquad\qquad t_6 = t_5 + t_4 = 8 + 5 = 13$$

Hence we have $\langle 1, 2, 3, 5, 8, 13, \ldots \rangle$.

19. Consider the sequence $\langle a_n = (-1)^{n-1}(2n-1) \rangle$:

$$\langle 1, -3, 5, -7, 9, -11, 13, -15, \ldots \rangle$$

Determine whether or not each of the following sequences is a subsequence of $\langle a_n \rangle$.

(i) $\langle b_n \rangle = \langle 1, 5, -3, -7, 9, 13, -11, -15, \ldots \rangle$

(ii) $\langle c_n \rangle = \langle 1, 3, 5, 7, 9, 11, 13, \ldots \rangle$

(iii) $\langle d_n \rangle = \langle -3, -7, -11, -15, -19, -23, \ldots \rangle$

Solution:

(i)   Note that 5 appears before -3 in $\langle b_n \rangle$, but -3 appears before 5 in $\langle a_n \rangle$. Hence $\langle b_n \rangle$ is not a subsequence of $\langle a_n \rangle$.

(ii)   The terms 3, 7 and 11 do not even appear in $\langle a_n \rangle$; hence $\langle c_n \rangle$ is not a subsequence of $\langle a_n \rangle$.

(iii)   The sequence $\langle d_n \rangle$ is a subsequence of $\langle a_n \rangle$, for $\langle i_n = 2n \rangle = \langle 2, 4, 6, \ldots \rangle$ is a sequence of positive integers such that $i_1 < i_2 < i_3 < \cdots$; so

$$\langle a_{i_1}, a_{i_2}, \ldots \rangle = \langle a_2, a_4, a_6, \ldots \rangle = \langle -3, -7, -11, \ldots \rangle$$

is a subsequence of $\langle a_n \rangle$.

20. Determine the range of each sequence:

(i) $\langle 1, \frac{1}{2}, 1, \frac{1}{3}, 1, \frac{1}{4}, 1, \frac{1}{5}, \ldots \rangle$ (iii) $\langle 2, 4, 6, 8, 10, \ldots \rangle$

(ii) $\langle 1, 0, -1, 0, 1, 0, -1, 0, 1, 0, -1, 0, \ldots \rangle$

Solution:

The range of a sequence is the set of image points. Hence the ranges of the sequences are

(i) $\{1, \frac{1}{2}, \frac{1}{3}, \frac{1}{4}, \ldots\}$,   (ii) $\{1, 0, -1\}$,   (iii) $\{2, 4, 6, 8, \ldots\}$

21. Prove: If the range of a sequence $\langle a_n \rangle$ is finite, then the sequence has a convergent subsequence.

Solution:

If the range $\{a_n\}$ of $\langle a_n \rangle$ is finite, then one of the image points, say b, appears an infinite number of times in the sequence. Hence $\langle b, b, b, b, \ldots \rangle$ is a subsequence of $\langle a_n \rangle$ and it converges.

22. Prove: If $\lim a_n = b$ and $\lim a_n = c$, then $b = c$.

Solution:

Suppose that b and c are distinct. Let $\delta = |b - c| > 0$. Then the open intervals $B = (b - \frac{1}{2}\delta, b + \frac{1}{2}\delta)$ and $C = (c - \frac{1}{2}\delta, c + \frac{1}{2}\delta)$, containing b and c respectively, are disjoint. Since $\langle a_n \rangle$ converges to b, B must contain all except a finite number of the terms of the sequence. Hence C can only contain a finite number of the terms of the sequence. But this contradicts the fact that $\langle a_n \rangle$ converges to c. Accordingly, b and c are not distinct.

23. Prove: If the range $\{a_n\}$ of a sequence $\langle a_n \rangle$ contains an accumulation point b, then the sequence $\langle a_n \rangle$ contains a subsequence $\langle a_{i_n} \rangle$ which converges to b.

Solution:

Since b is a limit point of $\{a_n\}$, each of the open intervals

$$S_1 = (b - 1, b + 1), \quad S_2 = (b - \tfrac{1}{2}, b + \tfrac{1}{2}), \quad S_3 = (b - \tfrac{1}{3}, b + \tfrac{1}{3}), \quad \ldots$$

contains an infinite number of elements of the set $\{a_n\}$ and, hence, an infinite number of the terms of the sequence $\langle a_n \rangle$. We choose a sequence $\langle a_{i_n} \rangle$ as follows:

Choose a_{i_1} to be a point in S_1.

Choose a_{i_2} to be a point in S_2 such that $i_2 > i_1$, i.e. such that a_{i_2} appears after a_{i_1} in the sequence $\langle a_n \rangle$.

Choose a_{i_3} to be a point in S_3 such that $i_3 > i_2$.

We continue in the same manner.

Observe that we are always able to choose the next term in the sequence $\langle a_{i_n} \rangle$ since there are an infinite number of the terms of the original sequence $\langle a_n \rangle$ in each interval S_n.

We claim that $\langle a_{i_n} \rangle$ satisfies the conditions of the theorem. Recall that we choose the terms of the sequence $\langle a_{i_n} \rangle$ so that $i_1 < i_2 < i_3 < \cdots$; hence $\langle a_{i_n} \rangle$ is a subsequence of $\langle a_n \rangle$. We need to show that $\lim a_{i_n} = b$. Let G be an open set containing b. Then G contains an open interval (d_1, d_2) containing b; so $d_1 < b < d_2$. Let $\delta = \min (b - d_1, d_2 - b) > 0$; then

$$\exists n_0 \in N \quad \text{such that} \quad 1/n_0 < \delta$$

Hence $S_{n_0} \subset (d_1, d_2) \subset G$, and so

$$n > n_0 \quad \text{implies} \quad a_{i_n} \in S_n \subset S_{n_0} \subset (d_1, d_2) \subset G$$

Thus G contains almost all the terms of the sequence $\langle a_{i_n} \rangle$; that is, $\lim a_{i_n} = b$.

24. Prove Theorem 4.6: Every bounded sequence $\langle a_n \rangle$ of real numbers contains a convergent subsequence.

Solution:

Consider the range $\{a_n\}$ of the sequence $\langle a_n \rangle$. If the range is finite, then by Problem 21 the sequence contains a convergent subsequence. On the other hand, if the range is infinite, then, by the Bolzano-Weierstrass Theorem, the bounded infinite set $\{a_n\}$ contains a limit point. But then, by the previous problem, the sequence in this case also contains a convergent subsequence.

25. Prove: Every Cauchy sequence $\langle a_n \rangle$ of real numbers is bounded.

Solution:

Let $\epsilon = 1$. Then, by definition of a Cauchy sequence,

$$\exists n_0 \in \mathbf{N} \quad \text{such that} \quad n, m \geq n_0 \quad \text{implies} \quad |a_n - a_m| < 1$$

In particular, $m \geq n_0$ implies $|a_{n_0} - a_m| < 1$, or, $a_{n_0} - 1 < a_m < a_{n_0} + 1$

Let
$$\alpha = \max (a_1, a_2, \ldots, a_{n_0}, a_{n_0} + 1)$$
$$\beta = \min (a_1, a_2, \ldots, a_{n_0}, a_{n_0} - 1)$$

Then α is an upper bound for the range $\{a_n\}$ of the sequence $\langle a_n \rangle$ and β is a lower bound. Accordingly, $\langle a_n \rangle$ is a bounded sequence.

26. Prove: Let $\langle a_n \rangle$ be a Cauchy sequence. If a subsequence $\langle a_{i_n} \rangle$ of $\langle a_n \rangle$ converges to a point b, then the Cauchy sequence itself converges to b.

Solution:

Let $\epsilon > 0$. We need to find a positive integer n_0 such that

$$n > n_0 \quad \text{implies} \quad |a_n - b| < \epsilon$$

Since $\langle a_n \rangle$ is a Cauchy sequence,

$$\exists n_0 \in \mathbf{N} \quad \text{such that} \quad n, m > n_0 \quad \text{implies} \quad |a_n - a_m| < \tfrac{1}{2}\epsilon$$

Also, since the subsequence $\langle a_{i_n} \rangle$ converges to b,

$$\exists i_m \in \mathbf{N} \quad \text{such that} \quad |a_{i_m} - b| < \tfrac{1}{2}\epsilon$$

Observe that we can choose i_m so that $i_m > n_0$. Accordingly,

$$
\begin{aligned}
n > n_0 \quad \text{implies} \quad |a_n - b| &= |a_n - a_{i_m} + a_{i_m} - b| \\
&\leq |a_n - a_{i_m}| + |a_{i_m} - b| \\
&< \tfrac{1}{2}\epsilon + \tfrac{1}{2}\epsilon = \epsilon
\end{aligned}
$$

Hence $\langle a_n \rangle$ converges to b.

Observe that we need $i_m > n_0$ in order to state that: $n > n_0$ implies $|a_n - a_{i_m}| < \tfrac{1}{2}\epsilon$.

27. Prove Theorem (Cauchy) 4.7: Every Cauchy sequence $\langle a_n \rangle$ of real numbers converges to a real number.

Solution:

By Problem 25, the Cauchy sequence $\langle a_n \rangle$ is bounded. Hence, by Theorem 4.6, the bounded sequence $\langle a_n \rangle$ contains a convergent subsequence $\langle a_{i_n} \rangle$. But, by the preceding problem, the Cauchy sequence $\langle a_n \rangle$ converges to the same limit as its subsequence $\langle a_{i_n} \rangle$. In other words, the Cauchy sequence $\langle a_n \rangle$ converges to a real number.

28. Determine whether or not each of the following subsets of $\mathbf{R}$ is complete:

(i) $\mathbf{N}$, the set of positive integers;   (ii) $\mathbf{Q}^c$, the set of irrational numbers.

Solution:

(i)   Let $\langle a_n \rangle$ be a Cauchy sequence of positive integers. If $\epsilon = \frac{1}{2}$, then

$$|a_n - a_m| < \epsilon = \tfrac{1}{2} \quad \text{implies} \quad a_n = a_m$$

Therefore, the Cauchy sequence $\langle a_n \rangle$ is of the form $\langle a_1, a_2, \ldots, a_{n_0}, b, b, b, \ldots \rangle$ which converges to the positive integer b. Hence $\mathbf{N}$ is complete.

(ii)   Observe that each of the open intervals

$$(-1, 1), \quad (-\tfrac{1}{2}, \tfrac{1}{2}), \quad (-\tfrac{1}{3}, \tfrac{1}{3}), \quad \ldots$$

contains irrational points. Hence there exists a sequence $\langle a_n \rangle$ of irrational numbers such that a_n belongs to the open interval $(-1/n, 1/n)$. The sequence $\langle a_n \rangle$ will be a Cauchy sequence of points in $\mathbf{Q}^c$ and it will converge to the rational number 0. Hence $\mathbf{Q}^c$ is not complete.

CONTINUITY

29. Prove: If the function $f : \mathbf{R} \to \mathbf{R}$ is constant, say $f(x) = a$ for every $x \in \mathbf{R}$, then f is continuous.

Solution:

Method 1. The function f is continuous iff the inverse $f^{-1}[G]$ of any open set G is also open. Since $f(x) = a$ for every $x \in \mathbf{R}$,

$$f^{-1}[G] = \begin{cases} \emptyset & \text{if } a \notin G \\ \mathbf{R} & \text{if } a \in G \end{cases}$$

for any open set G. In either case, $f^{-1}[G]$ is open since both $\mathbf{R}$ and $\emptyset$ are open sets.

Method 2. We show that f is continuous at any point x_0 using the $\epsilon - \delta$ definition of continuity. Let $\epsilon > 0$. Then for any $\delta > 0$, say $\delta = 1$,

$$|x - x_0| < 1 \quad \text{implies} \quad |f(x) - f(x_0)| = |a - a| = 0 < \epsilon$$

Hence f is continuous.

30. Prove: The identity function $f : \mathbf{R} \to \mathbf{R}$, that is, the function defined by $f(x) = x$, is continuous.

Solution:

Method 1. Let G be any open set. Then $f^{-1}[G] = G$ is also an open set. Accordingly, f is continuous.

Method 2. We show that f is continuous at any point x_0 using the $\epsilon - \delta$ definition of continuity. Let $\epsilon > 0$. Then choosing $\epsilon = \delta$,

$$|x - x_0| < \delta \quad \text{implies} \quad |f(x) - f(x_0)| = |x - x_0| < \delta = \epsilon$$

Accordingly, f is continuous.

31. Prove: Let the functions $f : \mathbf{R} \to \mathbf{R}$ and $g : \mathbf{R} \to \mathbf{R}$ be continuous. Then the composition function $g \circ f : \mathbf{R} \to \mathbf{R}$ is also continuous.

Solution:

We show that the inverse $(g \circ f)^{-1}[G]$ of any open set G is also open. Since g is continuous, the inverse $g^{-1}[G]$ is an open set. But since f is continuous, the inverse $f^{-1}[g^{-1}[G]]$ of $g^{-1}[G]$ is also open. Recall that

$$(g \circ f)^{-1} = f^{-1} \circ g^{-1}$$

Hence $$(g \circ f)^{-1}[G] = (f^{-1} \circ g^{-1})[G] = f^{-1}[g^{-1}[G]]$$

is an open set. Thus the composition function $g \circ f : \mathbf{R} \to \mathbf{R}$ is continuous.

32. Prove: Let $f : \mathbf{R} \to \mathbf{R}$ be continuous and let $f(q) = 0$ for every rational number $q \in Q$. Then $f(x) = 0$ for every real number $x \in \mathbf{R}$.

Solution:

Suppose $f(p)$ is not zero for some real number $p \in \mathbf{R}$, i.e. suppose

$$\exists p \in \mathbf{R} \quad \text{such that} \quad f(p) = \gamma, \ |\gamma| > 0$$

Choose $\epsilon = \frac{1}{2}|\gamma|$. Since f is continuous,

$$\exists \delta > 0 \quad \text{such that} \quad |x - p| < \delta \quad \text{implies} \quad |f(x) - f(p)| < \epsilon = \tfrac{1}{2}|\gamma|$$

Now there are rational points in every open interval. In particular,

$$\exists q \in \mathbf{Q} \quad \text{such that} \quad q \in \{x : |x - p| < \delta\}$$

which implies $$|f(q) - f(p)| = |f(p)| = |\gamma| < \epsilon = \tfrac{1}{2}|\gamma|$$

an impossibility. Hence $f(x) = 0$ for every $x \in \mathbf{R}$.

33. Prove Theorem 4.8:   A function $f : \mathbf{R}^2 \to \mathbf{R}^2$ is continuous if and only if the inverse image of every open set is open.

Solution:

Let $f : \mathbf{R}^2 \to \mathbf{R}^2$ be continuous and let V be an open subset of $\mathbf{R}^2$. We want to show that $f^{-1}[V]$ is also an open set. Let $p \in f^{-1}[V]$. Then $f(p) \in V$. By definition of continuity, there exists an open set U_p containing p such that $f[U_p] \subset V$. Hence (as indicated in the diagram below)

$$U_p \subset f^{-1}[f[U_p]] \subset f^{-1}[V]$$

We have shown that, for every point $p \in f^{-1}[V]$, there exists an open set U_p such that

$$p \in U_p \subset f^{-1}[V]$$

Accordingly, $$f^{-1}[V] = \mathbf{U}\{U_p : p \in f^{-1}[V]\}$$

So $f^{-1}[V]$ is the union of open sets and is, therefore, open itself.

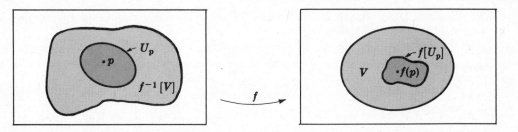

*On the other hand, suppose the inverse of every open set is open. We want to show that f is continuous at any point $p \in \mathbf{R}$. Let V be an open set containing $f(p)$, i.e. $f(p) \in V$. Then $f^{-1}[V]$ is an open set containing p with the property that $f[f^{-1}[V]] \subset V$. Hence f is continuous at p.

34. Give an example of two functions $f : \mathbf{R} \to \mathbf{R}$ and $g : \mathbf{R} \to \mathbf{R}$ such that f and g are each discontinuous (not continuous) at every point and such that the sum $f + g$ is continuous at every point in R.

Solution:
Consider the functions f and g defined by

$$f(x) = \begin{cases} 0 & \text{if } x \text{ is rational} \\ 1 & \text{if } x \text{ is irrational} \end{cases}, \qquad g(x) = \begin{cases} 1 & \text{if } x \text{ is rational} \\ 0 & \text{if } x \text{ is irrational} \end{cases}$$

The functions f and g are discontinuous at every point in $\mathbf{R}$, but the sum $f + g$ is the constant function $(f + g)(x) = 1$ which is continuous.

35. Prove: Let the function $f : \mathbf{R} \to \mathbf{R}$ be continuous at a point $p \in \mathbf{R}$.

(i)   If $f(p)$ is positive, i.e. $f(p) > 0$, then there exists an open interval S containing p such that f is positive at every point in S.

(ii) If $f(p)$ is negative, i.e. $f(p) < 0$, then there exists an open interval S containing p such that f is negative at every point in S.

Solution:
We prove (i). The proof of (ii) is similar and will be omitted. Suppose $f(p) = \epsilon > 0$. Since f is continuous at p,

$$\exists \delta > 0 \quad \text{such that} \quad |x - p| < \delta \quad \text{implies} \quad |f(x) - f(p)| < \epsilon$$

or, equivalently,

$$x \in (p - \delta, p + \delta) \quad \text{implies} \quad f(x) \in (f(p) - \epsilon, f(p) + \epsilon) = (0, 2\epsilon)$$

Thus for every point x in the open interval $(p - \delta, p + \delta)$, $f(x)$ is positive.

36. Prove: Let $f : \mathbf{R} \to \mathbf{R}$ be continuous at every point in a closed interval $[a, b]$, and let $f(a) < 0 < f(b)$. Then there exists a point $p \in [a, b]$ such that $f(p) = 0$. (In other words, the graph of a continuous function defined on a closed interval which lies both below and above the x-axis must cross the x-axis at at least one point, as indicated in the diagram.)

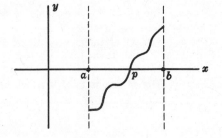

Solution:
Let A be the set of points in $[a, b]$ at which f is negative, i.e.,

$$A = \{x : x \in [a, b], \ f(x) < 0\}$$

Observe that A is not empty since, for example, $a \in A$. Let $p = \sup(A)$ be the least upper bound for A. Since $a \in A$, $a \le p$; and since b is an upper bound for A, $p \le b$. So p belongs to the interval $[a, b]$.

We claim that $f(p) = 0$. If $f(p) < 0$, then, by the preceding problem, there is an open interval $(p - \delta,\ p + \delta)$ in which f is negative, i.e.,

$$(p - \delta,\ p + \delta) \subset A$$

So p cannot be an upper bound for A. On the other hand, if $f(p) > 0$, then there exists an interval $(p - \delta,\ p + \delta)$ in which f is positive; so

$$(p - \delta,\ p + \delta) \cap A \ =\ \varnothing$$

which implies that p cannot be a least upper bound for A. Thus $f(p)$ can only be zero, i.e. $f(p) = 0$.

Remark.   The theorem is also true and proved similarly in the case $f(b) < 0 < f(a)$.

37. Prove Theorem (Weierstrass) 4.9:   Let $f : \mathbf{R} \to \mathbf{R}$ be continuous on a closed interval $[a, b]$.   Then the function assumes every value between $f(a)$ and $f(b)$.

Solution:

Suppose $f(a) < f(b)$ and let y_0 be a real number such that $f(a) < y_0 < f(b)$. We want to prove that there is a point p such that $f(p) = y_0$. Consider the function $g(x) = f(x) - y_0$ which is also continuous. Observe that $g(a) < 0 < g(b)$.

By the preceding problem, there exists a point p such that $g(p) = f(p) - y_0 = 0$. Hence $f(p) = y_0$.

The case when $f(b) < f(a)$ is proved similarly.

Supplementary Problems

OPEN SETS, CLOSED SETS, ACCUMULATION POINTS

38. Prove: If A is a finite subset of $\mathbf{R}$, then the derived set A' of A is empty, i.e. $A' = \varnothing$.

39. Prove: Every finite subset of $\mathbf{R}$ is closed.

40. Prove: If $A \subset B$, then $A' \subset B'$.

41. Prove: A subset B of $\mathbf{R}^2$ is closed if and only if $d(p, B) = 0$ implies $p \in B$, where $d(p, B) = \inf \{d(p, q) : q \in B\}$.

42. Prove: $A \cup A'$ is closed for any set A.

43. Prove: $A \cup A'$ is the smallest closed set containing A, i.e. if F is closed and $A \subset F \subset A \cup A'$ then $F = A \cup A'$.

44. Prove: The set of interior points of any set A, written $\text{int}\,(A)$, is an open set.

45. Prove: The set of interior points of A is the largest open set contained in A, i.e. if G is open and $\text{int}\,(A) \subset G \subset A$, then $\text{int}\,(A) = G$.

46. Prove: The only subsets of $\mathbf{R}$ which are both open and closed are $\varnothing$ and $\mathbf{R}$.

SEQUENCES

47. Prove: If the sequence $\langle a_n \rangle$ converges to $b \in \mathbf{R}$, then the sequence $\langle\, |a_n - b|\, \rangle$ converges to 0.

48. Prove: If the sequence $\langle a_n \rangle$ converges to 0, and the sequence $\langle b_n \rangle$ is bounded, then the sequence $\langle a_n b_n \rangle$ also converges to 0.

49. Prove: If $a_n \to a$ and $b_n \to b$, then the sequence $\langle a_n + b_n \rangle$ converges to $a + b$.

50. Prove: If $a_n \to a$ and $b_n \to b$, then the sequence $\langle a_n b_n \rangle$ converges to ab.

51. Prove: If $a_n \to a$ and $b_n \to b$ where $b_n \neq 0$ and $b \neq 0$, then the sequence $\langle a_n/b_n \rangle$ converges to a/b.

52. Prove: If the sequence $\langle a_n \rangle$ converges to b, then every subsequence $\langle a_{i_n} \rangle$ of $\langle a_n \rangle$ also converges to b.

53. Prove: If the sequence $\langle a_n \rangle$ converges to b, then either the range $\{a_n\}$ of the sequence $\langle a_n \rangle$ is finite, or b is an accumulation point of the range $\{a_n\}$.

54. Prove: If the sequence $\langle a_n \rangle$ of distinct elements is bounded and the range $\{a_n\}$ of $\langle a_n \rangle$ has exactly one limit point b, then the sequence $\langle a_n \rangle$ converges to b.

(*Remark*: The sequence $\langle 1, \frac{1}{2}, 2, \frac{1}{3}, 3, \frac{1}{4}, 4, \ldots \rangle$ shows that the condition of boundedness cannot be removed from this theorem.)

CONTINUITY

55. Prove: A function $f : \mathbf{R} \to \mathbf{R}$ is continuous at $a \in \mathbf{R}$ if and only if for every sequence $\langle a_n \rangle$ converging to a, the sequence $\langle f(a_n) \rangle$ converges to $f(a)$.

56. Prove: Let the function $f : \mathbf{R} \to \mathbf{R}$ be continuous at $p \in \mathbf{R}$. Then there exists an open interval S containing p such that f is bounded on the open interval S.

57. Give an example of a function $f : \mathbf{R} \to \mathbf{R}$ which is continuous at every point in the open interval $S = (0, 1)$ but which is not bounded on the open interval S.

58. Prove: Let $f : \mathbf{R} \to \mathbf{R}$ be continuous at every point in a closed interval $A = [a, b]$. Then f is bounded on A. (*Remark*: By the preceding problem, this theorem is not true if A is not closed.)

59. Prove: Let $f : \mathbf{R} \to \mathbf{R}$ and $g : \mathbf{R} \to \mathbf{R}$ be continuous. Then the sum $(f + g) : \mathbf{R} \to \mathbf{R}$ is continuous, where $f + g$ is defined by $(f + g)(x) \equiv f(x) + g(x)$.

60. Prove: Let $f : \mathbf{R} \to \mathbf{R}$ be continuous, and let k be any real number. Then the function $(kf) : \mathbf{R} \to \mathbf{R}$ is continuous, where kf is defined by $(kf)(x) \equiv k(f(x))$.

61. Prove: Let $f : \mathbf{R} \to \mathbf{R}$ and $g : \mathbf{R} \to \mathbf{R}$ be continuous. Then $\{x \in \mathbf{R} : f(x) = g(x)\}$ is a closed set.

62. Prove: The projection $\pi_x : \mathbf{R}^2 \to \mathbf{R}$ is continuous where π_x is defined by $\pi_x(\langle a, b \rangle) = a$.

63. Consider the functions $f : \mathbf{R} \to \mathbf{R}$ and $g : \mathbf{R} \to \mathbf{R}$ defined by

$$f(x) = \begin{cases} \sin(1/x) & \text{if } x \neq 0 \\ 0 & \text{if } x = 0 \end{cases}, \qquad g(x) = \begin{cases} x \sin(1/x) & \text{if } x \neq 0 \\ 0 & \text{if } x = 0 \end{cases}$$

Prove g is continuous at 0 but f is not continuous at 0.

64. Recall that every rational number $q \in \mathbf{Q}$ can be written uniquely in the form $q = a/b$ where $a \in \mathbf{Z}$, $b \in \mathbf{N}$, and a and b are relatively prime. Consider the function $f : \mathbf{R} \to \mathbf{R}$ defined by

$$f(x) = \begin{cases} 0 & \text{if } x \text{ is irrational} \\ 1/b & \text{if } x \text{ is rational and } x = a/b \text{ as above} \end{cases}$$

Prove that f is continuous at every irrational point, but f is discontinuous at every rational point.

Answers to Supplementary Problems

57. Consider the function

$$f(x) = \begin{cases} -x & \text{if } x \leq 0 \\ 1/x & \text{if } x > 0 \end{cases}$$

The function f is continuous at every point in $\mathbf{R}$ except at 0 as indicated in the adjacent graph of f. Hence f is continuous at every point in the open interval $(0, 1)$. But f is not bounded on $(0, 1)$.

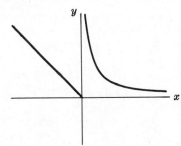

58. *Hint.* Use the result stated in Problem 56 and the Heine-Borel Theorem.

Chapter 5

Topological Spaces: Definitions

TOPOLOGICAL SPACES

Let X be a non-empty set. A class T of subsets of X is a *topology* on X iff T satisfies the following axioms.

[O₁] X and $\emptyset$ belong to T.

[O₂] The union of any number of sets in T belongs to T.

[O₃] The intersection of any two sets in T belongs to T.

The members of T are then called *T-open sets*, or simply *open sets*, and X together with T, i.e. the pair (X, T) is called a *topological space*.

Example 1.1: Let $\mathcal{U}$ denote the class of all open sets of real numbers discussed in Chapter 4. Then $\mathcal{U}$ is a topology on $\mathbf{R}$; it is called the *usual topology* on $\mathbf{R}$. Similarly, the class $\mathcal{U}$ of all open sets in the plane $\mathbf{R}^2$ is a topology, and also called the *usual topology*, on $\mathbf{R}^2$. We shall always assume the usual topology on $\mathbf{R}$ and $\mathbf{R}^2$ unless otherwise specified.

Example 1.2: Consider the following classes of subsets of $X = \{a, b, c, d, e\}$.
$$T_1 = \{X,\ \emptyset,\ \{a\},\ \{c,d\},\ \{a,c,d\},\ \{b,c,d,e\}\}$$
$$T_2 = \{X,\ \emptyset,\ \{a\},\ \{c,d\},\ \{a,c,d\},\ \{b,c,d\}\}$$
$$T_3 = \{X,\ \emptyset,\ \{a\},\ \{c,d\},\ \{a,c,d\},\ \{a,b,d,e\}\}$$

Observe that T_1 is a topology on X since it satisfies the necessary three axioms **[O₁]**, **[O₂]** and **[O₃]**. But T_2 is not a topology on X since the union
$$\{a,c,d\}\ \cup\ \{b,c,d\}\ =\ \{a,b,c,d\}$$
of two members of T_2 does not belong to T_2, i.e. T_2 does not satisfy the axiom **[O₂]**.

Also, T_3 is not a topology on X since the intersection
$$\{a,c,d\}\ \cap\ \{a,b,d,e\}\ =\ \{a,d\}$$
of two sets in T_3 does not belong to T_3, i.e. T_3 does not satisfy the axiom **[O₃]**.

Example 1.3: Let $\mathcal{D}$ denote the class of all subsets of X. Observe that $\mathcal{D}$ satisfies the axioms for a topology on X. This topology is called the *discrete topology*; and X together with its discrete topology, i.e. the pair $(X, \mathcal{D})$, is called a *discrete topological space* or simply a *discrete space*.

Example 1.4: As seen by axiom **[O₁]**, a topology on X must contain the sets X and $\emptyset$. The class $\mathcal{J} = \{X, \emptyset\}$, consisting of X and $\emptyset$ alone, is itself a topology on X. It is called the *indiscrete topology*; and X together with its indiscrete topology, i.e. $(X, \mathcal{J})$, is called an *indiscrete topological space* or simply an *indiscrete space*.

Example 1.5: Let T denote the class of all subsets of X whose complements are finite together with the empty set $\emptyset$. This class T is also a topology on X. It is called the *cofinite topology* or the *T_1-topology* on X. (The significance of the T_1 will appear in a later chapter.)

Example 1.6: The intersection $T_1 \cap T_2$ of any two topologies T_1 and T_2 on X is also a topology on X. For, by **[O$_1$]**, X and $\emptyset$ each belongs to both T_1 and T_2; hence X and $\emptyset$ each belongs to the intersection $T_1 \cap T_2$, i.e. $T_1 \cap T_2$ satisfies **[O$_1$]**. Furthermore, if $G, H \in T_1 \cap T_2$ then, in particular, $G, H \in T_1$ and $G, H \in T_2$. But since T_1 and T_2 are topologies, $G \cap H \in T_1$ and $G \cap H \in T_2$. Accordingly,

$$G \cap H \in T_1 \cap T_2$$

In other words $T_1 \cap T_2$ satisfies **[O$_3$]**. Similarly, $T_1 \cap T_2$ satisfies **[O$_2$]**.

The statement in the preceding example can, in fact, be generalized to any collection of topologies. Namely,

Theorem 5.1: Let $\{T_i : i \in I\}$ be any collection of topologies on a set X. Then the intersection $\cap_i T_i$ is also a topology on X.

In our last example, we show that the union of topologies need not be a topology.

Example 1.7: Each of the classes

$$T_1 = \{X, \emptyset, \{a\}\} \quad \text{and} \quad T_2 = \{X, \emptyset, \{b\}\}$$

is a topology on $X = \{a, b, c\}$. But the union

$$T_1 \cup T_2 = \{X, \emptyset, \{a\}, \{b\}\}$$

is not a topology on X since it violates **[O$_2$]**. That is, $\{a\} \in T_1 \cup T_2$, $\{b\} \in T_1 \cup T_2$ but $\{a\} \cup \{b\} = \{a, b\}$ does not belong to $T_1 \cup T_2$.

If G is an open set containing a point $p \in X$, then G is called an *open neighborhood* of p. Also, G without p, i.e. $G \setminus \{p\}$, is called a *deleted open neighborhood* of p.

Remark: The axioms **[O$_1$]**, **[O$_2$]** and **[O$_3$]** are equivalent to the following two axioms:

[O$_1^*$] The union of any number of sets in T belongs to T.

[O$_2^*$] The intersection of any finite number of sets in T belongs to T.

For **[O$_1^*$]** implies that $\emptyset$ belongs to T since

$$\cup \{G \in T : G \in \emptyset\} = \emptyset$$

i.e. the empty union of sets is the empty set. Furthermore, **[O$_2^*$]** implies that X belongs to T since

$$\cap \{G \in T : G \in \emptyset\} = X$$

i.e. the empty intersection of subsets of X is X itself.

ACCUMULATION POINTS

Let X be a topological space. A point $p \in X$ is an *accumulation point* or *limit point* (also called *cluster point* or *derived point*) of a subset A of X iff every open set G containing p contains a point of A different from p, i.e.,

$$G \text{ open}, \ p \in G \quad \text{implies} \quad (G \setminus \{p\}) \cap A \neq \emptyset$$

The set of accumulation points of A, denoted by A', is called the *derived set* of A.

Example 2.1: The class $\quad T = \{X, \emptyset, \{a\}, \{c, d\}, \{a, c, d\}, \{b, c, d\, e\}\}$

defines a topology on $X = \{a, b, c, d, e\}$. Consider the subset $A = \{a, b, c\}$ of X. Observe that $b \in X$ is a limit point of A since the open sets containing b are $\{b, c, d, e\}$ and X, and each contains a point of A different from b, i.e. c. On the other hand, the point $a \in X$ is not a limit point of A since the open set $\{a\}$, which contains a, does not contain a point of A different from a. Similarly, the points d and e are limit points of A and the point c is not a limit point of A. So $A' = \{b, d, e\}$ is the derived set of A.

Example 2.2:   Let X be an indiscrete topological space, i.e. X and $\emptyset$ are the only open subsets of X. Then X is the only open set containing any point $p \in X$. Hence p is an accumulation point of every subset of X except the empty set $\emptyset$ and the set consisting of p alone, i.e. the singleton set $\{p\}$. Accordingly, the derived set A' of any subset A of X is as follows:

$$A' = \begin{cases} \emptyset & \text{if } A = \emptyset \\ \{p\}^c = X \setminus \{p\} & \text{if } A = \{p\} \\ X & \text{if } A \text{ contains two or more points} \end{cases}$$

Observe that, for the usual topology on the line $\mathbf{R}$ and the plane $\mathbf{R}^2$, the above definition of an accumulation point is the same as that given in Chapter 4.

CLOSED SETS

Let X be a topological space. A subset A of X is a *closed set* iff its complement A^c is an open set.

Example 3.1:   The class $\mathcal{T} = \{X, \emptyset, \{a\}, \{c,d\}, \{a,c,d\}, \{b,c,d,e\}\}$

defines a topology on $X = \{a,b,c,d,e\}$. The closed subsets of X are

$$\emptyset, \ X, \ \{b,c,d,e\}, \ \{a,b,e\}, \ \{b,e\}, \ \{a\}$$

that is, the complements of the open subsets of X. Note that there are subsets of X, such as $\{b,c,d,e\}$, which are both open and closed, and there are subsets of X, such as $\{a,b\}$, which are neither open nor closed.

Example 3.2:   Let X be a discrete topological space, i.e. every subset of X is open. Then every subset of X is also closed since its complement is always open. In other words, all subsets of X are both open and closed.

Recall that $A^{cc} = A$, for any subset A of a space X. Hence

Proposition 5.2:   In a topological space X, a subset A of X is open if and only if its complement is closed.

The axioms [O₁], [O₂] and [O₃] of a topological space and DeMorgan's Laws give

Theorem 5.3:   Let X be a topological space. Then the class of closed subsets of X possesses the following properties:

(i) X and $\emptyset$ are closed sets.

(ii)   The intersection of any number of closed sets is closed.

(iii)   The union of any two closed sets is closed.

Closed sets can also be characterized in terms of their limit points as follows:

Theorem 5.4:   A subset A of a topological space X is closed if and only if A contains each of its accumulation points.

In other words, a set A is closed if and only if the derived set A' of A is a subset of A, i.e. $A' \subset A$.

CLOSURE OF A SET

Let A be a subset of a topological space X. The *closure of A*, denoted by

$$\bar{A} \quad \text{or} \quad A^-$$

is the intersection of all closed supersets of A. In other words, if $\{F_i : i \in I\}$ is the class of all closed subsets of X containing A, then

$$\bar{A} = \cap_i F_i$$

Observe first that $\bar{A}$ is a closed set since it is the intersection of closed sets. Furthermore, $\bar{A}$ is the smallest closed superset of A, that is, if F is a closed set containing A, then

$$A \subset \bar{A} \subset F$$

Accordingly, a set A is closed if and only if $A = \bar{A}$. We state these results formally:

Proposition 5.5: Let $\bar{A}$ be the closure of a set A. Then: (i) $\bar{A}$ is closed; (ii) if F is a closed superset of A, then $A \subset \bar{A} \subset F$; and (iii) A is closed iff $A = \bar{A}$.

> **Example 4.1:** Consider the topology $\mathcal{T}$ on $X = \{a, b, c, d, e\}$ of Example 3.1 where the closed subsets of X are
>
> $$\emptyset,\ X,\ \{b, c, d, e\},\ \{a, b, e\},\ \{b, e\},\ \{a\}$$
>
> Accordingly, $\quad \overline{\{b\}} = \{b, e\}, \quad \overline{\{a, c\}} = X, \quad \overline{\{b, d\}} = \{b, c, d, e\}$

> **Example 4.2:** Let X be a cofinite topological space, i.e. the complements of finite sets and $\emptyset$ are the open sets. Then the closed sets are precisely the finite subsets of X together with X. Hence if $A \subset X$ is finite, its closure $\bar{A}$ is A itself since A is closed. On the other hand, if $A \subset X$ is infinite then X is the only closed superset of A; so $\bar{A}$ is X. More concisely, for any subset A of a cofinite space X,
>
> $$\bar{A} \;=\; \begin{cases} A & \text{if } A \text{ is finite} \\ X & \text{if } A \text{ is infinite} \end{cases}$$

The closure of a set can be completely described in terms of its limit points as follows:

Theorem 5.6: Let A be a subset of a topological space X. Then the closure of A is the union of A and its set of accumulation points, i.e.,

$$\bar{A} \;=\; A \cup A'$$

A point $p \in X$ is called a *closure point* or *adherent point* of $A \subset X$ iff p belongs to the closure of A, i.e. $p \in \bar{A}$. In view of the preceding theorem, $p \in X$ is a closure point of $A \subset X$ iff $p \in A$ or p is a limit point of A.

> **Example 4.3:** Consider the set $\mathbf{Q}$ of rational numbers. As seen previously, in the usual topology for $\mathbf{R}$, every real number $a \in \mathbf{R}$ is a limit point of $\mathbf{Q}$. Hence the closure of $\mathbf{Q}$ is the entire set $\mathbf{R}$ of real numbers, i.e. $\bar{\mathbf{Q}} = \mathbf{R}$.

A subset A of a topological space X is said to be *dense* in $B \subset X$ if B is contained in the closure of A, i.e. $B \subset \bar{A}$. In particular, A is dense in X or is a dense subset of X iff $\bar{A} = X$.

> **Example 4.4:** Observe in Example 4.1 that
>
> $$\overline{\{a, c\}} = X \quad \text{and} \quad \overline{\{b, d\}} = \{b, c, d, e\}$$
>
> where $X = \{a, b, c, d, e\}$. Hence the set $\{a, c\}$ is a dense subset of X but the set $\{b, d\}$ is not.

> **Example 4.5:** As noted in Example 4.3, $\bar{\mathbf{Q}} = \mathbf{R}$. In other words, in the usual topology, the set $\mathbf{Q}$ of rational numbers is dense in $\mathbf{R}$.

The operator "closure", assigning to each subset A of X its closure $\bar{A} \subset X$ satisfies the four properties appearing in the proposition below, called the Kuratowski Closure Axioms. In fact, these axioms may be used to define a topology on X, as we shall prove subsequently.

Proposition 5.7: (i) $\bar{\emptyset} = \emptyset$; (ii) $A \subset \bar{A}$; (iii) $\overline{A \cup B} = \bar{A} \cup \bar{B}$; and (iv) $(A^-)^- = \bar{A}$.

INTERIOR, EXTERIOR, BOUNDARY

Let A be a subset of a topological space X. A point $p \in A$ is called an *interior point* of A if p belongs to an open set G contained in A:

$$p \in G \subset A \qquad \text{where } G \text{ is open}$$

The set of interior points of A, denoted by

$$\text{int}(A), \quad \overset{\circ}{A} \quad \text{or} \quad A°$$

is called the *interior* of A. The interior of A can also be characterized as follows:

Proposition 5.8:   The interior of a set A is the union of all open subsets of A. Furthermore: (i) $A°$ is open; (ii) $A°$ is the largest open subset of A, i.e. if G is an open subset of A then $G \subset A° \subset A$; and (iii) A is open iff $A = A°$.

The *exterior* of A, written $\text{ext}(A)$, is the interior of the complement of A, i.e. $\text{int}(A^c)$. The *boundary* of A, written $\text{b}(A)$, is the set of points which do not belong to the interior or the exterior of A. Next follows an important relationship between interior, exterior and closure.

Theorem 5.9:   Let A be any subset of a topological space X. Then the closure of A is the union of the interior and boundary of A, i.e. $\bar{A} = A° \cup \text{b}(A)$.

> **Example 5.1:**   Consider the four intervals $[a, b]$, (a, b), $(a, b]$ and $[a, b)$ whose endpoints are a and b. The interior of each is the open interval (a, b) and the boundary of each is the set of endpoints, i.e. $\{a, b\}$.

> **Example 5.2:**   Consider the topology
> $$\mathcal{T} = \{X, \emptyset, \{a\}, \{c, d\}, \{a, c, d\}, \{b, c, d, e\}\}$$
> on $X = \{a, b, c, d, e\}$, and the subset $A = \{b, c, d\}$ of X. The points c and d are each interior points of A since
> $$c, d \in \{c, d\} \subset A$$
> where $\{c, d\}$ is an open set. The point $b \in A$ is not an interior point of A; so $\text{int}(A) = \{c, d\}$. Only the point $a \in X$ is exterior to A, i.e. interior to the complement $A^c = \{a, e\}$ of A; hence $\text{int}(A^c) = \{a\}$. Accordingly the boundary of A consists of the points b and e, i.e. $\text{b}(A) = \{b, e\}$.

> **Example 5.3:**   Consider the set $\mathbf{Q}$ of rational numbers. Since every open subset of $\mathbf{R}$ contains both rational and irrational points, there are no interior or exterior points of $\mathbf{Q}$; so $\text{int}(\mathbf{Q}) = \emptyset$ and $\text{int}(\mathbf{Q}^c) = \emptyset$. Hence the boundary of $\mathbf{Q}$ is the entire set of real numbers, i.e. $\text{b}(\mathbf{Q}) = \mathbf{R}$.

A subset A of a topological space X is said to be *nowhere dense* in X if the interior of the closure of A is empty, i.e. $\text{int}(\bar{A}) = \emptyset$.

> **Example 5.4:**   Consider the subset $A = \{1, \frac{1}{2}, \frac{1}{3}, \frac{1}{4}, \ldots\}$ of $\mathbf{R}$. As noted previously, A has exactly one limit point, 0. Hence $\bar{A} = \{0, 1, \frac{1}{2}, \frac{1}{3}, \frac{1}{4}, \ldots\}$. Observe that $\bar{A}$ has no interior points; so A is nowhere dense in $\mathbf{R}$.

> **Example 5.5:**   Let A consist of the rational points between 0 and 1, i.e. $A = \{x : x \in Q, 0 < x < 1\}$. Observe that the interior of A is empty, i.e. $\text{int}(A) = \emptyset$. But A is not nowhere dense in $\mathbf{R}$; for the closure of A is $[0, 1]$, and so
> $$\text{int}(\bar{A}) = \text{int}([0, 1]) = (0, 1)$$
> is not empty.

NEIGHBORHOODS AND NEIGHBORHOOD SYSTEMS

Let p be a point in a topological space X. A subset N of X is a *neighborhood* of p iff N is a superset of an open set G containing p:

$$p \in G \subset N \qquad \text{where } G \text{ is an open set}$$

In other words, the relation "N is a neighborhood of a point p" is the inverse of the relation "p is an interior point of N". The class of all neighborhoods of $p \in X$, denoted by $\mathcal{N}_p$, is called the *neighborhood system* of p.

> **Example 6.1:**   Let a be any real number, i.e. $a \in \mathbf{R}$. Then each closed interval $[a - \delta, a + \delta]$, with center a, is a neighborhood of a since it contains the open interval $(a - \delta, a + \delta)$ containing a. Similarly, if p is a point in the plane $\mathbf{R}^2$, then every closed disc $\{q \in \mathbf{R}^2 : d(p, q) < \delta \neq 0\}$, with center p, is a neighborhood of p since it contains the open disc with center p.

The central facts about the neighborhood system $\mathcal{N}_p$ of any point $p \in X$ are the four properties appearing in the proposition below, called the Neighborhood Axioms. In fact, these axioms may be used to define a topology on X, as we shall note subsequently.

Proposition 5.10:   (i) $\mathcal{N}_p$ is not empty and p belongs to each member of $\mathcal{N}_p$.

  (ii)   The intersection of any two members of $\mathcal{N}_p$ belongs to $\mathcal{N}_p$.

  (iii)   Every superset of a member of $\mathcal{N}_p$ belongs to $\mathcal{N}_p$.

  (iv)   Each member $N \in \mathcal{N}_p$ is a superset of a member $G \in \mathcal{N}_p$ where G is a neighborhood of each of its points, i.e. $G \in \mathcal{N}_g$ for every $g \in G$.

CONVERGENT SEQUENCES

A sequence $\langle a_1, a_2, \ldots \rangle$ of points in a topological space X *converges* to a point $b \in X$, or b is the *limit* of the sequence $\langle a_n \rangle$, denoted by

$$\lim_{n \to \infty} a_n = b, \quad \lim a_n = b \quad \text{or} \quad a_n \to b$$

iff for each open set G containing b there exists a positive integer $n_0 \in N$ such that

$$n > n_0 \quad \text{implies} \quad a_n \in G$$

that is, if G contains almost all, i.e. all except a finite number, of the terms of the sequence.

Example 7.1:   Let $\langle a_1, a_2, \ldots \rangle$ be a sequence of points in an indiscrete topological space $(X, \mathcal{J})$. Note that: (i) X is the only open set containing any point $b \in X$; and (ii) X contains every term of the sequence $\langle a_n \rangle$. Accordingly, the sequence $\langle a_1, a_2, \ldots \rangle$ converges to every point $b \in X$.

Example 7.2:   Let $\langle a_1, a_2, \ldots \rangle$ be a sequence of points in a discrete topological space $(X, \mathcal{D})$. Now for every point $b \in X$, the singleton set $\{b\}$ is an open set containing b. So, if $a_n \to b$, then the set $\{b\}$ must contain almost all of the terms of the sequence. In other words, the sequence $\langle a_n \rangle$ converges to a point $b \in X$ iff the sequence is of the form $\langle a_1, a_2, \ldots, a_{n_0}, b, b, b, \ldots \rangle$.

Example 7.3:   Let T be the topology on an infinite set X which consists of $\emptyset$ and the complements of countable sets (see Problem 56). We claim that a sequence $\langle a_1, a_2, \ldots \rangle$ in X converges to $b \in X$ iff the sequence is also of the form $\langle a_1, a_2, \ldots, a_{n_0}, b, b, b, \ldots \rangle$, i.e. the set A consisting of the terms of $\langle a_n \rangle$ different from b is finite. Now A is countable and so A^c is an open set containing b. Hence if $a_n \to b$ then A^c contains all except a finite number of the terms of the sequence, and so A is finite.

COARSER AND FINER TOPOLOGIES

Let T_1 and T_2 be topologies on a non-empty set X. Suppose that each T_1-open subset of X is also a T_2-open subset of X. That is, suppose that T_1 is a subclass of T_2, i.e. $T_1 \subset T_2$. Then we say that T_1 is *coarser* or *smaller* (sometimes called *weaker*) than T_2 or that T_2 is *finer* or *larger* than T_1. Observe that the collection $\mathbf{T} = \{T_i\}$ of all topologies on X is partially ordered by class inclusion; so we shall also write

$$T_1 \precsim T_2 \quad \text{for} \quad T_1 \subset T_2$$

and we shall say that two topologies on X are *not comparable* if neither is coarser than the other.

Example 8.1:   Consider the discrete topology $\mathcal{D}$, the indiscrete topology $\mathcal{J}$ and any other topology T on any set X. Then T is coarser than $\mathcal{D}$ and T is finer than $\mathcal{J}$. That is, $\mathcal{J} \precsim T \precsim \mathcal{D}$.

Example 8.2:   Consider the cofinite topology T and the usual topology $\mathcal{U}$ on the plane $\mathbf{R}^2$. Recall that every finite subset of $\mathbf{R}^2$ is a $\mathcal{U}$-closed set; hence the complement of any finite subset of $\mathbf{R}^2$, i.e. any member of T, is also a $\mathcal{U}$-open set. In other words, T is coarser than $\mathcal{U}$, i.e. $T \precsim \mathcal{U}$.

SUBSPACES, RELATIVE TOPOLOGIES

Let A be a non-empty subset of a topological space $(X, \mathcal{T})$. The class $\mathcal{T}_A$ of all intersections of A with $\mathcal{T}$-open subsets of X is a topology on A; it is called the *relative topology* on A or the *relativization* of $\mathcal{T}$ to A, and the topological space $(A, \mathcal{T}_A)$ is called a *subspace* of $(X, \mathcal{T})$. In other words, a subset H of A is a $\mathcal{T}_A$-open set, i.e. open relative to A, if and only if there exists a $\mathcal{T}$-open subset G of X such that

$$H = G \cap A$$

Example 9.1:   Consider the topology

$$\mathcal{T} = \{X, \varnothing, \{a\}, \{c, d\}, \{a, c, d\}, \{b, c, d, e\}\}$$

on $X = \{a, b, c, d, e\}$, and the subset $A = \{a, d, e\}$ of X. Observe that

$$X \cap A = A, \quad \{a\} \cap A = \{a\}, \quad \{a, c, d\} \cap A = \{a, d\}$$

$$\varnothing \cap A = \varnothing, \quad \{c, d\} \cap A = \{d\}, \quad \{b, c, d, e\} \cap A = \{d, e\}$$

Hence the relativization of $\mathcal{T}$ to A is

$$\mathcal{T}_A = \{A, \varnothing, \{a\}, \{d\}, \{a, d\}, \{d, e\}\}$$

Example 9.2:   Consider the usual topology $\mathcal{U}$ on $\mathbf{R}$ and the relative topology $\mathcal{T}_A$ on the closed interval $A = [3, 8]$. Note that the closed-open interval $[3, 5)$ is open in the relative topology on A, i.e. is a $\mathcal{T}_A$-open set, since

$$[3, 5) = (2, 5) \cap A$$

where $(2, 5)$ is a $\mathcal{T}$-open subset of $\mathbf{R}$. Thus we see that a set may be open relative to a subspace but be neither open nor closed in the entire space.

EQUIVALENT DEFINITIONS OF TOPOLOGIES

Our definition of a topological space gave axioms for the open sets in the topological space, that is, we used the open set as the primitive notion for the topology. We now state two theorems which exhibit alternate methods of defining a topology on a set, using as primitives the notions of "neighborhood of a point" and "closure of a set".

Theorem 5.11:   Let X be a non-empty set and let there be assigned to each point $p \in X$ a class $\mathcal{A}_p$ of subsets of X satisfying the following axioms:

[A_1] $\mathcal{A}_p$ is not empty and p belongs to each member of $\mathcal{A}_p$.

[A_2]   The intersection of any two members of $\mathcal{A}_p$ belongs to $\mathcal{A}_p$.

[A_3]   Every superset of a member of $\mathcal{A}_p$ belongs to $\mathcal{A}_p$.

[A_4]   Each member $N \in \mathcal{A}_p$ is a superset of a member $G \in \mathcal{A}_p$ such that $G \in \mathcal{A}_g$ for every $g \in G$.

Then there exists one and only one topology $\mathcal{T}$ on X such that $\mathcal{A}_p$ is the $\mathcal{T}$-neighborhood system of the point $p \in X$.

Theorem 5.12:   Let X be a non-empty set and let k be an operation which assigns to each subset A of X the subset A^k of X, satisfying the following axioms, called the Kuratowski Closure Axioms:

[K_1] $\varnothing^k = \varnothing$

[K_2] $A \subset A^k$

[K_3] $(A \cup B)^k = A^k \cup B^k$

[K_4] $(A^k)^k = A^k$

Then there exists one and only one topology $\mathcal{T}$ on X such that A^k will be the $\mathcal{T}$-closure of the subset A of X.

Solved Problems

TOPOLOGIES, OPEN SETS

1. Let $X = \{a, b, c, d, e\}$. Determine whether or not each of the following classes of subsets of X is a topology on X.

   (i) $T_1 = \{X, \emptyset, \{a\}, \{a, b\}, \{a, c\}\}$

   (ii) $T_2 = \{X, \emptyset, \{a, b, c\}, \{a, b, d\}, \{a, b, c, d\}\}$

   (iii) $T_3 = \{X, \emptyset, \{a\}, \{a, b\}, \{a, c, d\}, \{a, b, c, d\}\}$

   Solution:

   (i) T_1 is not a topology on X since
   $$\{a, b\}, \{a, c\} \in T_1 \quad \text{but} \quad \{a, b\} \cup \{a, c\} = \{a, b, c\} \notin T_1$$

   (ii) T_2 is not a topology on X since
   $$\{a, b, c\}, \{a, b, d\} \in T_2 \quad \text{but} \quad \{a, b, c\} \cap \{a, b, d\} = \{a, b\} \notin T_2$$

   (iii) T_3 is a topology on X since it satisfies the necessary axioms.

2. Let T be the class consisting of $\mathbf{R}$, $\emptyset$ and all infinite open intervals $A_q = (q, \infty)$ with $q \in \mathbf{Q}$, the rationals. Show that T is not a topology on $\mathbf{R}$.

   Solution:
   Observe that
   $$A = \cup \{A_q : q \in \mathbf{Q}, \, q > \sqrt{2}\} = (\sqrt{2}, \infty)$$

   is the union of members of T, but $A \notin T$ since $\sqrt{2}$ is irrational. Hence T violates $[\mathbf{O_2}]$ and is therefore not a topology on $\mathbf{R}$.

3. Let T be a topology on a set X consisting of four sets, i.e.
   $$T = \{X, \emptyset, A, B\}$$

   where A and B are non-empty distinct proper subsets of X. What conditions must A and B satisfy?

   Solution:
   Since $A \cap B$ must also belong to T, there are two possibilities:

   Case I. $A \cap B = \emptyset$
   Then $A \cup B$ cannot be A or B; hence $A \cup B = X$. Thus the class $\{A, B\}$ is a partition of X.

   Case II. $A \cap B = A$ or $A \cap B = B$
   In either case, one of the sets is a subset of the other, and the members of T are totally ordered by inclusion: $\emptyset \subset A \subset B \subset X$ or $\emptyset \subset B \subset A \subset X$.

4. List all topologies on $X = \{a, b, c\}$ which consist of exactly four members.

   Solution:
   Each topology T on X with four members is of the form $T = \{X, \emptyset, A, B\}$ where A and B correspond to Case I or Case II of the preceding problem.

   Case I. $\{A, B\}$ is a partition of X.
   The topologies in this case are the following:
   $$T_1 = \{X, \emptyset, \{a\}, \{b, c\}\}, \quad T_2 = \{X, \emptyset, \{b\}, \{a, c\}\}, \quad T_3 = \{X, \emptyset, \{c\}, \{a, b\}\}$$

   Case II. The members of T are totally ordered by inclusion.
   The topologies in this case are the following:

   $$T_4 = \{X, \emptyset, \{a\}, \{a, b\}\} \qquad T_7 = \{X, \emptyset, \{b\}, \{a, b\}\}$$
   $$T_5 = \{X, \emptyset, \{a\}, \{a, c\}\} \qquad T_8 = \{X, \emptyset, \{c\}, \{a, c\}\}$$
   $$T_6 = \{X, \emptyset, \{b\}, \{b, c\}\} \qquad T_9 = \{X, \emptyset, \{c\}, \{b, c\}\}$$

5. Let $f : X \to Y$ be a function from a non-empty set X into a topological space $(Y, \mathcal{U})$. Furthermore, let $\mathcal{T}$ be the class of inverses of open subsets of Y:
$$\mathcal{T} \;=\; \{f^{-1}[G] \,:\, G \in \mathcal{U}\}$$
Show that $\mathcal{T}$ is a topology on X.

Solution:

Since $\mathcal{U}$ is a topology, $Y, \emptyset \in \mathcal{U}$. But $X = f^{-1}[Y]$ and $\emptyset = f^{-1}[\emptyset]$, so $X, \emptyset \in \mathcal{T}$ and $\mathcal{T}$ satisfies $[\mathbf{O_1}]$.

Let $\{A_i\}$ be a class of sets in $\mathcal{T}$. By definition, there exist $G_i \in \mathcal{U}$ for which $A_i = f^{-1}[G_i]$. But
$$\cup_i A_i \;=\; \cup_i f^{-1}[G_i] \;=\; f^{-1}[\cup_i G_i]$$
Since $\mathcal{U}$ is a topology, $\cup_i G_i \in \mathcal{U}$, so $\cup_i A_i \in \mathcal{T}$, and $\mathcal{T}$ satisfies $[\mathbf{O_2}]$.

Lastly, let $A_1, A_2 \in \mathcal{T}$. Then
$$\exists\, G_1, G_2 \in \mathcal{U} \quad \text{such that} \quad A_1 = f^{-1}[G_1], \;\; A_2 = f^{-1}[G_2]$$
But
$$A_1 \cap A_2 \;=\; f^{-1}[G_1] \cap f^{-1}[G_2] \;=\; f^{-1}[G_1 \cap G_2]$$
and $G_1 \cap G_2 \in \mathcal{U}$. Thus $A_1 \cap A_2 \in \mathcal{T}$ and $[\mathbf{O_3}]$ is also satisfied.

6. Consider the second axiom for a topology $\mathcal{T}$ on a set X:

[$\mathbf{O_2}$] The union of any number of sets in $\mathcal{T}$ belongs to $\mathcal{T}$.

Show that $[\mathbf{O_2}]$ can be replaced by the following weaker axiom:

[$\mathbf{O_2'}$] The union of any number of sets in $\mathcal{T} \setminus \{X, \emptyset\}$ belongs to $\mathcal{T}$.

In other words, show that the axioms $[\mathbf{O_1}]$, $[\mathbf{O_2'}]$ and $[\mathbf{O_3}]$ are equivalent to the axioms $[\mathbf{O_1}]$, $[\mathbf{O_2}]$ and $[\mathbf{O_3}]$.

Solution:

Let $\mathcal{T}$ be a class of subsets of X satisfying $[\mathbf{O_1}]$, $[\mathbf{O_2'}]$ and $[\mathbf{O_3}]$, and let $\mathcal{A}$ be a subclass of $\mathcal{T}$. We want to show that $\mathcal{T}$ also satisfies $[\mathbf{O_2}]$, i.e. that $\bigcup\{E : E \in \mathcal{A}\} \in \mathcal{T}$.

Case I. $X \in \mathcal{A}$.

Then $\bigcup\{E : E \in \mathcal{A}\} = X$ and therefore belongs to $\mathcal{T}$ by $[\mathbf{O_1}]$.

Case II. $X \notin \mathcal{A}$.

Then
$$\bigcup\{E : E \in \mathcal{A}\} \;=\; \bigcup\{E : E \in \mathcal{A} \setminus \{X\}\}$$
But the empty set $\emptyset$ does not contribute any elements to a union of sets; hence
$$\bigcup\{E : E \in \mathcal{A}\} \;=\; \bigcup\{E : E \in \mathcal{A} \setminus \{X\}\} \;=\; \bigcup\{E : E \in \mathcal{A} \setminus \{X, \emptyset\}\} \qquad (1)$$
Since $\mathcal{A}$ is a subclass of $\mathcal{T}$, $\mathcal{A} \setminus \{X, \emptyset\}$ is a subclass of $\mathcal{T} \setminus \{X, \emptyset\}$, so by $[\mathbf{O_2'}]$ the union in (1) belongs to $\mathcal{T}$.

7. Prove: Let A be a subset of a topological space X with the property that each point $p \in A$ belongs to an open set G_p contained in A. Then A is open.

Solution:

For each point $p \in A$, $p \in G_p \subset A$. Hence $\bigcup\{G_p : p \in A\} = A$ and so A is a union of open sets and, by $[\mathbf{O_2}]$, is open.

8. Let $\mathcal{T}$ be a class of subsets of X totally ordered by set inclusion. Show that $\mathcal{T}$ satisfies $[\mathbf{O_3}]$, i.e. the intersection of any two members of $\mathcal{T}$ belongs to $\mathcal{T}$.

Solution:

Let $A, B \in \mathcal{T}$. Since $\mathcal{T}$ is totally ordered by set inclusion,
$$\text{either} \quad A \cap B = A \quad \text{or} \quad A \cap B = B$$
In either case $A \cap B \in \mathcal{T}$, and so $\mathcal{T}$ satisfies $[\mathbf{O_3}]$.

9. Let $\mathcal{T}$ be the class of subsets of $\mathbf{R}$ consisting of $\mathbf{R}$, $\emptyset$ and all open infinite intervals $E_a = (a, \infty)$ with $a \in \mathbf{R}$. Show that $\mathcal{T}$ is a topology on $\mathbf{R}$.

Solution:

Since $\mathbf{R}$ and $\emptyset$ belong to $\mathcal{T}$, $\mathcal{T}$ satisfies $[\mathbf{O_1}]$. Observe that $\mathcal{T}$ is totally ordered by set inclusion; hence $\mathcal{T}$ satisfies $[\mathbf{O_3}]$.

Now let $\mathcal{A}$ be a subclass of $\mathcal{T} \setminus \{X, \emptyset\}$, that is $\mathcal{A} = \{E_i : i \in I\}$ where I is some set of real numbers. We want to show that $\cup_i E_i$ belongs to $\mathcal{T}$. If I is not bounded from below, i.e. if $\inf(I) = -\infty$, then $\cup_i E_i = \mathbf{R}$. If I is bounded from below, say $\inf(I) = i_0$, then $\cup_i E_i = (i_0, \infty) = E_{i_0}$. In either case, $\cup_i E_i \in \mathcal{T}$, and $\mathcal{T}$ satisfies $[\mathbf{O_2'}]$.

10. Let $\mathcal{T}$ be the class of subsets of $\mathbf{N}$ consisting of $\emptyset$ and all subsets of $\mathbf{N}$ of the form $E_n = \{n, n+1, n+2, \ldots\}$ with $n \in \mathbf{N}$.

(i) Show that $\mathcal{T}$ is a topology on $\mathbf{N}$.

(ii) List the open sets containing the positive integer 6.

Solution:

(i) Since $\emptyset$ and $E_1 = \{1, 2, 3, \ldots\} = \mathbf{N}$ belong to $\mathcal{T}$, $\mathcal{T}$ satisfies $[\mathbf{O_1}]$. Furthermore, since $\mathcal{T}$ is totally ordered by set inclusion, $\mathcal{T}$ also satisfies $[\mathbf{O_3}]$.

Now let $\mathcal{A}$ be a subclass of $\mathcal{T} \setminus \{\mathbf{N}, \emptyset\}$, that is, $\mathcal{A} = \{E_n : n \in I\}$ where I is some set of positive integers. Note that I contains a smallest positive integer n_0 and

$$\cup \{E_n : n \in I\} = \{n_0, n_0 + 1, n_0 + 2, \ldots\} = E_{n_0}$$

which belongs to $\mathcal{T}$. Hence $\mathcal{T}$ satisfies $[\mathbf{O_2'}]$, and so $\mathcal{T}$ is a topology on $\mathbf{N}$.

(ii) Since the non-empty open sets are of the form

$$E_n = \{n, n+1, n+2, \ldots\}$$

with $n \in \mathbf{N}$, the open sets containing the positive integer 6 are the following:

$$E_1 = \mathbf{N} = \{1, 2, 3, \ldots\} \qquad E_4 = \{4, 5, 6, \ldots\}$$

$$E_2 = \{2, 3, 4, \ldots\} \qquad E_5 = \{5, 6, 7, \ldots\}$$

$$E_3 = \{3, 4, 5, \ldots\} \qquad E_6 = \{6, 7, 8, \ldots\}$$

ACCUMULATION POINTS, DERIVED SETS

11. Let $\mathcal{T}$ be the topology on $\mathbf{N}$ which consists of $\emptyset$ and all subsets of $\mathbf{N}$ of the form $E_n = \{n, n+1, n+2, \ldots\}$ where $n \in \mathbf{N}$ as in Problem 10.

(i) Find the accumulation points of the set $A = \{4, 13, 28, 37\}$.

(ii) Determine those subsets E of $\mathbf{N}$ for which $E' = \mathbf{N}$.

Solution:

(i) Observe that the open sets containing any point $p \in \mathbf{N}$ are the sets E_i where $i \le p$. If $n_0 \le 36$, then every open set containing n_0 also contains $37 \in A$ which is different from n_0; hence $n_0 \le 36$ is a limit point of A. On the other hand, if $n_0 > 36$ then the open set $E_{n_0} = \{n_0, n_0 + 1, n_0 + 2, \ldots\}$ contains no point of A different from n_0. So $n_0 > 36$ is not a limit point of A. Accordingly, the derived set of A is $A' = \{1, 2, 3, \ldots, 34, 35, 36\}$.

(ii) If E is an infinite subset of $\mathbf{N}$ then E is not bounded from above. So every open set containing any point $p \in \mathbf{N}$ will contain points of E other than p. Hence $E' = \mathbf{N}$.

On the other hand, if E is finite then E is bounded from above, say, by $n_0 \in \mathbf{N}$. Then the open set $E_{n_0 + 1}$ contains no point of E. Hence $n_0 + 1 \in \mathbf{N}$ is not a limit point of E, and so $E' \ne \mathbf{N}$.

12. Let A be a subset of a topological space $(X, \mathcal{T})$. When will a point $p \in X$ not be a limit point of A?

   Solution:

   The point $p \in X$ is a limit point of A iff every open neighborhood of p contains a point of A other than p, i.e.,

   $$p \in G \text{ and } G \in \mathcal{T} \quad \text{implies} \quad (G \setminus \{p\}) \cap A \neq \emptyset$$

   So p is not a limit point of A if there exists an open set G such that

   $$p \in G \quad \text{and} \quad (G \setminus \{p\}) \cap A = \emptyset$$

   or, equivalently, $p \in G \quad \text{and} \quad G \cap A = \emptyset \text{ or } G \cap A = \{p\}$

   or, equivalently, $p \in G \quad \text{and} \quad G \cap A \subset \{p\}$

13. Let A be any subset of a discrete topological space X. Show that the derived set A' of A is empty.

   Solution:

   Let p be any point in X. Recall that every subset of a discrete space is open. Hence, in particular, the singleton set $G = \{p\}$ is an open subset of X. But

   $$p \in G \quad \text{and} \quad G \cap A = (\{p\} \cap A) \subset \{p\}$$

   Hence, by the above problem, $p \notin A'$ for every $p \in X$, i.e. $A' = \emptyset$.

14. Consider the topology

   $$\mathcal{T} = \{X, \emptyset, \{a\}, \{a, b\}, \{a, c, d\}, \{a, b, c, d\}, \{a, b, e\}\}$$

   on $X = \{a, b, c, d, e\}$. Determine the derived sets of (i) $A = \{c, d, e\}$ and (ii) $B = \{b\}$.

   Solution:

   (i)   Note that $\{a, b\}$ and $\{a, b, e\}$ are open subsets of X and that

   $$a, b \in \{a, b\} \quad \text{and} \quad \{a, b\} \cap A = \emptyset$$

   $$e \in \{a, b, e\} \quad \text{and} \quad \{a, b, e\} \cap A = \{e\}$$

   Hence a, b and e are not limit points of A. On the other hand, every other point in X is a limit point of A since every open set containing it also contains a point of A different from it. Accordingly, $A' = \{c, d\}$.

   (ii)   Note that $\{a\}$, $\{a, b\}$ and $\{a, c, d\}$ are open subsets of X and that

   $$a \in \{a\} \quad \text{and} \quad \{a\} \cap B = \emptyset$$

   $$b \in \{a, b\} \quad \text{and} \quad \{a, b\} \cap B = \{b\}$$

   $$c, d \in \{a, c, d\} \quad \text{and} \quad \{a, c, d\} \cap B = \emptyset$$

   Hence a, b, c and d are not limit points of $B = \{b\}$. But e is a limit point of B since the open sets containing e are $\{a, b, e\}$ and X and each contains the point $b \in B$ different from e. Thus $B' = \{e\}$.

15. Prove: If A is a subset of B, then every limit point of A is also a limit point of B, i.e., $A \subset B$ implies $A' \subset B'$.

   Solution:

   Recall that $p \in A'$ iff $(G \setminus \{p\}) \cap A \neq \emptyset$ for every open set G containing p. But $B \supset A$; hence

   $$(G \setminus \{p\}) \cap B \supset (G \setminus \{p\}) \cap A \neq \emptyset$$

   So $p \in A'$ implies $p \in B'$, i.e. $A' \subset B'$.

16. Let $\mathcal{T}_1$ and $\mathcal{T}_2$ be topologies on X such that $\mathcal{T}_1 \subset \mathcal{T}_2$, i.e. every $\mathcal{T}_1$-open subset of X is also a $\mathcal{T}_2$-open subset of X. Furthermore, let A be any subset of X.

    (i)   Show that every $\mathcal{T}_2$-limit point of A is also a $\mathcal{T}_1$-limit point of A.

    (ii) Construct a space in which a $\mathcal{T}_1$-limit point is not a $\mathcal{T}_2$-limit point.

Solution:

(i) Let p be a $\mathcal{T}_2$-limit point of A; i.e. $(G \setminus \{p\}) \cap A \neq \emptyset$ for every $G \in \mathcal{T}_2$ such that $p \in G$. But $\mathcal{T}_1 \subset \mathcal{T}_2$; so, in particular, $(G \setminus \{p\}) \cap A \neq \emptyset$ for every $G \in \mathcal{T}_1$ such that $p \in G$, i.e. p is a $\mathcal{T}_1$-limit point of A.

(ii) Consider the usual topology $\mathcal{U}$ and the discrete topology $\mathcal{D}$ on $\mathbf{R}$. Note that $\mathcal{U} \subset \mathcal{D}$ since $\mathcal{D}$ contains every subset of $\mathbf{R}$. By Problem 13, 0 is not a $\mathcal{D}$-limit point of the set $A = \{1, \frac{1}{2}, \frac{1}{3}, \ldots\}$ since A' is empty. But 0 is a limit point of A with respect to the usual topology on $\mathbf{R}$.

17. Prove: Let A and B be subsets of a topological space $(X, \mathcal{T})$. Then $(A \cup B)' = A' \cup B'$.

Solution:

Utilizing Problem 15, $A \subset A \cup B$ implies $A' \subset (A \cup B)'$

$\qquad\qquad\qquad\qquad\qquad B \subset A \cup B$ implies $B' \subset (A \cup B)'$

So $A' \cup B' \subset (A \cup B)'$, and we need only show that
$$(A \cup B)' \subset A' \cup B'$$

Assume $p \notin A' \cup B'$; thus $\exists G, H \in \mathcal{T}$ such that
$$p \in G \text{ and } G \cap A \subset \{p\} \quad \text{and} \quad p \in H \text{ and } H \cap B \subset \{p\}$$

But $G \cap H \in \mathcal{T}$, $p \in G \cap H$ and
$$(G \cap H) \cap (A \cup B) = (G \cap H \cap A) \cup (G \cap H \cap B) \subset (G \cap A) \cup (H \cap B) \subset \{p\} \cup \{p\} = \{p\}$$

Thus $p \notin (A \cup B)'$, and so $(A \cup B)' \subset (A' \cup B')$.

CLOSED SETS, CLOSURE OPERATION, DENSE SETS

18. Consider the following topology on $X = \{a, b, c, d, e\}$:
$$\mathcal{T} = \{X, \emptyset, \{a\}, \{a, b\}, \{a, c, d\}, \{a, b, c, d\}, \{a, b, e\}\}$$

    (i)   List the closed subsets of X.

    (ii) Determine the closure of the sets $\{a\}$, $\{b\}$ and $\{c, e\}$.

    (iii) Which sets in (ii) are dense in X?

Solution:

(i) A set is closed iff its complement is open. Hence write the complement of each set in $\mathcal{T}$:
$$\emptyset, \ X, \ \{b, c, d, e\}, \ \{c, d, e\}, \ \{b, e\}, \ \{e\}, \ \{c, d\}$$

(ii) The closure $\bar{A}$ of any set A is the intersection of all closed supersets of A. The only closed superset of $\{a\}$ is X; the closed supersets of $\{b\}$ are $\{b, e\}$, $\{b, c, d, e\}$ and X; and the closed supersets of $\{c, e\}$ are $\{c, d, e\}$, $\{b, c, d, e\}$ and X. Thus,
$$\overline{\{a\}} = X, \quad \overline{\{b\}} = \{b, e\}, \quad \overline{\{c, e\}} = \{c, d, e\}$$

(iii) A set A is dense in X iff $\bar{A} = X$; so $\{a\}$ is the only dense set.

19. Let $\mathcal{T}$ be the topology on $\mathbf{N}$ which consists of $\emptyset$ and all subsets of $\mathbf{N}$ of the form $E_n = \{n, n+1, n+2, \ldots\}$ where $n \in \mathbf{N}$ as in Problem 10.

    (i)   Determine the closed subsets of $(\mathbf{N}, \mathcal{T})$.

    (ii) Determine the closure of the sets $\{7, 24, 47, 85\}$ and $\{3, 6, 9, 12, \ldots\}$.

    (iii) Determine those subsets of $\mathbf{N}$ which are dense in $\mathbf{N}$.

Solution:

(i) A set is closed iff its complement is open. Hence the closed subsets of $\mathbf{N}$ are as follows:
$$\mathbf{N}, \ \emptyset, \ \{1\}, \ \{1, 2\}, \ \{1, 2, 3\}, \ \ldots, \ \{1, 2, \ldots, m\}, \ \ldots$$

(ii)   The closure of a set is the smallest closed superset. So

$$\overline{\{7, 24, 47, 85\}} = \{1, 2, \ldots, 84, 85\}, \qquad \overline{\{3, 6, 9, 12, \ldots\}} = \{1, 2, 3, \ldots\} = \mathbf{N}$$

(iii) If a subset A of $\mathbf{N}$ is infinite, or equivalently unbounded, then $\bar{A} = \mathbf{N}$, i.e. A is dense in $\mathbf{N}$. If A is finite then its closure is not $\mathbf{N}$, i.e. A is not dense in $\mathbf{N}$.

20. Let $\mathcal{T}$ be the topology on $\mathbf{R}$ consisting of $\mathbf{R}$, $\emptyset$ and all open infinite intervals $E_a = (a, \infty)$ where $a \in \mathbf{R}$.

(i)   Determine the closed subsets of $(\mathbf{R}, \mathcal{T})$.

(ii) Determine the closure of the sets $[3, 7)$, $\{7, 24, 47, 85\}$ and $\{3, 6, 9, 12, \ldots\}$.

Solution:

(i)   A set is closed iff its complement is open. Hence the closed subsets of $(\mathbf{R}, \mathcal{T})$ are $\emptyset$, $\mathbf{R}$ and all closed infinite intervals $E_a^c = (-\infty, a]$.

(ii) The closure of a set is the smallest closed superset. Hence

$$\overline{[3, 7)} = (-\infty, 7], \qquad \overline{\{7, 24, 47, 85\}} = (-\infty, 85], \qquad \overline{\{3, 6, 9, 12, \ldots\}} = (-\infty, \infty) = \mathbf{R}$$

21. Let X be a discrete topological space. (i) Determine the closure of any subset A of X. (ii) Determine the dense subsets of X.

Solution:

(i)   Recall that in a discrete space X any $A \subset X$ is closed; hence $\bar{A} = A$.

(ii) A is dense in X iff $\bar{A} = X$. But $\bar{A} = A$, so X is the only dense subset of X.

22. Let X be an indiscrete space. (i) Determine the closed subsets of X. (ii) Determine the closure of any subset A of X. (iii) Determine the dense subsets of X.

Solution:

(i)   Recall that the only open subsets of an indiscrete space X are X and $\emptyset$; hence the closed subsets of X are also X and $\emptyset$.

(ii) If $A = \emptyset$, then $\bar{A} = \emptyset$. If $A \neq \emptyset$, then X is the only closed superset of A; so $\bar{A} = X$. That is, for any $A \subset X$,

$$\bar{A} = \begin{cases} \emptyset & \text{if } A = \emptyset \\ X & \text{if } A \neq \emptyset \end{cases}$$

(iii) $A \subset X$ is dense in X iff $\bar{A} = X$; hence every non-empty subset of X is dense in X.

23. Prove Theorem 5.4: A subset A of a topological space X is closed if and only if A contains each of its accumulation points, i.e. $A' \subset A$.

Solution:

Suppose A is closed, and let $p \notin A$, i.e. $p \in A^c$. But A^c, the complement of a closed set, is open; hence $p \notin A'$ for A^c is an open set such that

$$p \in A^c \quad \text{and} \quad A^c \cap A = \emptyset$$

Thus $A' \subset A$ if A is closed.

Now assume $A' \subset A$; we show that A^c is open. Let $p \in A^c$; then $p \notin A'$, so $\exists$ an open set G such that

$$p \in G \quad \text{and} \quad (G \setminus \{p\}) \cap A = \emptyset$$

But $p \notin A$; hence

$$G \cap A = (G \setminus \{p\}) \cap A = \emptyset$$

So $G \subset A^c$. Thus p is an interior point of A^c, and so A^c is open.

24. Prove: If F is a closed superset of any set A, then $A' \subset F$.

Solution:

By Problem 15, $A \subset F$ implies $A' \subset F'$. But $F' \subset F$, by Theorem 5.4, since F is closed. Thus $A' \subset F' \subset F$, which implies $A' \subset F$.

25. Prove: $A \cup A'$ is a closed set.

Solution:

Let $p \in (A \cup A')^c$. Since $p \notin A'$, $\exists$ an open set G such that

$$p \in G \quad \text{and} \quad G \cap A = \emptyset \text{ or } \{p\}$$

However, $p \notin A$; hence, in particular, $G \cap A = \emptyset$.

We also claim that $G \cap A' = \emptyset$. For if $g \in G$, then

$$g \in G \quad \text{and} \quad G \cap A = \emptyset$$

where G is an open set. So $g \notin A'$ and thus $G \cap A' = \emptyset$. Accordingly,

$$G \cap (A \cup A') = (G \cap A) \cup (G \cap A') = \emptyset \cup \emptyset = \emptyset$$

and so $G \subset (A \cup A')^c$. Thus p is an interior point of $(A \cup A')^c$ which is therefore an open set. Hence $A \cup A'$ is closed.

26. Prove Theorem 5.6: $\bar{A} = A \cup A'$.

Solution:

Since $A \subset \bar{A}$ and $\bar{A}$ is closed, $A' \subset (\bar{A})' \subset \bar{A}$ and hence $A \cup A' \subset \bar{A}$. But $A \cup A'$ is a closed set containing A, so $A \subset \bar{A} \subset A \cup A'$. Thus $\bar{A} = A \cup A'$.

27. Prove: If $A \subset B$ then $\bar{A} \subset \bar{B}$.

Solution:

If $A \subset B$, then by Problem 15, $A' \subset B'$. So $A \cup A' \subset B \cup B'$ or, by the preceding problem, $\bar{A} \subset \bar{B}$.

28. Prove: $\overline{A \cup B} = \bar{A} \cup \bar{B}$.

Solution:

Utilizing the preceding problem, $\bar{A} \subset \overline{A \cup B}$ and $\bar{B} \subset \overline{A \cup B}$; hence $(\bar{A} \cup \bar{B}) \subset \overline{A \cup B}$. But $(A \cup B) \subset (\bar{A} \cup \bar{B})$, a closed set since it is the union of two closed sets. Then (Proposition 5.5) $(A \cup B) \subset \overline{A \cup B} \subset (\bar{A} \cup \bar{B})$ and therefore $\overline{A \cup B} = \bar{A} \cup \bar{B}$.

29. Prove Proposition 5.7: (i) $\overline{\emptyset} = \emptyset$; (ii) $A \subset \bar{A}$; (iii) $\overline{A \cup B} = \bar{A} \cup \bar{B}$; and (iv) $(A^-)^- = A^-$.

Solution:

(i) and (iv): $\emptyset$ and $\bar{A}$ are closed; hence they are equal to their closures.   (ii) $A \subset A \cup A' = \bar{A}$ (Problem 26).   (iii) Preceding problem.

INTERIOR, EXTERIOR, BOUNDARY

30. Consider the following topology on $X = \{a, b, c, d, e\}$:

$$\mathcal{T} = \{X, \emptyset, \{a\}, \{a, b\}, \{a, c, d\}, \{a, b, c, d\}, \{a, b, e\}\}$$

(i) Find the interior points of the subset $A = \{a, b, c\}$ of X.   (ii) Find the exterior points of A.   (iii) Find the boundary points of A.

Solution:

(i)   The points a and b are interior points of A since

$$a, b \in \{a, b\} \subset A = \{a, b, c\}$$

where $\{a, b\}$ is an open set, i.e. since each belongs to an open set contained in A. Note that c is not an interior point of A since c does not belong to any open set contained in A. Hence $\text{int}(A) = \{a, b\}$ is the interior of A.

(ii)   The complement of A is $A^c = \{d, e\}$. Neither d nor e are interior points of A^c since neither belongs to any open subset of $A^c = \{d, e\}$. Hence $\text{int}(A^c) = \emptyset$, i.e. there are no exterior points of A.

(iii)   The boundary $b(A)$ of A consists of those points which are neither interior nor exterior to A. So $b(A) = \{c, d, e\}$.

31. Prove Proposition 5.8: The interior of a set A is the union of all open subsets of A. Furthermore: (i) $A°$ is open; (ii) $A°$ is the largest open subset of A, i.e. if G is an open subset of A then $G \subset A° \subset A$; and (iii) A is open iff $A = A°$.

Solution:

Let $\{G_i\}$ be the class of all open subsets of A. If $x \in A°$, then x belongs to an open subset of A, i.e.,

$$\exists i_0 \quad \text{such that} \quad x \in G_{i_0}$$

Hence $x \in \cup_i G_i$ and so $A° \subset \cup_i G_i$. On the other hand, if $y \in \cup_i G_i$, then $y \in G_{i_0}$ for some i_0. Thus $y \in A°$, and $\cup_i G_i \subset A°$. Accordingly, $A° = \cup_i G_i$.

(i) $A° = \cup_i G_i$ is open since it is the union of open sets.

(ii)   If G is an open subset of A then $G \in \{G_i\}$; so $G \subset \cup_i G_i = A° \subset A$.

(iii)   If A is open then $A \subset A° \subset A$ or $A = A°$. If $A = A°$ then A is open since $A°$ is open.

32. Let A be a non-empty proper subset of an indiscrete space X. Find the interior, exterior and boundary of A.

Solution:

X and $\emptyset$ are the only open subsets of X. Since $X \neq A$, $\emptyset$ is the only open subset of A; hence $\text{int}(A) = \emptyset$. Similarly, $\text{int}(A^c) = \emptyset$, i.e. the exterior of A is empty. Thus $\text{b}(A) = X$.

33. Let $\mathcal{T}$ be the topology on $\mathbf{R}$ consisting of $\mathbf{R}$, $\emptyset$ and all open infinite intervals $E_a = (a, \infty)$ where $a \in \mathbf{R}$. Find the interior, exterior and boundary of the closed infinite interval $A = [7, \infty)$.

Solution:

Since the interior of A is the largest open subset of A, $\text{int}(A) = (7, \infty)$. Note that $A^c = (-\infty, 7)$ contains no open set except $\emptyset$; so $\text{int}(A^c) = \text{ext}(A) = \emptyset$. The boundary consists of those points which do not belong to $\text{int}(A)$ or $\text{ext}(A)$; hence $\text{b}(A) = (-\infty, 7]$.

34. Prove Theorem 5.9: $\bar{A} = \text{int}(A) \cup \text{b}(A)$

Solution:

Since $X = \text{int}(A) \cup \text{b}(A) \cup \text{ext}(A)$, $(\text{int}(A) \cup \text{b}(A))^c = \text{ext}(A)$ and it suffices to show $(\bar{A})^c = \text{ext}(A)$.

Let $p \in \text{ext}(A)$; then $\exists$ an open G such that

$$p \in G \subset A^c \quad \text{which implies} \quad G \cap A = \emptyset$$

So p is not a limit point of A, i.e. $p \notin A'$, and $p \notin A$. Hence $p \notin A' \cup A = \bar{A}$. In other words, $\text{ext}(A) \subset (\bar{A})^c$.

Now assume $p \in (\bar{A})^c = (A \cup A')^c$. Thus $p \notin A'$, so $\exists$ an open set G such that

$$p \in G \quad \text{and} \quad (G \setminus \{p\}) \cap A = \emptyset$$

But also $p \notin A$, so $G \cap A = \emptyset$ and $p \in G \subset A^c$. Thus $p \in \text{ext}(A)$, and $(\bar{A})^c = \text{ext}(A)$.

35. Show by a counterexample that the function f which assigns to each set its interior, i.e. $f(A) = \text{int}(A)$, does not commute with the function g which assigns to each set its closure, i.e. $g(A) = \bar{A}$.

Solution:

Consider $\mathbf{Q}$, the set of rational numbers, as a subset of $\mathbf{R}$ with the usual topology. Recall (Example 5.3) that the interior of $\mathbf{Q}$ is empty; hence

$$(g \circ f)(\mathbf{Q}) = g(f(\mathbf{Q})) = g(\text{int}(\mathbf{Q})) = g(\emptyset) = \bar{\emptyset} = \emptyset$$

On the other hand, $\bar{\mathbf{Q}} = \mathbf{R}$ and the interior of $\mathbf{R}$ is $\mathbf{R}$ itself. So

$$(f \circ g)(\mathbf{Q}) = f(g(\mathbf{Q})) = f(\bar{\mathbf{Q}}) = f(\mathbf{R}) = \mathbf{R}$$

Thus $g \circ f \neq f \circ g$, or f and g do not commute.

NEIGHBORHOODS, NEIGHBORHOOD SYSTEMS

36. Consider the following topology on $X = \{a, b, c, d, e\}$:

$$\mathcal{T} = \{X, \emptyset, \{a\}, \{a, b\}, \{a, c, d\}, \{a, b, c, d\}, \{a, b, e\}\}$$

List the neighborhoods (i) of the point e, (ii) of the point c.

Solution:

(i)   A neighborhood of e is any superset of an open set containing e. The open sets containing e are $\{a, b, e\}$ and X. The supersets of $\{a, b, e\}$ are $\{a, b, e\}$, $\{a, b, c, e\}$, $\{a, b, d, e\}$ and X; the only superset of X is X. Accordingly, the class of neighborhoods of e, i.e. the neighborhood system of e, is

$$\mathcal{N}_e = \{\{a, b, e\}, \{a, b, c, e\}, \{a, b, d, e\}, X\}$$

(ii)   The open sets containing c are $\{a, c, d\}$, $\{a, b, c, d\}$ and X. Hence the neighborhood system of c is

$$\mathcal{N}_c = \{\{a, c, d\}, \{a, b, c, d\}, \{a, c, d, e\}, X\}$$

37. Determine the neighborhood system of a point p in an indiscrete space X.

Solution:

X and $\emptyset$ are the only open subsets of X; hence X is the only open set containing p. In addition, X is the only superset of X. Hence $\mathcal{N}_p = \{X\}$.

38. Prove: The intersection $N \cap M$ of any two neighborhoods N and M of a point p is also a neighborhood of p.

Solution:

N and M are neighborhoods of p, so $\exists$ open sets G, H such that

$$p \in G \subset N \quad \text{and} \quad p \in H \subset M$$

Hence $p \in G \cap H \subset N \cap M$, and $G \cap H$ is open, or $N \cap M$ is a neighborhood of p.

39. Prove: Any superset M of a neighborhood N of a point p is also a neighborhood of p.

Solution:

N is a neighborhood of p, so $\exists$ an open set G such that $p \in G \subset N$. By hypothesis, $N \subset M$, so

$$p \in G \subset N \subset M \quad \text{which implies} \quad p \in G \subset M$$

and hence M is a neighborhood of p.

40. Determine whether or not each of the following intervals is a neighborhood of 0 under the usual topology for the real line $\mathbf{R}$. (i) $(-\frac{1}{2}, \frac{1}{2}]$, (ii) $(-1, 0]$, (iii) $[0, \frac{1}{2})$, (iv) $(0, 1]$.

Solution:

(i)   Note that $0 \in (-\frac{1}{2}, \frac{1}{2}) \subset (-\frac{1}{2}, \frac{1}{2}]$ and $(-\frac{1}{2}, \frac{1}{2})$ is open; so $(-\frac{1}{2}, \frac{1}{2}]$ is a neighborhood of 0.

(ii)   and (iii)   Any $\mathcal{U}$-open set G containing 0 contains an open interval (a, b) containing 0, i.e. $a < 0 < b$; hence G contains points both greater and less than 0. So neither $(-1, 0]$ nor $[0, \frac{1}{2})$ is a neighborhood of 0.

(iv)   The interval $(0, 1]$ does not even contain 0 and hence is not a neighborhood of 0.

41. Prove: A set G is open if and only if it is a neighborhood of each of its points.

Solution:

Suppose G is open; then each point $p \in G$ belongs to the open set G contained in G. Hence G is a neighborhood of each of its points.

Conversely, suppose G is a neighborhood of each of its points. So, for each point $p \in G$, $\exists$ an open set G_p such that $p \in G_p \subset G$. Hence $G = \cup \{G_p : p \in G\}$ and is open since it is a union of open sets.

42. Prove Proposition 5.10: Let $\mathcal{N}_p$ be the neighborhood system of a point p in a topological space X. Then:

(i) $\mathcal{N}_p$ is not empty and p belongs to each member of $\mathcal{N}_p$.

(ii)   The intersection of any two members of $\mathcal{N}_p$ belongs to $\mathcal{N}_p$.

(iii) Every superset of a member of $\mathcal{N}_p$ belongs to $\mathcal{N}_p$.

(iv)   Each member $N \in \mathcal{N}_p$ is a superset of a member $G \in \mathcal{N}_p$ where G is a neighborhood of each of its points.

Solution:

(i)   If $N \in \mathcal{N}_p$, then ∃ an open set G such that $p \in G \subset N$; hence $p \in N$. Note $X \in \mathcal{N}_p$ since X is an open set containing p; so $\mathcal{N}_p \neq \emptyset$.

(ii)   Proven in Problem 38.   (iii) Proven in Problem 39.

(iv)   If $N \in \mathcal{N}_p$, then N is a neighborhood of p, so ∃ an open set G such that $p \in G \subset N$. But by the preceding problem $G \in \mathcal{N}_p$ and G is a neighborhood of each of its points.

SUBSPACES, RELATIVE TOPOLOGIES

43. Consider the following topology on $X = \{a, b, c, d, e\}$:

$$\mathcal{T} \;=\; \{X, \emptyset, \{a\}, \{a,b\}, \{a,c,d\}, \{a,b,c,d\}, \{a,b,e\}\}$$

List the members of the relative topology $\mathcal{T}_A$ on $A = \{a, c, e\}$.

Solution:

$\mathcal{T}_A = \{A \cap G : G \in \mathcal{T}\}$,   so the members of $\mathcal{T}_A$ are:

$A \cap X = A$	$A \cap \{a\} = \{a\}$	$A \cap \{a,c,d\} = \{a,c\}$	$A \cap \{a,b,e\} = \{a,e\}$
$A \cap \emptyset = \emptyset$	$A \cap \{a,b\} = \{a\}$	$A \cap \{a,b,c,d\} = \{a,c\}$	

In other words, $\mathcal{T}_A = \{A, \emptyset, \{a\}, \{a,c\}, \{a,e\}\}$.   Observe that $\{a,c\}$ is not open in X, but is relatively open in A, i.e. is $\mathcal{T}_A$-open.

44. Consider the usual topology $\mathcal{U}$ on the real line **R**.   Describe the relative topology $\mathcal{U}_N$ on the set **N** of positive integers.

Solution:

Observe that, for each positive integer $n_0 \in \mathbf{N}$,

$$\{n_0\} \;=\; \mathbf{N} \cap (n_0 - \tfrac{1}{2},\, n_0 + \tfrac{1}{2})$$

and $(n_0 - \tfrac{1}{2}, n_0 + \tfrac{1}{2})$ is a $\mathcal{U}$-open set; so every singleton subset $\{n_0\}$ of **N** is open relative to **N**. Hence every subset of **N** is open relative to **N** since it is a union of singleton sets. In other words, $\mathcal{U}_N$ is the discrete topology on **N**.

45. Let A be a $\mathcal{T}$-open subset of $(X, \mathcal{T})$ and let $A \subset Y \subset X$.   Show that A is also open relative to the relative topology on Y, i.e. A is a $\mathcal{T}_Y$-open subset of Y.

Solution:

$\mathcal{T}_Y = \{Y \cap G : G \in \mathcal{T}\}$.   But $A \subset Y$ and $A \in \mathcal{T}$; so $A = Y \cap A \in \mathcal{T}_Y$.

46. Consider the usual topology $\mathcal{U}$ on the real line **R**.   Determine whether or not each of the following subsets of $I = [0, 1]$ are open relative to I, i.e. $\mathcal{T}_I$-open: (i) $(\tfrac{1}{2}, 1]$, (ii) $(\tfrac{1}{2}, \tfrac{2}{3})$, (iii) $(0, \tfrac{1}{2}]$.

Solution:

(i)   Note that $(\tfrac{1}{2}, 1] = I \cap (\tfrac{1}{2}, 3)$ and $(\tfrac{1}{2}, 3)$ is open in **R**; hence $(\tfrac{1}{2}, 1]$ is open relative to I.

(ii)   Since $(\tfrac{1}{2}, \tfrac{2}{3})$ is open in **R**, i.e. $(\tfrac{1}{2}, \tfrac{2}{3}) \in \mathcal{U}$, it is open relative I by the preceding problem. In fact, $(\tfrac{1}{2}, \tfrac{2}{3}) = I \cap (\tfrac{1}{2}, \tfrac{2}{3})$.

(iii) Since $(0, \tfrac{1}{2}]$ is not the intersection of I with any $\mathcal{U}$-open subset of **R**, it is not $\mathcal{U}_I$-open.

47. Let A be a subset of a topological space (X, T). Show that the relative topology T_A is well-defined. In other words, show that $T_A = \{A \cap G : G \in T\}$ is a topology on A.

Solution:

Since T is a topology, X and $\emptyset$ belong to T. Hence $A \cap X = A$ and $A \cap \emptyset = \emptyset$ both belong to T_A, which then satisfies $[\mathbf{O_1}]$.

Now let $\{H_i : i \in I\}$ be a subclass of T_A. By definition of T_A, for each $i \in I$ $\exists$ a T-open set G_i such that $H_i = A \cap G_i$. By the distributive law of intersection over union,

$$\cup_i H_i = \cup_i (A \cap G_i) = A \cap (\cup_i G_i)$$

But $\cup_i G_i \in T$ as it is the union of T-open sets; hence $\cup_i H_i \in T_A$. Thus T_A satisfies $[\mathbf{O_2}]$.

Now suppose $H_1, H_2 \in T_A$. Then $\exists$ $G_1, G_2 \in T$ such that $H_1 = A \cap G_1$ and $H_2 = A \cap G_2$. But $G_1 \cap G_2 \in T$ since T is a topology. Hence

$$H_1 \cap H_2 = (A \cap G_1) \cap (A \cap G_2) = A \cap (G_1 \cap G_2)$$

belongs to T_A. Accordingly, T_A satisfies $[\mathbf{O_3}]$ and is a topology on A.

48. Let (X, T) be a subspace of (Y, T^*) and let (Y, T^*) be a subspace of (Z, T^{**}). Show that (X, T) is also a subspace of (Z, T^{**}).

Solution:

Since $X \subset Y \subset Z$, (X, T) is a subspace of (Z, T^{**}) if and only if $T_X^{**} = T$. Let $G \in T$; now $T_X^* = T$, so $\exists G^* \in T_X^*$ for which $G = X \cap G^*$. But $T^* = T_Y^{**}$, so $\exists G^{**} \in T^{**}$ such that $G^* = Y \cap G^{**}$. Thus

$$G = X \cap G^* = X \cap Y \cap G^{**} = X \cap G^{**}$$

since $X \subset Y$; so $G \in T_X^{**}$. Accordingly, $T \subset T_X^{**}$.

Now assume $G \in T_X^{**}$, i.e. $\exists H \in T^{**}$ such that $G = X \cap H$. But $Y \cap H \in T_Y^{**} = T^*$ so $X \cap (Y \cap H) \in T_X^* = T$. Since

$$X \cap (Y \cap H) = X \cap H = G$$

we have $G \in T$. Accordingly, $T_X^{**} \subset T$ and the theorem is proved.

MISCELLANEOUS PROBLEMS

49. Let $\mathcal{P}(X)$ be the power set, i.e. class of subsets, of a non-empty set X. Furthermore, let $k : \mathcal{P}(X) \to \mathcal{P}(X)$ be the identity mapping, i.e. for each $A \subset X$, $k(A) = A$.

  (i)   Verify that k satisfies the Kuratowski Closure Axioms of Theorem 5.12.

  (ii) Determine the topology on X induced by k.

Solution:

(i) $k(\emptyset) = \emptyset$, so $[\mathbf{K_1}]$ is satisfied. $k(A \cup B) = A \cup B = k(A) \cup k(B)$, so $[\mathbf{K_3}]$ is satisfied.

      $k(A) = A \supset A$, so $[\mathbf{K_2}]$ is satisfied. $k(k(A)) = k(A)$, so $[\mathbf{K_4}]$ is satisfied.

(ii) A subset $F \subset X$ is closed in the topology induced by k if and only if $k(F) = F$. But $k(A) = A$ for every $A \subset X$, so every set is closed and k induces the discrete topology.

50. Let T be the cofinite topology on the real line $\mathbf{R}$, and let $\langle a_1, a_2, \ldots \rangle$ be a sequence in $\mathbf{R}$ with distinct terms. Show that $\langle a_n \rangle$ converges to every real number $p \in \mathbf{R}$.

Solution:

Let G be any open set containing $p \in \mathbf{R}$. By definition of the cofinite topology, G^c is a finite set and hence can contain only a finite number of the terms of the sequence $\langle a_n \rangle$ since the terms are distinct. Thus G contains almost all of the terms of $\langle a_n \rangle$, and so $\langle a_n \rangle$ converges to p.

51. Let $\mathbf{T}$ be the collection of all topologies on a non-empty set X, partially ordered by class inclusion. Show that $\mathbf{T}$ is a complete lattice, i.e. if $\mathbf{S}$ is a non-empty subcollection of $\mathbf{T}$ then $\sup(\mathbf{S})$ and $\inf(\mathbf{S})$ exist.

Solution:

Let $T_1 = \cap \{T : T \in \mathbf{S}\}$. By Theorem 5.1, T_1 is a topology; so $T_1 \in \mathbf{T}$ and $T_1 = \inf(\mathbf{S})$.

Now let $\mathbf{B}$ be the collection of all upper bounds of $\mathbf{S}$. Observe that $\mathbf{B}$ is non-empty since, for example, the discrete topology $\mathcal{D}$ on X belongs to $\mathbf{B}$. Let $T_2 = \cap \{T : T \in \mathbf{B}\}$. Again by Theorem 5.1, T_2 is a topology on X and, furthermore, $T_2 = \sup(\mathbf{S})$.

52. Let X be a non-empty set and, for each point $p \in X$, let $\mathcal{A}_p$ denote the class of subsets of X containing p.

(i)   Verify that $\mathcal{A}_p$ satisfies the Neighborhood Axioms of Theorem 5.11.

(ii)   Determine the induced topology on X.

Solution:

(i)   Since $p \in X$, $X \in \mathcal{A}_p$ and, so, $\mathcal{A}_p \neq \emptyset$. By hypothesis, p belongs to each member of $\mathcal{A}_p$. Hence [$\mathbf{A}_1$] is satisfied.

      If $M, N \in \mathcal{A}_p$, then $p \in M$ and $p \in N$, and so $p \in M \cap N$. Hence $M \cap N \in \mathcal{A}_p$ and so [$\mathbf{A}_2$] is satisfied.

      If $N \in \mathcal{A}_p$ and $N \subset M$, i.e. if $p \in N \subset M$, then $p \in M$. Hence $M \in \mathcal{A}_p$ and so [$\mathbf{A}_3$] is satisfied.

      By definition of $\mathcal{A}_p$, every $A \subset X$ has the property that $A \in \mathcal{A}_p$ for every $p \in A$. Hence [$\mathbf{A}_4$] is satisfied.

(ii)   A subset $A \subset X$ is open in the induced topology if and only if $A \in \mathcal{A}_p$ for every $p \in A$. Since every subset of X has this property, the induced topology on X is the discrete topology.

Supplementary Problems

TOPOLOGICAL SPACES

53. List all possible topologies on the set $X = \{a, b\}$.

54. Prove Theorem 5.1: Let $\{\mathcal{T}_i : i \in I\}$ be any collection of topologies on a set X. Then the intersection $\cap_i \mathcal{T}_i$ is also a topology on X.

55. Let X be an infinite set and let $\mathcal{T}$ be a topology on X in which all infinite subsets of X are open. Show that $\mathcal{T}$ is the discrete topology on X.

56. Let X be an infinite set and let $\mathcal{T}$ consist of $\emptyset$ and all subsets of X whose complements are countable.

(i)   Prove that $(X, \mathcal{T})$ is a topological space.

(ii)   If X is countable, describe the topology determined by $\mathcal{T}$.

57. Let $\mathcal{T} = \{\mathbf{R}^2, \emptyset\} \cup \{G_k : k \in \mathbf{R}\}$ be the class of subsets of the plane $\mathbf{R}^2$ where

$$G_k = \{\langle x, y \rangle : x, y \in \mathbf{R}, \ x > y + k\}$$

(i)   Prove that $\mathcal{T}$ is a topology on $\mathbf{R}^2$.

(ii)   Is $\mathcal{T}$ a topology on $\mathbf{R}^2$ if "$k \in \mathbf{R}$" is replaced by "$k \in \mathbf{N}$"? by "$k \in \mathbf{Q}$"?

58. Prove that $(\mathbf{R}^2, \mathcal{T})$ is a topological space where the elements of $\mathcal{T}$ are $\emptyset$ and the complements of finite sets of lines and points.

59. Let $\{p\}$ be an arbitrary singleton set such that $p \notin \mathbf{R}$; e.g. $\{\mathbf{R}\}$. Furthermore, let $\mathbf{R}^* = \mathbf{R} \cup \{p\}$ and let $\mathcal{T}$ be the class of subsets of $\mathbf{R}^*$ consisting of all $\mathcal{U}$-open subsets of $\mathbf{R}$ and the complements (relative to $\mathbf{R}^*$) of all bounded $\mathcal{U}$-closed subsets of $\mathbf{R}$. Prove that $\mathcal{T}$ is a topology on $\mathbf{R}^*$.

60. Let $\{p\}$ be an arbitrary singleton set such that $p \notin \mathbf{R}$; and let $\mathbf{R}^* = \mathbf{R} \cup \{p\}$. Furthermore, let $\mathcal{T}$ be the class of subsets of $\mathbf{R}^*$ consisting of all subsets of $\mathbf{R}$ and the complements (relative to $\mathbf{R}^*$) of all finite subsets of $\mathbf{R}$. Prove that $\mathcal{T}$ is a topology on $\mathbf{R}^*$.

ACCUMULATION POINTS, DERIVED SETS

61. Prove: $A' \cup B' = (A \cup B)'$.

62. Prove: If p is a limit point of the set A, then p is also a limit point of $A \setminus \{p\}$.

63. Prove: Let X be a cofinite topological space. Then A' is closed for any subset A of X.

64. Consider the topological space $(\mathbf{R}, \mathcal{T})$ where $\mathcal{T}$ consists of $\mathbf{R}$, $\emptyset$ and all open infinite intervals $E_a = (a, \infty)$, $a \in \mathbf{R}$. Find the derived set of: (i) the interval $[4, 10]$; (ii) $\mathbf{Z}$, the set of integers.

65. Let $\mathcal{T}$ be the topology on $\mathbf{R}^* = \mathbf{R} \cup \{p\}$ defined in Problem 59.
   (i)   Determine the accumulation points of the following sets:
        (1) open interval (a, b), $a, b \in \mathbf{R}$ (2) infinite open interval (a, ∞), $a \in \mathbf{R}$ (3) $\mathbf{R}$.
   (ii) Determine those subsets of $\mathbf{R}^*$ which have p as a limit point.

66. Let $\mathcal{T}_1$ and $\mathcal{T}_2$ be topologies on a set X with $\mathcal{T}_1$ coarser than $\mathcal{T}_2$, i.e. $\mathcal{T}_1 \subset \mathcal{T}_2$.
   (i)   Show that every $\mathcal{T}_2$-accumulation point of a subset A of X is also a $\mathcal{T}_1$-accumulation point.
   (ii) Construct an example in which the converse of (i) does not hold.

CLOSED SETS, CLOSURE OF A SET, DENSE SUBSETS

67. Construct a non-discrete topological space in which the closed sets are identical to the open sets.

68. Prove: $\overline{A \cap B} \subset \bar{A} \cap \bar{B}$. Construct an example in which equality does not hold.

69. Prove: $\bar{A} \setminus \bar{B} \subset \overline{(A \setminus B)}$. Construct an example in which equality does not hold.

70. Prove: If A is open, then $A \cap \bar{B} \subset \overline{A \cap B}$.

71. Prove: Let A be a dense subset of $(X, \mathcal{T})$, and let B be a non-empty open subset of X. Then $A \cap B \neq \emptyset$.

72. Let $\mathcal{T}_1$ and $\mathcal{T}_2$ be topologies on X with $\mathcal{T}_1$ coarser than $\mathcal{T}_2$. Show that the $\mathcal{T}_2$-closure of any subset A of X is contained in the $\mathcal{T}_1$-closure of A.

73. Show that every non-finite subset of an infinite cofinite space X is dense in X.

74. Show that every non-empty open subset of an indiscrete space X is dense in X.

INTERIOR, EXTERIOR, BOUNDARY

75. Let X be a discrete space and let $A \subset X$. Find (i) int (A), (ii) ext (A), and (iii) b (A).

76. Prove: (i)   b $(A) \subset A$ if and only if A is closed.
          (ii) b $(A) \cap A = \emptyset$ if and only if A is open.
          (iii) b $(A) = \emptyset$ if and only if A is both open and closed.

77. Prove: If $\bar{A} \cap \bar{B} = \emptyset$, then b $(A \cup B) =$ b $(A) \cup$ b (B).

78. Prove: (i) $A^\circ \cap B^\circ = (A \cap B)^\circ$; (ii) $A^\circ \cup B^\circ \subset (A \cup B)^\circ$. Construct an example in which equality in (ii) does not hold.

79. Prove: b $(A^\circ) \subset$ b (A). Construct an example in which equality does not hold.

80. Show that int $(A) \cup$ ext (A) need not be dense in a space X. (It is true if $X = \mathbf{R}$.)

81. Prove: Let $\mathcal{T}_1$ and $\mathcal{T}_2$ be topologies on X with $\mathcal{T}_1$ coarser than $\mathcal{T}_2$, i.e. $\mathcal{T}_1 \subset \mathcal{T}_2$, and let $A \subset X$. Then:
   (i)   The $\mathcal{T}_1$-interior of A is a subset of the $\mathcal{T}_2$-interior of A.
   (ii) The $\mathcal{T}_2$-boundary of A is a subset of the $\mathcal{T}_1$-boundary of A.

NEIGHBORHOODS, NEIGHBORHOOD SYSTEMS

82. Let X be a cofinite topological space. Show that every neighborhood of a point $p \in X$ is an open set.

83. Let X be an indiscrete space. Determine the neighborhood system $\mathcal{N}_p$ of any point $p \in X$.

84. Show that if $\mathcal{N}_p$ is finite, then $\bigcap \{N : N \in \mathcal{N}_p\}$ belongs to $\mathcal{N}_p$.

SUBSPACES, RELATIVE TOPOLOGIES

85. Show that every subspace of a discrete space is also discrete.

86. Show that every subspace of an indiscrete space is indiscrete.

87. Let $(Y, \mathcal{T}_Y)$ be a subspace of $(X, \mathcal{T})$. Show that $E \subset Y$ is $\mathcal{T}_Y$-closed if and only if $E = Y \cap F$, where F is a $\mathcal{T}$-closed subset of X.

88. Let $(A, \mathcal{T}_A)$ be a subspace of $(X, \mathcal{T})$. Prove that $\mathcal{T}_A$ consists of the members of $\mathcal{T}$ contained in A, i.e. $\mathcal{T}_A = \{G : G \subset A, G \in \mathcal{T}\}$, if and only if A is a $\mathcal{T}$-open subset of X.

89. Let $(Y, \mathcal{T}_Y)$ be a subspace of $(X, \mathcal{T})$. For any subset A of Y, let $\bar{A}$ and A° be the closure and interior of A with respect to $\mathcal{T}$ and let $(\bar{A})_Y$ and $(A^\circ)_Y$ be the closure and interior of A with respect to $\mathcal{T}_Y$. Prove   (i) $(\bar{A})_Y = \bar{A} \cap Y$,   (ii) $A^\circ = (A^\circ)_Y \cap Y^\circ$.

90. Let A, B and C be subsets of a topological space X with $C \subset A \cup B$. If A, B and $A \cup B$ are given the relative topologies, prove that C is open with respect to $A \cup B$ if and only if $C \cap A$ is open with respect to A and $C \cap B$ is open with respect to B.

EQUIVALENT DEFINITIONS OF TOPOLOGIES

91. Prove Theorem 5.11: Let X be a non-empty set and let there be assigned to each point $p \in X$ a class $\mathcal{A}_p$ of subsets of X satisfying the following axioms:

[A_1] $\mathcal{A}_p$ is not empty and p belongs to each member of $\mathcal{A}_p$.
[A_2]   The intersection of any two members of $\mathcal{A}_p$ belongs to $\mathcal{A}_p$.
[A_3]   Every superset of a member of $\mathcal{A}_p$ belongs to $\mathcal{A}_p$.
[A_4]   Each member $N \in \mathcal{A}_p$ is a superset of a member $G \in \mathcal{A}_p$ such that $G \in \mathcal{A}_g$ for every $g \in G$.

Then there exists one and only one topology $\mathcal{T}$ on X such that $\mathcal{A}_p$ is the $\mathcal{T}$-neighborhood system of the point $p \in X$.

92. Prove Theorem 5.12: Let X be a non-empty set and let $k : \mathcal{P}(X) \to \mathcal{P}(X)$ satisfy the following Kuratowski Closure Axioms:

$$[\textbf{K}_1] \ k(\emptyset) = \emptyset, \quad [\textbf{K}_2] \ A \subset k(A), \quad [\textbf{K}_3] \ k(A \cup B) = k(A) \cup k(B), \quad [\textbf{K}_4] \ k(k(A)) = k(A)$$

Then there exists one and only one topology $\mathcal{T}$ on X such that $k(A)$ will be the $\mathcal{T}$-closure of $A \subset X$.

93. Prove: Let X be a non-empty set and let $i : \mathcal{P}(X) \to \mathcal{P}(X)$ satisfy the following properties:

$$\text{(i) } i(X) = X, \quad \text{(ii) } i(A) \subset A, \quad \text{(iii) } i(A \cup B) = i(A) \cup i(B), \quad \text{(iv) } i(i(A)) = i(A)$$

Then there exists one and only one topology $\mathcal{T}$ on X such that $i(A)$ will be the $\mathcal{T}$-interior of $A \subset X$.

94. Prove: Let X be a non-empty set and let $\mathcal{F}$ be a class of subsets of X satisfying the following properties:
    (i) X and $\emptyset$ belong to $\mathcal{F}$.
    (ii)   The intersection of any number of members of $\mathcal{F}$ belongs to $\mathcal{F}$.
    (iii) The union of any two members of $\mathcal{F}$ belongs to $\mathcal{F}$.

Then there exists one and only one topology $\mathcal{T}$ on X such that the members of $\mathcal{F}$ are precisely the $\mathcal{T}$-closed subsets of X.

95. Let a neighborhood of a real number $p \in \mathbf{R}$ be any set containing p and containing all the rational numbers of some open interval (a, b) where $a < p < b$.
    (i)   Show that these neighborhoods actually satisfy the neighborhood axioms and hence define a topology on the real line $\mathbf{R}$.
    (ii)   Show that any set of irrational numbers does not contain any accumulation points.
    (iii) Show that any sequence of irrational numbers, such as $\langle \pi/2, \pi/3, \pi/4, \ldots \rangle$, does not converge.

Answers to Supplementary Problems

53. $\{X, \emptyset\}$, $\{X, \{a\}, \emptyset\}$, $\{X, \{b\}, \emptyset\}$ and $\{X, \{a\}, \{b\}, \emptyset\}$.

56. (ii) Discrete topology.

64. (i) $(-\infty, 10]$ (ii) $\mathbf{R}$

65. (i): (1) $[a, b]$,   (2) $[a, \infty) \cup \{p\}$,   (3) $\mathbf{R}^*$.   (ii) Unbounded subsets of $\mathbf{R}$.

67. $X = \{a, b, c\}$, $\mathcal{T} = \{X, \emptyset, \{a, b\}, \{c\}\}$

75. (i) A, (ii) A^c, (iii) $\emptyset$

80. Let $X = \{a, b\}$ be an indiscrete space and let $A = \{a\}$.

Chapter 6

Bases and Subbases

BASE FOR A TOPOLOGY

Let $(X, \mathcal{T})$ be a topological space. A class $\mathcal{B}$ of open subsets of X, i.e. $\mathcal{B} \subset \mathcal{T}$, is a *base* for the topology $\mathcal{T}$ iff

    (i)   every open set $G \in \mathcal{T}$ is the union of members of $\mathcal{B}$.

Equivalently, $\mathcal{B} \subset \mathcal{T}$ is a base for $\mathcal{T}$ iff

    (ii)   for any point p belonging to an open set G, there exists $B \in \mathcal{B}$ with $p \in B \subset G$.

Example 1.1:   The open intervals form a base for the usual topology on the line $\mathbf{R}$. For if $G \subset \mathbf{R}$ is open and $p \in G$, then by definition, $\exists$ an open interval (a, b) with $p \in (a, b) \subset G$.

                Similarly, the open discs form a base for the usual topology on the plane $\mathbf{R}^2$.

Example 1.2:   The open rectangles in the plane $\mathbf{R}^2$, bounded by sides parallel to the x-axis and y-axis, also form a base $\mathcal{B}$ for the usual topology on $\mathbf{R}^2$. For, let $G \subset \mathbf{R}^2$ be open and $p \in G$. Hence there exists an open disc D_p centered at p with $p \in D_p \subset G$. Then any rectangle $B \in \mathcal{B}$ whose vertices lie on the boundary of D_p satisfies

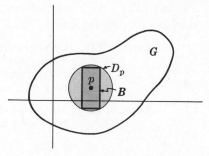

$$p \in B \subset D_p \subset G \quad \text{or} \quad p \in B \subset G$$

as indicated in the diagram. In other words, $\mathcal{B}$ satisfies (ii) above.

Example 1.3:   Consider any discrete space $(X, \mathcal{D})$. Then the class $\mathcal{B} = \{\{p\} : p \in X\}$ of all singleton subsets of X is a base for the discrete topology $\mathcal{D}$ on X. For each singleton set $\{p\}$ is $\mathcal{D}$-open, since every $A \subset X$ is $\mathcal{D}$-open; furthermore, every set is the union of singleton sets. In fact any other class $\mathcal{B}^*$ of subsets of X is a base for $\mathcal{D}$ if and only if it is a superclass of $\mathcal{B}$, i.e. $\mathcal{B}^* \supset \mathcal{B}$.

We now ask the following question: Given a class $\mathcal{B}$ of subsets of a set X, when will the class $\mathcal{B}$ be a base for some topology on X? Clearly $X = \bigcup\{B : B \in \mathcal{B}\}$ is necessary since X is open in every topology on X. The next example shows that other conditions are also needed.

Example 1.4:   Let $X = \{a, b, c\}$. We show that the class $\mathcal{B}$ consisting of $\{a, b\}$ and $\{b, c\}$, i.e. $\mathcal{B} = \{\{a, b\}, \{b, c\}\}$, cannot be a base for any topology on X. For then $\{a, b\}$ and $\{b, c\}$ would themselves be open and therefore their intersection $\{a, b\} \cap \{b, c\} = \{b\}$ would also be open; but $\{b\}$ is not the union of members of $\mathcal{B}$.

The next theorem gives both necessary and sufficient conditions for a class of sets to be a base for some topology.

Theorem 6.1:   Let $\mathcal{B}$ be a class of subsets of a non-empty set X. Then $\mathcal{B}$ is a base for some topology on X if and only if it possesses the following two properties:

        (i) $X = \bigcup\{B : B \in \mathcal{B}\}$.

        (ii)   For any $B, B^* \in \mathcal{B}$, $B \cap B^*$ is the union of members of $\mathcal{B}$, or, equivalently, if $p \in B \cap B^*$ then $\exists B_p \in \mathcal{B}$ such that $p \in B_p \subset B \cap B^*$.

Example 1.5: Let $\mathcal{B}$ be the class of open-closed intervals in the real line $\mathbf{R}$:

$$\mathcal{B} \;=\; \{(a,b] \,:\, a,b \in \mathbf{R},\; a < b\}$$

Clearly $\mathbf{R}$ is the union of members of $\mathcal{B}$ since every real number belongs to some open-closed intervals. In addition, the intersection $(a,b] \cap (c,d]$ of any two open-closed intervals is either empty or another open-closed interval. For example,

$$\text{if}\quad a < c < b < d \quad \text{then}\quad (a,b] \cap (c,d] = (c,b]$$

as indicated in the diagram below.

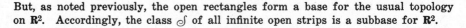

$$a \qquad\qquad c \qquad\qquad b \qquad\qquad d$$

Thus the class $\mathcal{T}$ consisting of unions of open-closed intervals is a topology on $\mathbf{R}$, i.e. $\mathcal{B}$ is a base for a topology $\mathcal{T}$ on $\mathbf{R}$. This topology $\mathcal{T}$ is called the *upper limit* topology on $\mathbf{R}$. Observe that $\mathcal{T} \neq \mathcal{U}$.

Similarly, the class of closed-open intervals,

$$\mathcal{B}^* \;=\; \{[a,b) \,:\, a,b \in \mathbf{R},\; a < b\}$$

is a base for a topology $\mathcal{T}^*$ on $\mathbf{R}$ called the *lower limit* topology on $\mathbf{R}$.

SUBBASES

Let $(X, \mathcal{T})$ be a topological space. A class $\mathcal{S}$ of open subsets of X, i.e. $\mathcal{S} \subset \mathcal{T}$, is a *subbase* for the topology $\mathcal{T}$ on X iff finite intersections of members of $\mathcal{S}$ form a base for $\mathcal{T}$.

Example 2.1: Observe that every open interval (a,b) in the line $\mathbf{R}$ is the intersection of two infinite open intervals (a,∞) and $(-\infty,b)$: $(a,b) = (a,\infty) \cap (-\infty,b)$. But the open intervals form a base for the usual topology on $\mathbf{R}$; hence the class $\mathcal{S}$ of all infinite open intervals is a subbase for $\mathbf{R}$.

Example 2.2: The intersection of a vertical and a horizontal infinite open strip in the plane $\mathbf{R}^2$ is an open rectangle as indicated in the diagram below.

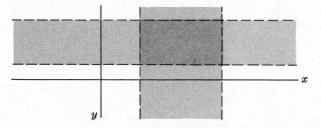

But, as noted previously, the open rectangles form a base for the usual topology on $\mathbf{R}^2$. Accordingly, the class $\mathcal{S}$ of all infinite open strips is a subbase for $\mathbf{R}^2$.

TOPOLOGIES GENERATED BY CLASSES OF SETS

Let $\mathcal{A}$ be any class of subsets of a non-empty set X. As seen previously, $\mathcal{A}$ may not be a base for a topology on X. However, $\mathcal{A}$ always *generates* a topology on X in the following sense:

Theorem 6.2: Any class $\mathcal{A}$ of subsets of a non-empty set X is the subbase for a unique topology $\mathcal{T}$ on X. That is, finite intersections of members of $\mathcal{A}$ form a base for the topology $\mathcal{T}$ on X.

Example 3.1: Consider the following class of subsets of $X = \{a,b,c,d\}$:

$$\mathcal{A} \;=\; \{\{a,b\},\, \{b,c\},\, \{d\}\}$$

Finite intersections of members of $\mathcal{A}$ gives the class

$$\mathcal{B} \;=\; \{\{a,b\},\, \{b,c\},\, \{d\},\, \{b\},\, \emptyset,\, X\}$$

(Note $X \in \mathcal{B}$ since, by definition, it is the empty intersection of members of $\mathcal{A}$.) Taking unions of members of $\mathcal{B}$ gives the class

$$\mathcal{T} \;=\; \{\{a,b\},\ \{b,c\},\ \{d\},\ \{b\},\ \varnothing,\ X,\ \{a,b,d\},\ \{b,c,d\},\ \{b,d\},\ \{a,b,c\}\}$$

which is the topology on X generated by the class $\mathcal{A}$.

Example 3.2: Let $(X, \preceq)$ be any non-empty totally ordered set. The topology on X generated by the subsets of X of the form

$$\{x \in X \,:\, x \prec p,\ p \in X\} \quad \text{or} \quad \{x \in X \,:\, p \prec x,\ p \in X\}$$

is called the *order topology* on X. Observe, by Example 2.1, that the usual topology on **R** is, in fact, identical to the (natural) order topology on **R**.

The topology generated by a class of sets can also be characterized as follows:

Proposition 6.3: Let $\mathcal{A}$ be a class of subsets of a non-empty set X. Then the topology $\mathcal{T}$ on X generated by $\mathcal{A}$ is the intersection of all topologies on X which contain $\mathcal{A}$.

LOCAL BASES

Let p be any arbitrary point in a topological space X. A class $\mathcal{B}_p$ of open sets containing p is called a *local base at* p iff for each open set G containing p, $\exists G_p \in \mathcal{B}_p$ with the property $p \in G_p \subset G$.

Example 4.1: Consider the usual topology on the plane $\mathbf{R}^2$ and any point $p \in \mathbf{R}^2$. Then the class $\mathcal{B}_p$ of all open discs centered at p is a local base at p. For, as proven previously, any open set G containing p also contains an open disc D_p whose center is p.

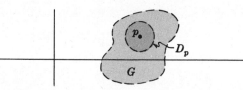

Similarly, the class of all open intervals $(a - \delta,\ a + \delta)$ in the line **R** with center $a \in \mathbf{R}$ is a local base at the point a.

The following relationship between a base ("in the large") for a topology and a local base ("in the small") at a point clearly holds:

Proposition 6.4: Let $\mathcal{B}$ be a base for a topology $\mathcal{T}$ on X and let $p \in X$. Then the members of the base $\mathcal{B}$ which contain p form a local base at the point p.

Some concepts previously defined in terms of the open sets containing a point p can also be defined merely in terms of the members of a local base at p. For example,

Proposition 6.5: A point p in a topological space X is an accumulation point of $A \subset X$ iff each member of some local base $\mathcal{B}_p$ at p contains a point of A different from p.

Proposition 6.6: A sequence $\langle a_1, a_2, \ldots \rangle$ of points in a topological space X converges to $p \in X$ iff each member of some local base $\mathcal{B}_p$ at p contains almost all of the terms of the sequence.

The previous three propositions imply the following useful corollary.

Corollary 6.7: Let $\mathcal{B}$ be a base for a topology $\mathcal{T}$ on X. Then:

    (i) $p \in X$ is an accumulation point of $A \subset X$ iff each open base set $B \in \mathcal{B}$ containing p contains a point of A different from p;

    (ii) a sequence $\langle a_1, a_2, \ldots \rangle$ of points in X converges to $p \in X$ iff each open base set $B \in \mathcal{B}$ containing p contains almost all of the terms of the sequence.

Example 4.2:   Consider the lower limit topology T on the real line $\mathbf{R}$ which has as a base the class of closed-open intervals $[a, b)$, and let $A = (0, 1)$. Note that $G = [1, 2)$ is a T-open set containing $1 \in \mathbf{R}$ for which $G \cap A = \emptyset$; hence 1 is not a limit point of A. On the other hand, $0 \in \mathbf{R}$ is a limit point of A since any open base set $[a, b)$ containing 0, i.e. for which $a \leqq 0 < b$, contains points of A other than 0.

Solved Problems

BASES

1. Show the equivalence of both definitions of a base for a topology, that is, if $\mathcal{B}$ is a subclass of T then the following statements are equivalent:

    (i)   Each $G \in T$ is the union of members of $\mathcal{B}$.

    (ii) For any point p belonging to an open set G, $\exists B_p \in \mathcal{B}$ such that $p \in B_p \subset G$.

    Solution:

    If $G = \cup_i B_i$ where $B_i \in \mathcal{B}$, then each point $p \in G = \cup_i B_i$ belongs to at least one member B_{i_0} in the union; so

    $$p \in B_{i_0} \subset \cup_i B_i = G$$

    On the other hand, if for each $p \in G$, $\exists B_p \in \mathcal{B}$ such that $p \in B_p \subset G$, then

    $$G = \mathbf{U}\{B_p : p \in G\}$$

    and G is the union of members of $\mathcal{B}$.

2. Determine whether or not each of the following classes of subsets of the plane $\mathbf{R}^2$ is a base for the usual topology on $\mathbf{R}^2$:   (i) the class of open equilateral triangles;   (ii) the class of open squares with horizontal and vertical sides.

    Solution:

    Both of the above classes are a base for the usual topology on $\mathbf{R}^2$. For let G be an open subset of $\mathbf{R}^2$ and let $p \in G$. Then $\exists$ an open disc D_p centered at p such that $p \in D_p \subset G$. Observe that either an equilateral triangle or a square can be inscribed in D_p as indicated in the diagrams below.

    Hence each class satisfies the second definition of a base for a topology.

3. Let $\mathcal{B}$ be a base for a topology T on X and let $\mathcal{B}^*$ be a class of open sets containing $\mathcal{B}$, i.e. $\mathcal{B} \subset \mathcal{B}^* \subset T$. Show that $\mathcal{B}^*$ is also a base for T.

    Solution:

    Let G be an open subset of X. Since $\mathcal{B}$ is a base for (X, T), G is the union of members of $\mathcal{B}$, i.e. $G = \cup_i B_i$ where $B_i \in \mathcal{B}$. But $\mathcal{B} \subset \mathcal{B}^*$; hence each $B_i \in \mathcal{B}$ also belongs to $\mathcal{B}^*$. So G is the union of members of $\mathcal{B}^*$, and therefore $\mathcal{B}^*$ is also a base for (X, T).

4. Let X be a discrete space and let $\mathcal{B}$ be the class of all singleton subsets of X, i.e.
    $\mathcal{B} = \{\{p\} : p \in X\}$. Show that any class $\mathcal{B}^*$ of subsets of X is a base for X if and
    only if it is a superclass of $\mathcal{B}$.

    Solution:

    Suppose $\mathcal{B}^*$ is a base for X. Since any singleton set $\{p\}$ is open in a discrete space, $\{p\}$ must be
    a union of members of $\mathcal{B}^*$. But a singleton set can only be the union of itself or itself with the
    empty set $\emptyset$. Hence $\{p\}$ must be a member of $\mathcal{B}^*$, so $\mathcal{B} \subset \mathcal{B}^*$.

    On the other hand, since $\mathcal{B}$ is a base for the discrete space X (see Example 1.3), any superset of $\mathcal{B}$
    is also a base for X.

5. Prove Theorem 6.1: Let $\mathcal{B}$ be a class of subsets of a non-empty set X. Then $\mathcal{B}$ is a
    base for some topology on X if and only if it satisfies the following two properties:

    (i) $X = \bigcup \{B : B \in \mathcal{B}\}$.

    (ii) For any $B, B^* \in \mathcal{B}$, $B \cap B^*$ is the union of members of $\mathcal{B}$, or, equivalently, if
          $p \in B \cap B^*$ then $\exists B_p \in \mathcal{B}$ such that $p \in B_p \subset B \cap B^*$.

    Solution:

    Suppose $\mathcal{B}$ is a base for a topology $\mathcal{T}$ on X. Since X is open, X is the union of members of $\mathcal{B}$.
    Hence X is the union of all the members of $\mathcal{B}$, i.e. $X = \bigcup \{B : B \in \mathcal{B}\}$. Furthermore, if $B, B^* \in \mathcal{B}$
    then, in particular, B and B^* are open. Hence the intersection $B \cap B^*$ is also open and, since $\mathcal{B}$ is
    a base for $\mathcal{T}$, it is the union of members of $\mathcal{B}$. Thus (i) and (ii) are satisfied.

    Conversely, suppose $\mathcal{B}$ is a class of subsets of X which satisfy (i) and (ii) above. Let $\mathcal{T}$ be the
    class of all subsets of X which are unions of members of $\mathcal{B}$. We claim that $\mathcal{T}$ is a topology on X.
    Observe that $\mathcal{B} \subset \mathcal{T}$ will be a base for this topology.

    By (i), $X = \bigcup \{B : B \in \mathcal{B}\}$; so $X \in \mathcal{T}$. Note that $\emptyset$ is the union of the empty subclass of $\mathcal{B}$, i.e.
    $\emptyset = \bigcup \{B : B \in \emptyset \subset \mathcal{B}\}$; hence $\emptyset \in \mathcal{T}$, and so $\mathcal{T}$ satisfies $[\mathbf{O_1}]$.

    Now let $\{G_i\}$ be a class of members of $\mathcal{T}$. By definition of $\mathcal{T}$, each G_i is the union of members
    of $\mathcal{B}$; hence the union $\cup_i G_i$ is also the union of members of $\mathcal{B}$ and so belongs to $\mathcal{T}$. Thus $\mathcal{T}$
    satisfies $[\mathbf{O_2}]$.

    Lastly, suppose $G, H \in \mathcal{T}$. We need to show that $G \cap H$ also belongs to $\mathcal{T}$. By definition
    of $\mathcal{T}$, there exist two subclasses $\{B_i : i \in I\}$ and $\{B_j : j \in J\}$ of $\mathcal{B}$ such that $G = \cup_i B_i$ and
    $H = \cup_j B_j$. Then, by the distributive laws,

    $$G \cap H = (\cup_i B_i) \cap (\cup_j B_j) = \bigcup \{B_i \cap B_j : i \in I, j \in J\}$$

    But by (ii), $B_i \cap B_j$ is the union of members of $\mathcal{B}$; hence $G \cap H = \bigcup \{B_i \cap B_j : i \in I, j \in J\}$ is also
    the union of members of $\mathcal{B}$ and so belongs to $\mathcal{T}$ which therefore satisfies $[\mathbf{O_3}]$. Hence $\mathcal{T}$ is a topology
    on X with base $\mathcal{B}$.

6. Let $\mathcal{B}$ and $\mathcal{B}^*$ be bases, respectively, for topologies $\mathcal{T}$ and $\mathcal{T}^*$ on a set X. Suppose
    that each $B \in \mathcal{B}$ is the union of members of $\mathcal{B}^*$. Show that $\mathcal{T}$ is coarser than $\mathcal{T}^*$, i.e.
    $\mathcal{T} \subset \mathcal{T}^*$.

    Solution:

    Let G be a $\mathcal{T}$-open set. Then G is the union of members of $\mathcal{B}$, i.e. $G = \cup_i B_i$ where $B_i \in \mathcal{B}$.
    But, by hypothesis, each $B_i \in \mathcal{B}$ is the union of members of $\mathcal{B}^*$, and so $G = \cup_i B_i$ is also the union
    of members of $\mathcal{B}^*$ which are $\mathcal{T}^*$-open sets. Hence G is also a $\mathcal{T}^*$-open set, and so $\mathcal{T} \subset \mathcal{T}^*$.

7. Show that the usual topology $\mathcal{U}$ on the real line $\mathbf{R}$ is coarser than the upper limit
    topology $\mathcal{T}$ on $\mathbf{R}$ which has as a base the class of open-closed intervals $(a, b]$.

    Solution:

    Note first that any open interval is the union of open-closed intervals. For example,

    $$(a, b) = \bigcup \{(a, b - 1/n] : n \in \mathbf{N}\}$$

    Since the class of open intervals is a base for $\mathcal{U}$, by the preceding problem, $\mathcal{U} \subset \mathcal{T}$, i.e. any $\mathcal{U}$-open
    set is also $\mathcal{T}$-open.

8. Consider the upper limit topology T on the real line **R** which has as a base the class of open-closed intervals $(a, b]$.   (i) Show that the open infinite interval $(4, \infty)$ and the closed infinite interval $(-\infty, 2]$ are T-open sets.   (ii) Show that any open infinite interval (a, ∞) and any closed infinite interval $(-\infty, b]$ are T-open sets.   (iii) Show that any open-closed interval $(a, b]$ is both T-open and T-closed.

Solution:

(i)   Observe that $(4, \infty) \;=\; (4, 5] \,\cup\, (4, 6] \,\cup\, (4, 7] \,\cup\, (4, 8] \,\cup\, \cdots$

$$(-\infty, 2] \;=\; (0, 2] \,\cup\, (-1, 2] \,\cup\, (-2, 2] \,\cup\, \cdots$$

Hence each is T-open since each is the union of members of the base for T.

(ii)   Similarly, $(a, \infty) \;=\; (a, a+1] \,\cup\, (a, a+2] \,\cup\, (a, a+3] \,\cup\, \cdots$

$$(-\infty, b] \;=\; (b-1, b] \,\cup\, (b-2, b] \,\cup\, (b-3, b] \,\cup\, (b-4, b] \,\cup\, \cdots$$

Hence each is T-open.

(iii) $(a, b]^c = (-\infty, a] \,\cup\, (b, \infty)$,   and the two intervals on the right are open, so their union is open and therefore $(a, b]$ is closed.   But $(a, b]$ belongs to the base for T and so is also open.

SUBBASES, TOPOLOGIES GENERATED BY CLASSES OF SETS

9. Let $X = \{a, b, c, d, e\}$ and let $\mathcal{A} = \{\{a, b, c\}, \{c, d\}, \{d, e\}\}$.   Find the topology on X generated by $\mathcal{A}$.

Solution:

First compute the class $\mathcal{B}$ of all finite intersections of sets in $\mathcal{A}$:

$$\mathcal{B} \;=\; \{X, \{a, b, c\}, \{c, d\}, \{d, e\}, \{c\}, \{d\}, \varnothing\}$$

(Note that $X \in \mathcal{B}$, since by definition X is the empty intersection of members of $\mathcal{A}$.)   Taking unions of members of $\mathcal{B}$ gives the class

$$T \;=\; \{X, \{a, b, c\}, \{c, d\}, \{d, e\}, \{c\}, \{d\}, \varnothing, \{a, b, c, d\}, \{c, d, e\}\}$$

which is the topology on X generated by $\mathcal{A}$.

10. Determine the topology T on the real line **R** generated by the class $\mathcal{A}$ of all closed intervals $[a, a+1]$ with length one.

Solution:

Let p be any point in **R**.   Note that the closed intervals $[p-1, p]$ and $[p, p+1]$ belong to $\mathcal{A}$ as they have length one.   Hence

$$[p-1, p] \,\cap\, [p, p+1] \;=\; \{p\}$$

belongs to the topology T, i.e. all singleton sets $\{p\}$ are T-open, and so T is the discrete topology on X.

11. Let $\mathcal{A}$ be the class of all open half-planes H in the plane **R**^{2} of the form

$$H \;=\; \{\langle x, y \rangle : x < a, \text{ or } x > a, \text{ or } y < a, \text{ or } y > a\}$$

(See diagrams below.)

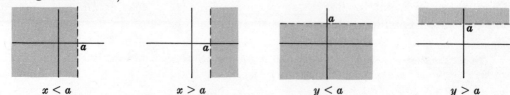

$x < a$ $x > a$ $y < a$ $y > a$

Find the topology on **R**^{2} generated by $\mathcal{A}$.

Solution:

Observe that every open rectangle $B = \{\langle x, y \rangle : a < x < b,\ c < y < d\}$ is the intersection of the four half-planes

$$H_1 = \{\langle x, y \rangle : a < x\} \qquad H_3 = \{\langle x, y \rangle : c < y\}$$
$$H_2 = \{\langle x, y \rangle : x < b\} \qquad H_4 = \{\langle x, y \rangle : y < d\}$$

Since each $H \in \mathcal{A}$ is $\mathcal{U}$-open, and since the class of all open rectangles B is a base for the usual topology $\mathcal{U}$ on $\mathbf{R}^2$, the class $\mathcal{A}$ is a subbase for $\mathcal{U}$. That is, $\mathcal{A}$ generates the usual topology on the plane $\mathbf{R}^2$.

12. Consider the discrete topology $\mathcal{D}$ on $X = \{a, b, c, d, e\}$. Find a subbase $\mathcal{S}$ for $\mathcal{D}$ which does not contain any singleton sets.

Solution:

Recall that any class $\mathcal{B}$ of subsets of X is a base for the discrete topology $\mathcal{D}$ on X iff it contains all singleton subsets of X. Hence $\mathcal{S}$ is a subbase for $\mathcal{D}$ iff finite intersections of members of $\mathcal{S}$ gives $\{a\}, \{b\}, \{c\}, \{d\}$ and $\{e\}$. So $\mathcal{S} = \{\{a, b\}, \{b, c\}, \{c, d\}, \{d, e\}, \{e, a\}\}$ is a subbase for $\mathcal{D}$.

13. Let $\mathcal{S}$ be a subbase for a topology $\mathcal{T}$ on X and let A be a subset of X. Show that the class $\mathcal{S}_A = \{A \cap S : S \in \mathcal{S}\}$ is a subbase for the relative topology $\mathcal{T}_A$ on A.

Solution:

Let H be a $\mathcal{T}_A$-open subset of A. Then $H = A \cap G$ where G is a $\mathcal{T}$-open subset of X. By hypothesis, $\mathcal{S}$ is a subbase for $\mathcal{T}$; so

$$G = \cup_i (S_{i_1} \cap S_{i_2} \cap \cdots \cap S_{i_{n_i}}) \qquad \text{where } S_{i_k} \in \mathcal{S}$$

Hence
$$H = A \cap G = A \cap [\cup_i (S_{i_1} \cap \cdots \cap S_{i_{n_i}})]$$
$$= \cup_i [(A \cap S_{i_1}) \cap \cdots \cap (A \cap S_{i_{n_i}})]$$

Thus H is the union of finite intersections of members of $\mathcal{S}_A$ and therefore $\mathcal{S}_A$ is a subbase for $\mathcal{T}_A$.

14. Show that all intervals $(a, 1]$ and $[0, b)$, where $0 < a, b < 1$, form a subbase for the relative usual topology on the unit interval $I = [0, 1]$.

Solution:

Recall that the infinite open intervals (a, ∞) and $(-\infty, b)$ form a subbase for the usual topology on the real line $\mathbf{R}$. The intersection of these infinite open intervals with $I = [0, 1]$ are the sets $\emptyset, I, (a, 1]$ and $[0, b)$ which, by the preceding problem, form a subbase for $I = [0, 1]$. But we can exclude the empty set $\emptyset$ and the whole space I from any subbase; so the intervals $(a, 1]$ and $[0, b)$ form a subbase for I.

15. Show that if $\mathcal{S}$ is a subbase for topologies $\mathcal{T}$ and $\mathcal{T}^*$ on X, then $\mathcal{T} = \mathcal{T}^*$.

Solution:

Suppose $G \in \mathcal{T}$. Since $\mathcal{S}$ is a subbase for $\mathcal{T}$, $G = \cup_i (S_{i_1} \cap \cdots \cap S_{i_{n_i}})$ where $S_{i_k} \in \mathcal{S}$.

But $\mathcal{S}$ is also a subbase for $\mathcal{T}^*$ and so $\mathcal{S} \subset \mathcal{T}^*$; hence each $S_{i_k} \in \mathcal{T}^*$. Since $\mathcal{T}^*$ is a topology, $S_{i_1} \cap \cdots \cap S_{i_{n_i}} \in \mathcal{T}^*$ and hence $G \in \mathcal{T}^*$. Thus $\mathcal{T} \subset \mathcal{T}^*$. Similarly $\mathcal{T}^* \subset \mathcal{T}$, and so $\mathcal{T} = \mathcal{T}^*$.

16. Prove Theorem 6.2: Any class $\mathcal{A}$ of subsets of a non-empty set X is the subbase for a unique topology on X. That is, finite intersections of members of $\mathcal{A}$ form a base for a topology $\mathcal{T}$ on X.

Solution:

We show that the class $\mathcal{B}$ of finite intersections of members of $\mathcal{A}$ satisfies the two conditions in Theorem 6.1 for it to be a base for a topology on X:

(i) $X = \cup\{B : B \in \mathcal{B}\}$.

(ii) For any $G, H \in \mathcal{B}$, $G \cap H$ is the union of members of $\mathcal{B}$.

Note $X \in \mathcal{B}$, since X by definition is the empty intersection of members of $\mathcal{A}$; so

$$X = \bigcup\{B : B \in \mathcal{B}\}$$

Furthermore, if $G, H \in \mathcal{B}$, then G and H are finite intersections of members of $\mathcal{A}$. Hence $G \cap H$ is also a finite intersection of members of $\mathcal{A}$ and therefore belongs to $\mathcal{B}$. Accordingly, $\mathcal{B}$ is a base for a topology $\mathcal{T}$ on X for which $\mathcal{A}$ is a subbase. The preceding problem shows that $\mathcal{T}$ is unique.

17. Prove Proposition 6.3:   Let $\mathcal{A}$ be a class of subsets of a non-empty set X. Then the topology $\mathcal{T}$ on X generated by $\mathcal{A}$ is the intersection of all topologies on X which contain $\mathcal{A}$.

Solution:

Let $\{T_i\}$ be the collection of topologies on X containing $\mathcal{A}$, and let $\mathcal{T}^* = \cap_i T_i$. Note that $\mathcal{A} \subset \mathcal{T}^*$. We want to prove that $\mathcal{T} = \mathcal{T}^*$. Since $\mathcal{T}$ is a topology containing $\mathcal{A}$, and $\mathcal{T}^*$ is the intersection of all such topologies, we have $\mathcal{T}^* \subset \mathcal{T}$.

On the other hand, suppose $G \in \mathcal{T}$. Then by the definition of $\mathcal{T}$,

$$G = \cup_i (S_{i_1} \cap S_{i_2} \cap \cdots \cap S_{i_{n_i}}) \qquad \text{where} \qquad S_{i_k} \in \mathcal{A}$$

But $\mathcal{A} \subset \mathcal{T}^*$, hence each $S_{i_k} \in \mathcal{T}^*$. Accordingly, $S_{i_1} \cap \cdots \cap S_{i_{n_i}} \in \mathcal{T}^*$ and so

$$G = \cup_i (S_{i_1} \cap \cdots \cap S_{i_{n_i}}) \in \mathcal{T}^*$$

We have shown that $G \in \mathcal{T}$ implies $G \in \mathcal{T}^*$; hence $\mathcal{T} \subset \mathcal{T}^*$. Consequently $\mathcal{T} = \mathcal{T}^*$.

LOCAL BASES

18. Prove Proposition 6.5:   A point p in a topological space $(X, \mathcal{T})$ is an accumulation point of $A \subset X$ iff each member of some local base $\mathcal{B}_p$ at p contains a point of A different from p.

Solution:

Recall $p \in X$ is an accumulation point of A iff $(G \setminus \{p\}) \cap A \neq \emptyset$ for all $G \in \mathcal{T}$ such that $p \in G$. But $\mathcal{B}_p \subset \mathcal{T}$, so in particular $(B \setminus \{p\}) \cap A \neq \emptyset$ for all $B \in \mathcal{B}_p$.

Conversely, suppose $(B \setminus \{p\}) \cap A \neq \emptyset$ for all $B \in \mathcal{B}_p$, and let G be any open subset of X containing p. Then $\exists B_0 \in \mathcal{B}_p$ for which $p \in B_0 \subset G$. But then

$$(G \setminus \{p\}) \cap A \supset (B_0 \setminus \{p\}) \cap A \neq \emptyset$$

So $(G \setminus \{p\}) \cap A \neq \emptyset$, or p is an accumulation point of A.

19. Prove Proposition 6.6:   A sequence $\langle a_1, a_2, \ldots \rangle$ of points in a topological space $(X, \mathcal{T})$ converges to $p \in X$ iff each member of some local base $\mathcal{B}_p$ at p contains almost all of the terms of the sequence.

Solution:

Recall that $a_n \to p$ iff every open set $G \in \mathcal{T}$ containing p contains almost all the terms of the sequence. But $\mathcal{B}_p \subset \mathcal{T}$, so in particular each $B \in \mathcal{B}_p$ contains almost all the terms of the sequence.

On the other hand, suppose every $B \in \mathcal{B}_p$ contains almost all the terms of the sequence, and let G be any open set containing p. Then $\exists B_0 \in \mathcal{B}_p$ for which $p \in B_0 \subset G$. Hence G also contains almost all the terms of the sequence, and so $\langle a_n \rangle$ converges to p.

20. Show that every point p in a discrete space X has a finite local base.

Solution:

Note that the singleton set $\{p\}$ is open since every subset of a discrete space is open. Accordingly the class $\mathcal{B}_p = \{\{p\}\}$, i.e. the class consisting of the singleton set $\{p\}$, is a local base at p since every open set G containing p must be a superset of $\{p\}$.

21. Consider the upper limit topology T on the real line **R** which has as a base the class of open-closed intervals $(a, b]$. Determine whether or not each of the following sequences converges to 0:

(i) $\langle 1, \frac{1}{2}, \frac{1}{3}, \ldots \rangle$ (ii) $\langle -1, -\frac{1}{2}, -\frac{1}{3}, \ldots \rangle$

Solution:

(i)   No. For the T-open set $(-2, 0]$ containing 0 does not contain any term of the sequence.

(ii) Yes. For any open basic set $(a, b]$ containing 0, i.e. for which $a < 0 \leqq b$, $\exists n_0 \in N$ such that $a < -1/n_0 < 0$. Hence $n > n_0$ implies $-1/n \in (a, b]$.

Supplementary Problems

BASES FOR TOPOLOGIES

22. Show that the class of closed intervals $[a, b]$, where a and b are rational and $a < b$, is not a base for a topology on the real line **R**.

23. Show that the class of closed intervals $[a, b]$, where a is rational and b is irrational and $a < b$, is a base for a topology on the real line **R**.

24. Let $\mathcal{B}$ be a base for a topology T on X and let $A \subset X$. Show that the class $\mathcal{B}_A = \{A \cap G : G \in \mathcal{B}\}$ is a base for the relative topology T_A on A.

25. Let $\mathcal{B}$ be the class of half-open rectangles in the plane **R**^{2} indicated in the diagram on the right, i.e. of the form

$$\{\langle x, y \rangle : a \leqq x < b, \ c \leqq y < d\}$$

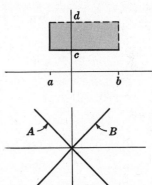

(i)   Show that $\mathcal{B}$ is a base for a topology T on **R**^{2}.

(ii) Show that the relative topology T_A on the line

$$A = \{\langle x, y \rangle : x + y = 0\}$$

is the discrete topology on A.

(iii) Show that the relative topology T_B on the line

$$B = \{\langle x, y \rangle : x = y\}$$

is not the discrete topology on B.

26. Let $\mathcal{B}$ be a class of subsets of a non-empty set X totally ordered by set inclusion. Show that $\mathcal{B}$ is a base for a topology on X provided that $X = \bigcup \{B : B \in \mathcal{B}\}$.

27. Show that a topology T on X is finite if and only if T has a finite base.

SUBBASES

28. Let $X = \{a, b, c, d, e\}$. Find the topology T on X generated by $\mathcal{A} = \{\{a\}, \{a, b, c\}, \{c, d\}\}$.

29. Determine the smallest subbase $\mathcal{S}$ for the discrete topology T on any non-empty set X.

30. Let $\mathcal{S}$ be the class of all closed intervals $[a, b]$ where a and b are rational, i.e. $a, b \in \mathbf{Q}$, and $a < b$. Show that $\mathcal{S} \cup \{\{p\} : p \in \mathbf{Q}\}$ is a base for the topology T on the real line **R** generated by $\mathcal{S}$.

31. Show that if $\mathcal{S}$ is a subbase for a topology T on X, then $\mathcal{S} \setminus \{X, \emptyset\}$ is also a subbase for T.

32. Let T and T^* be the topologies on X generated respectively by $\mathcal{A}$ and $\mathcal{A}^*$.
Show that: (i) $\mathcal{A} \subset \mathcal{A}^*$ implies $T \subset T^*$; and (ii) $\mathcal{A} \subset \mathcal{A}^* \subset T$ implies $T = T^*$.

33. Let $\mathcal{S}$ be a subbase for a topological space X and let $G \subset X$ be an open set containing a point $p \in X$. Show that there exists a finite number of members of $\mathcal{S}$, say $S_1, S_2, \ldots, S_m$, with the property that $p \in S_1 \cap S_2 \cap \cdots \cap S_m \subset G$.

LOCAL BASES

34. Let $(X, \mathcal{T})$ be a topological space and let $\mathcal{A}$ be a $\mathcal{T}$-local base at $p \in X$. Consider any subset A of X such that $p \in A \subset X$, and consider the relative topology $\mathcal{T}_A$ on A. Show that the following class of subsets of A is a $\mathcal{T}_A$-local base at $p \in A$: $\mathcal{A}_A = \{A \cap G : G \in \mathcal{A}\}$.

35. Let X be a topological space, let $p \in X$, let $\mathcal{N}_p$ be the neighborhood system at p and let $\mathcal{B}_p$ be a local base at p. Show that every neighborhood of p contains a member of the local base at p; i.e. for every $N \in \mathcal{N}_p$, $\exists G \in \mathcal{B}_p$ for which $G \subset N$.

36. Show that if a point p has a finite local base $\mathcal{B}_p$ then it also has a local base consisting of exactly one set.

37. Consider the upper limit topology $\mathcal{T}$ on the real line $\mathbf{R}$ which has as a base the class of open-closed intervals $(a, b]$. Determine whether or not each of the following sequences converges:
    (i) $\langle 1, \frac{1}{2}, \frac{1}{3}, \ldots \rangle$, (ii) $\langle -1, -\frac{1}{2}, -\frac{1}{3}, \ldots \rangle$, (iii) $\langle -1, \frac{1}{2}, -\frac{1}{3}, \frac{1}{4}, \ldots \rangle$.

38. Let $\mathcal{T}$ be the topology on the real line $\mathbf{R}$ generated by the class $\mathcal{S}$ of all closed intervals $[a, b]$ where a and b are rational (see Problem 30).
    (i)   Determine whether or not each of the following sequences converges:
          (a) $\langle 2 + \frac{1}{2}, 2 + \frac{1}{3}, 2 + \frac{1}{4}, \ldots \rangle$, (b) $\langle \sqrt{2} + \frac{1}{2}, \sqrt{2} + \frac{1}{3}, \sqrt{2} + \frac{1}{4}, \ldots \rangle$.
    (ii) Determine the closure of each of the following subsets of $\mathbf{R}$:
          (a) $(2, 4)$, (b) $(\sqrt{2}, 5]$, (c) $(-3, \pi)$, (d) $A = \{1, \frac{1}{2}, \frac{1}{3}, \ldots \}$.
    (iii) Show that any finite subset of $\mathbf{R}$ is $\mathcal{T}$-closed.

39. Let $\mathcal{S}$ be a subbase for a topological space X and let $p \in X$.
    (i)   Show by a counterexample that the class $\mathcal{S}_p = \{S \in \mathcal{S} : p \in S\}$ need not be a local base at p.
    (ii) Show that finite intersections of members of $\mathcal{S}_p$ do form a local base at p.
    (iii) Show that a sequence $\langle a_n \rangle$ in X converges to p if and only if every $S \in \mathcal{S}_p$ contains all except a finite number of the terms of the sequence.

Answers to Supplementary Problems

28. $\mathcal{T} = \{X, \emptyset, \{a\}, \{c\}, \{a, c\}, \{c, d\}, \{a, b, c\}, \{a, c, d\}, \{a, b, c, d\}\}$

37. (i) No   (ii) Yes   (iii) No.

38. (i): (a) No, (b) Yes.   (ii): (a) $(2, 4)$, (b) $[\sqrt{2}, 5]$, (c) $(-3, \pi]$, (d) A.

39. (ii) *Hint.* Use Problem 33.

Chapter 7

Continuity and Topological Equivalence

CONTINUOUS FUNCTIONS

Let (X, T) and (Y, T^*) be topological spaces. A function f from X into Y is *continuous relative to* T *and* T^*, or T-T^* *continuous*, or simply *continuous*, iff the inverse image $f^{-1}[H]$ of every T^*-open subset H of Y is a T-open subset of X, that is, iff

$$H \in T^* \quad \text{implies} \quad f^{-1}[H] \in T$$

We shall write $f : (X, T) \to (Y, T^*)$ for a function from X into Y when it is convenient to indicate the topologies involved.

Example 1.1: Consider the following topologies on $X = \{a, b, c, d\}$ and $Y = \{x, y, z, w\}$ respectively:

$$T = \{X, \emptyset, \{a\}, \{a, b\}, \{a, b, c\}\}, \quad T^* = \{Y, \emptyset, \{x\}, \{y\}, \{x, y\}, \{y, z, w\}\}$$

Also consider the functions $f : X \to Y$ and $g : X \to Y$ defined by the diagrams below:

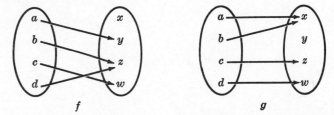

$$f \qquad\qquad\qquad g$$

The function f is continuous since the inverse of each member of the topology T^* on Y is a member of the topology T on X. The function g is not continuous since $\{y, z, w\} \in T^*$, i.e. is an open subset of Y, but its inverse image $g^{-1}[\{y, z, w\}] = \{c, d\}$ is not an open subset of X, i.e. does not belong to T.

Example 1.2: Consider any discrete space $(X, \mathcal{D})$ and any topological space (Y, T). Then every function $f : X \to Y$ is $\mathcal{D}$-T continuous. For if H is any open subset of Y, its inverse $f^{-1}[H]$ is an open subset of X since every subset of a discrete space is open.

Example 1.3: Let $f : X \to Y$ where X and Y are topological spaces, and let $\mathcal{B}$ be a base for the topology on Y. Suppose for each member $B \in \mathcal{B}$, $f^{-1}[B]$ is an open subset of X; then f is a continuous function. For let H be an open subset of Y; then $H = \cup_i B_i$, a union of members of $\mathcal{B}$. But

$$f^{-1}[H] = f^{-1}[\cup_i B_i] = \cup_i f^{-1}[B_i]$$

and each $f^{-1}[B_i]$ is open by hypothesis; hence $f^{-1}[H]$ is the union of open sets and is therefore open. Accordingly, f is continuous.

We formally state the result of the preceding example.

Proposition 7.1: A function $f : X \to Y$ is continuous iff the inverse of each member of a base $\mathcal{B}$ for Y is an open subset of X.

This proposition can in fact be strengthened as follows:

Theorem 7.2: Let $\mathcal{S}$ be a subbase for a topological space Y. Then a function $f : X \to Y$ is continuous iff the inverse of each member of $\mathcal{S}$ is an open subset of X.

Example 1.4: The projection mappings from the plane $\mathbf{R}^2$ into the line $\mathbf{R}$ are both continuous relative to the usual topologies. Consider, for example, the projection $\pi : \mathbf{R}^2 \to \mathbf{R}$ defined by $\pi(\langle x, y \rangle) = y$. Then the inverse of any open interval (a, b) is an infinite open strip as illustrated below:

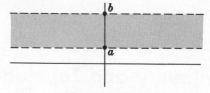

$\pi^{-1}[(a, b)]$ is shaded

Hence by Proposition 7.1, the inverse of every open subset of $\mathbf{R}$ is open in $\mathbf{R}^2$, i.e. π is continuous.

Example 1.5: The absolute value function f on $\mathbf{R}$, i.e. $f(x) = |x|$ for $x \in \mathbf{R}$, is continuous. For if $A = (a, b)$ is an open interval in $\mathbf{R}$, then

$$f^{-1}[A] = \begin{cases} \varnothing & \text{if } a < b \leq 0 \\ (-b, b) & \text{if } a < 0 < b \\ (-b, -a) \cup (a, b) & \text{if } 0 \leq a < b \end{cases}$$

as illustrated below. In each case $f^{-1}[A]$ is open; hence f is continuous.

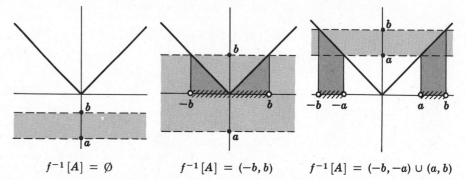

$f^{-1}[A] = \varnothing$ $f^{-1}[A] = (-b, b)$ $f^{-1}[A] = (-b, -a) \cup (a, b)$

Continuous functions can be characterized by their behavior with respect to closed sets, as follows:

Theorem 7.3: A function $f : X \to Y$ is continuous if and only if the inverse image of every closed subset of Y is a closed subset of X.

CONTINUOUS FUNCTIONS AND ARBITRARY CLOSENESS

Let X be a topological space. A point $p \in X$ is said to be *arbitrarily close* to a set $A \subset X$ if

     either   (i) $p \in A$ or   (ii) p is an accumulation point of A

Recall that $\bar{A} = A \cup A'$; so the closure of A consists precisely of those points in X which are arbitrarily close to A. Recall also that $\bar{A} = A^\circ \cup \mathrm{b}(A)$; hence p is arbitrarily close to A if p is either an interior or a boundary point of A.

Continuous functions can also be characterized as those functions which *preserve arbitrary closeness*, namely,

Theorem 7.4: A function $f : X \to Y$ is continuous if and only if, for any $p \in X$ and any $A \subset X$,

     p arbitrarily close to A $\Rightarrow$ $f(p)$ arbitrarily close to $f[A]$

     or $p \in \bar{A}$ $\Rightarrow$ $f(p) \in \overline{f[A]}$

     or $f[\bar{A}] \subset \overline{f[A]}$

CONTINUITY AT A POINT

Continuity as we have defined it is a *global* property, that is, it restricts the way in which a function behaves on the entire set X. There also exists a corresponding local concept of *continuity at a point*.

A function $f: X \to Y$ is continuous at $p \in X$ iff the inverse image $f^{-1}[H]$ of every open set $H \subset Y$ containing $f(p)$ is a superset of an open set $G \subset X$ containing p or, equivalently, iff the inverse image of every neighborhood of $f(p)$ is a neighborhood of p, i.e.,

$$N \in \mathcal{N}_{f(p)} \quad \Rightarrow \quad f^{-1}[N] \in \mathcal{N}_p$$

Notice that, with respect to the usual topology on the real line $\mathbf{R}$, this definition coincides with the $\epsilon - \delta$ definition of continuity at a point for functions $f: \mathbf{R} \to \mathbf{R}$. In fact, the relationship between local and global continuity for functions $f: \mathbf{R} \to \mathbf{R}$ holds true in general; namely,

Theorem 7.5:   Let X and Y be topological spaces. Then a function $f: X \to Y$ is continuous if and only if it is continuous at every point of X.

SEQUENTIAL CONTINUITY AT A POINT

A function $f: X \to Y$ is *sequentially continuous* at a point $p \in X$ iff for every sequence $\langle a_n \rangle$ in X converging to p, the sequence $\langle f(a_n) \rangle$ in Y converges to $f(p)$, i.e.,

$$a_n \to p \quad \text{implies} \quad f(a_n) \to f(p)$$

Sequential continuity and continuity at a point are related as follows:

Proposition 7.6:   If a function $f: X \to Y$ is continuous at $p \in X$, then it is sequentially continuous at p.

Remark: The converse of the previous proposition is not true. Consider, for example, the topology $\mathcal{T}$ on the real line $\mathbf{R}$ consisting of $\emptyset$ and the complements of countable sets. Recall (see Example 7.3 of Chapter 5) that a sequence $\langle a_n \rangle$ converges to p if and only if it has the form

$$\langle a_1, a_2, \ldots, a_{n_0}, p, p, p, \ldots \rangle$$

Then for any function $f: (\mathbf{R}, \mathcal{T}) \to (X, \mathcal{T}^*)$,

$$\langle f(a_n) \rangle = \langle f(a_1), \ldots, f(a_{n_0}), f(p), f(p), f(p), \ldots \rangle$$

converges to $f(p)$. In other words, every function on $(\mathbf{R}, \mathcal{T})$ is sequentially continuous. On the other hand, the function $f(\mathbf{R}, \mathcal{T}) \to (\mathbf{R}, \mathcal{U})$ defined by $f(x) = x$, i.e. the identity function, is not $\mathcal{T}$-$\mathcal{U}$ continuous since $f^{-1}[(0,1)] = (0,1)$ is not a $\mathcal{T}$-open subset of $\mathbf{R}$.

OPEN AND CLOSED FUNCTIONS

A continuous function has the property that the *inverse* image of every open set is open and the *inverse* image of every closed set is closed. It is natural then to ask about the following types of functions:

(1) A function $f: X \to Y$ is called an *open* (or *interior*) *function* if the image of every open set is open.

(2) A function $g: X \to Y$ is called a *closed function* if the image of every closed set is closed.

In general, functions which are open need not be closed and vice versa. In fact, the function in our first example is open and continuous but not closed.

Example 2.1: Consider the projection mapping $\pi : \mathbf{R}^2 \to \mathbf{R}$ of the plane $\mathbf{R}^2$ into the x-axis, i.e. $\pi(\langle x, y \rangle) = x$. Observe that the projection $\pi[D]$ of any open disc $D \subset \mathbf{R}^2$ is an open interval. Hence any point $\pi(p)$ in the image $\pi[G]$ of an open set $G \subset \mathbf{R}^2$ belongs to an open interval contained in $\pi[G]$, or $\pi[G]$ is open. Accordingly, π is an open function. On the other hand, π is not a closed function, for the set $A = \{\langle x, y \rangle : xy \geq 1,\ x > 0\}$ is closed, but its projection $\pi[A] = (0, \infty)$ is not closed. (See diagrams below.)

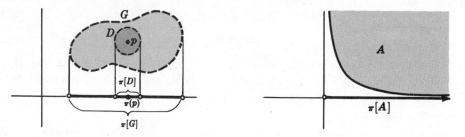

HOMEOMORPHIC SPACES

A topological space $(X, \mathcal{T})$ is, as we have seen, a set X together with a distinguished class $\mathcal{T}$ of subsets of X, satisfying certain axioms. Between any two such spaces $(X, \mathcal{T})$ and $(Y, \mathcal{T}^*)$ there are many functions $f : X \to Y$. We choose to discuss continuous, or open, or closed functions rather than arbitrary functions since it is these functions which preserve some aspect of the structure of the spaces $(X, \mathcal{T})$ and $(Y, \mathcal{T}^*)$.

Now suppose there is some bijective (i.e. one-one and onto) mapping $f : X \to Y$. Then f induces a bijective function $f : \mathcal{P}(X) \to \mathcal{P}(Y)$ from the power set of X, i.e. the class of subsets of X, into the power set of Y. If this induced function also takes $\mathcal{T}$ onto $\mathcal{T}^*$, i.e. defines a one-to-one correspondence between the open sets in X and the open sets in Y, then the spaces $(X, \mathcal{T})$ and $(Y, \mathcal{T}^*)$ are identical from the topological point of view. Specifically:

Definition: Two topological spaces X and Y are called *homeomorphic* or *topologically equivalent* if there exists a bijective (i.e. one-one, onto) function $f : X \to Y$ such that f and f^{-1} are continuous. The function f is called a *homeomorphism*.

A function f is called *bicontinuous* or *topological* if f is open and continuous. Thus $f : X \to Y$ is a homeomorphism iff f is bicontinuous and bijective.

Example 3.1: Let $X = (-1, 1)$. The function $f : X \to \mathbf{R}$ defined by $f(x) = \tan \frac{1}{2}\pi x$ is one-one, onto and continuous. Furthermore, the inverse function f^{-1} is also continuous. Hence the real line $\mathbf{R}$ and the open interval $(-1, 1)$ are homeomorphic.

Example 3.2: Let X and Y be discrete spaces. Then, as seen in Example 1.2, all functions from one to the other are continuous. Hence X and Y are homeomorphic iff there exists a one-one, onto function from one to the other, i.e. iff they are cardinally equivalent.

Proposition 7.7: The relation in any collection of topological spaces defined by "X is homeomorphic to Y" is an equivalence relation.

Thus, by the Fundamental Theorem on Equivalence Relations, any collection of topological spaces can be partitioned into classes of topologically equivalent spaces.

TOPOLOGICAL PROPERTIES

A property P of sets is called *topological* or a *topological invariant* if whenever a topological space $(X, \mathcal{T})$ has P then every space homeomorphic to $(X, \mathcal{T})$ also has P.

Example 4.1: As seen in Example 3.1, the real line $\mathbf{R}$ is homeomorphic to the open interval $X = (-1, 1)$. Hence *length* is not a topological property since X and $\mathbf{R}$ have different lengths, and *boundedness* is not a topological property since X is bounded but $\mathbf{R}$ is not.

Example 4.2: Let X be the set of positive real numbers, i.e. $X = (0, \infty)$. The function $f : X \to X$ defined by $f(x) = 1/x$ is a homeomorphism from X onto X. Observe that the sequence

$$\langle a_n \rangle \;=\; \langle 1, \tfrac{1}{2}, \tfrac{1}{3}, \ldots \rangle$$

corresponds, under the homeomorphism, to the sequence

$$\langle f(a_n) \rangle \;=\; \langle 1, 2, 3, \ldots \rangle$$

The sequence $\langle a_n \rangle$ is a Cauchy sequence; the sequence $\langle f(a_n) \rangle$ is not. Hence the property of being a Cauchy sequence is not topological.

Most of topology is an investigation of the consequences of certain topological properties as *compactness* and *connectedness*. In fact, formally topology is the study of topological invariants. In the next example, connectedness is defined and is shown to be a topological property.

Example 4.3: A topological space $(X, \mathcal{T})$ is *disconnected* iff X is the union of two open, non-empty, disjoint subsets, i.e.

$$X = G \cup H \quad \text{where} \quad G, H \in \mathcal{T}, \; G \cap H = \emptyset \quad \text{but} \quad G, H \neq \emptyset$$

If $f : X \to Y$ is a homeomorphism then $X = G \cup H$ if and only if $Y = f[G] \cup f[H]$ and so Y is disconnected if and only if X is.

The space $(X, \mathcal{T})$ is *connected* iff it is not disconnected.

TOPOLOGIES INDUCED BY FUNCTIONS

Let $\{(Y_i, \mathcal{T}_i)\}$ be any collection of topological spaces and for each Y_i let there be given a function $f_i : X \to Y_i$ defined on some arbitrary non-empty set X. We want to investigate those topologies on X with respect to which all the functions f_i are continuous. Recall that f_i is continuous relative to some topology on X provided the inverse image of each open subset of Y_i is an open subset of X. Thus we consider the following class of subsets of X:

$$\mathcal{S} \;=\; \bigcup_i \{f_i^{-1}[H] : H \in \mathcal{T}_i\}$$

That is, $\mathcal{S}$ consists of the inverse image of each open subset of every space Y_i. The topology $\mathcal{T}$ on X generated by $\mathcal{S}$ is called the topology *induced* (or *generated*) by the functions f_i. The main properties of $\mathcal{T}$ are listed in the next theorem.

Theorem 7.8:   (i)   All the functions f_i are continuous relative to $\mathcal{T}$.

(ii) $\mathcal{T}$ is the intersection of all the topologies on X with respect to which the functions f_i are continuous.

(iii) $\mathcal{T}$ is the smallest, i.e. coarsest, topology on X with respect to which the functions f_i are continuous.

(iv) $\mathcal{S}$ is a subbase for the topology $\mathcal{T}$.

We shall call $\mathcal{S}$ the *defining subbase* for the topology induced by the functions f_i, i.e. the coarsest topology on X with respect to which the functions f_i are continuous.

Example 5.1: Let π_1 and π_2 be the projections of the plane $\mathbf{R}^2$ into $\mathbf{R}$, i.e.,

$$\pi_1(\langle x, y \rangle) = x \quad \text{and} \quad \pi_2(\langle x, y \rangle) = y$$

Observe, as illustrated below, that the inverse image of an open interval (a, b) in $\mathbf{R}$ is an infinite open strip in $\mathbf{R}^2$.

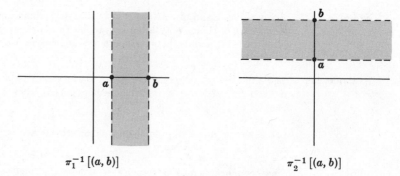

$$\pi_1^{-1}[(a,b)] \qquad\qquad \pi_2^{-1}[(a,b)]$$

Recall that these infinite open strips form a subbase for the usual topology on $\mathbf{R}^2$. Accordingly, the usual topology on $\mathbf{R}^2$ is the smallest topology on $\mathbf{R}^2$ with respect to which the projections π_1 and π_2 are continuous.

Solved Problems

CONTINUOUS FUNCTIONS

1. Prove: Let $f : X \to Y$ be a constant function, say $f(x) = p \in Y$, for every $x \in X$. Then f is continuous relative to any topology T on X and any topology T^* on Y.

    Solution:

    We need to show that the inverse image of any T^*-open subset of Y is a T-open subset of X. Let $H \in T^*$. Now $f(x) = p$ for all $x \in X$, so

    $$f^{-1}[H] \;=\; \begin{cases} X & \text{if } p \in H \\ \emptyset & \text{if } p \notin H \end{cases}$$

    In either case $f^{-1}[H]$ is an open subset of X since X and $\emptyset$ belong to every topology T on X.

2. Prove: Let $f : X \to Y$ be any function. If $(Y, \mathcal{J})$ is an indiscrete space, then $f : (X, T) \to (Y, \mathcal{J})$ is continuous for any T.

    Solution:

    We want to show that the inverse image of every open subset of Y is an open subset of X. Since $(Y, \mathcal{J})$ is an indiscrete space, Y and $\emptyset$ are the only open subsets of Y. But

    $$f^{-1}[Y] = X, \quad f^{-1}[\emptyset] = \emptyset$$

    and X and $\emptyset$ belong to any topology T on X. Hence f is continuous for any T.

3. Let $\mathcal{U}$ be the usual topology on the real line $\mathbf{R}$ and let T be the upper limit topology on $\mathbf{R}$ which is generated by the open-closed intervals $(a, b]$. Furthermore, let $f : \mathbf{R} \to \mathbf{R}$ be defined by

    $$f(x) \;=\; \begin{cases} x & \text{if } x \leq 1 \\ x+2 & \text{if } x > 1 \end{cases}$$

    (See diagram on the right.)

    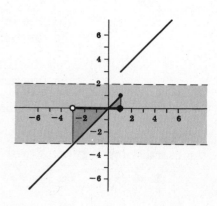

    (i)   Show that f is not $\mathcal{U}$-$\mathcal{U}$ continuous.

    (ii) Show that f is T-T continuous.

    Solution:

    (i)   Let $A = (-3, 2)$. Then $f^{-1}[A] = (-3, 1]$. Now $A \in \mathcal{U}$ but $f^{-1}[A] \notin \mathcal{U}$, so f is not $\mathcal{U}$-$\mathcal{U}$ continuous.

(ii)   Let $A = (a, b]$.   Then:

$$f^{-1}[A] \;=\; \begin{cases} (a, b] & \text{if } a < b \leq 1 \\ (a, 1] & \text{if } a < 1 < b \leq 3 \\ (a, b-2] & \text{if } a < 1 < 3 < b \\ \emptyset & \text{if } 1 \leq a < b \leq 3 \\ (1, b-2] & \text{if } 1 \leq a < 3 < b \\ (a-2, b-2] & \text{if } 3 \leq a < b \end{cases}$$

In each case, $f^{-1}[A]$ is a T-open set. Hence f is T-T continuous.

4. Suppose a function $f : (X, T_1) \to (Y, T_2)$ is not T_1-T_2 continuous.   Show that if T_1^* is a topology on X coarser than T_1 and if T_2^* is a topology on Y finer than T_2, i.e. $T_1^* \subset T_1$ and $T_2 \subset T_2^*$, then f is also not T_1^*-T_2^* continuous.

Solution:
Since $f : (X, T_1) \to (Y, T_2)$ is not continuous,

$$\exists G \in T_2 \quad \text{for which} \quad f^{-1}[G] \notin T_1$$

Now, $T_1^* \subset T_1$ and $T_2 \subset T_2^*$. Hence $G \in T_2$ implies $G \in T_2^*$, and $f^{-1}[G] \notin T_1$ implies $f^{-1}[G] \notin T_1^*$. Thus f is not continuous with respect to T_1^* and T_2^*.

5. Show that the identity function $i : (X, T) \to (X, T^*)$ is continuous if and only if T is finer than T^*, i.e. $T^* \subset T$.

Solution:
By definition, i is T-T^* continuous if and only if

$$G \in T^* \;\Rightarrow\; i^{-1}[G] \in T$$

But $i^{-1}[G] = G$, so i is T-T^* continuous, if and only if

$$G \in T^* \;\Rightarrow\; G \in T$$

that is, $T^* \subset T$.

6. Prove Theorem 7.2:   Let $f : (X, T) \to (Y, T^*)$, and let $\mathcal{S}$ be a subbase for the topology T^* on Y.   Then f is continuous if and only if the inverse of every member of the subbase $\mathcal{S}$ is an open subset of X, i.e. $f^{-1}[S] \in T$ for every $S \in \mathcal{S}$.

Solution:
Suppose $f^{-1}[S] \in T$ for every $S \in \mathcal{S}$. We want to show that f is continuous, i.e. $G \in T^*$ implies $f^{-1}[G] \in T$. Let $G \in T^*$. By definition of subbase,

$$G = \cup_i (S_{i_1} \cap \cdots \cap S_{i_{n_i}}) \quad \text{where} \quad S_{i_k} \in \mathcal{S}$$

Hence, $\quad f^{-1}[G] \;=\; f^{-1}[\cup_i (S_{i_1} \cap \cdots \cap S_{i_{n_i}})] \;=\; \cup_i f^{-1}[S_{i_1} \cap \cdots \cap S_{i_{n_i}}]$

$$= \;\cup_i (f^{-1}[S_{i_1}] \cap \cdots \cap f^{-1}[S_{i_{n_i}}])$$

But $S_{i_k} \in \mathcal{S}$ implies $f^{-1}[S_{i_k}] \in T$. Hence $f^{-1}[G] \in T$ since it is the union of finite intersections of open sets. Accordingly, f is continuous.

On the other hand, if f is continuous then the inverse of all open sets, including the members of $\mathcal{S}$ are open.

7. Let f be a function from a topological space X into the unit interval $[0, 1]$.   Show that if $f^{-1}[(a, 1]]$ and $f^{-1}[[0, b)]$ are open subsets of X for all $0 < a, b < 1$, then f is continuous.

Solution:
Recall that the intervals $(a, 1]$ and $[0, b)$ form a subbase for the unit interval $I = [0, 1]$. Hence f is continuous by the preceding problem, i.e. by Theorem 7.2.

8. **Prove:** Let the functions $f : X \to Y$ and $g : Y \to Z$ be continuous. Then the composition function $g \circ f : X \to Z$ is also continuous.

Solution:

Let G be an open subset of Z. Then $g^{-1}[G]$ is open in Y since g is continuous. But f is also continuous, so $f^{-1}[g^{-1}[G]]$ is open in X. Now

$$(g \circ f)^{-1}[G] \;=\; f^{-1}[g^{-1}[G]]$$

Thus $(g \circ f)^{-1}[G]$ is open in X for every open subset G of Z, or, $g \circ f$ is continuous.

9. **Prove:** Let $\{\mathcal{T}_i\}$ be a collection of topologies on a set X. If a function $f : X \to Y$ is continuous with respect to each $\mathcal{T}_i$, then f is continuous with respect to the intersection topology $\mathcal{T} = \cap_i \mathcal{T}_i$.

Solution:

Let G be an open subset of Y. Then, by hypothesis, $f^{-1}[G]$ belongs to each $\mathcal{T}_i$. Hence $f^{-1}[G]$ belongs to the intersection, i.e. $f^{-1}[G] \in \cap_i \mathcal{T}_i = \mathcal{T}$, and so f is continuous with respect to $\mathcal{T}$.

10. **Prove Theorem 7.3:** A function $f : X \to Y$ is continuous if and only if the inverse image of every closed subset of Y is a closed subset of X.

Solution:

Suppose $f : X \to Y$ is continuous, and let F be a closed subset of Y. Then F^c is open, and so $f^{-1}[F^c]$ is open in X. But $f^{-1}[F^c] = (f^{-1}[F])^c$; therefore $f^{-1}[F]$ is closed.

Conversely, assume F closed in Y implies $f^{-1}[F]$ closed in X. Let G be an open subset of Y. Then G^c is closed in Y, and so $f^{-1}[G^c] = (f^{-1}[G])^c$ is closed in X. Accordingly, $f^{-1}[G]$ is open and therefore f is continuous.

11. **Prove Theorem 7.4:** A function $f : X \to Y$ is continuous if and only if, for every subset $A \subset X$, $f[\bar{A}] \subset \overline{f[A]}$.

Solution:

Suppose $f : X \to Y$ is continuous. Now $f[A] \subset \overline{f[A]}$, so

$$A \subset f^{-1}[f[A]] \subset f^{-1}\big[\overline{f[A]}\big]$$

But $\overline{f[A]}$ is closed, and so $f^{-1}\big[\overline{f[A]}\big]$ is also closed; hence

$$A \subset \bar{A} \subset f^{-1}\big[\overline{f[A]}\big]$$

and therefore

$$f[\bar{A}] \subset \overline{f[A]} = f\big[f^{-1}\big[\overline{f[A]}\big]\big]$$

Conversely, assume $f[\bar{A}] \subset \overline{f[A]}$ for any $A \subset X$, and let F be a closed subset of Y. Set $A = f^{-1}[F]$; we wish to show that A is also closed or, equivalently, that $\bar{A} = A$. Now

$$f[\bar{A}] = f\big[\overline{f^{-1}[F]}\big] \subset \overline{f[f^{-1}[F]]} = \bar{F} = F$$

Hence

$$\bar{A} \subset f^{-1}[f[\bar{A}]] \subset f^{-1}[F] = A$$

But $A \subset \bar{A}$, so $\bar{A} = A$ and f is continuous.

12. **Prove:** Let $f : (X, \mathcal{T}) \to (Y, \mathcal{T}^*)$ be continuous. Then $f_A : (A, \mathcal{T}_A) \to (Y, \mathcal{T}^*)$ is continuous, where $A \subset X$ and f_A is the restriction of f to A.

Solution:

Observe that $f_A^{-1}[G] = A \cap f^{-1}[G]$ for any $G \subset Y$.

Let $G \in \mathcal{T}^*$. Then $f^{-1}[G] \in \mathcal{T}$, and so $A \cap f^{-1}[G] \in \mathcal{T}_A$ by definition of the induced topology. Thus $A \cap f^{-1}[G] = f_A^{-1}[G] \in \mathcal{T}_A$, so f_A is continuous.

CONTINUITY AT A POINT

13. Under what conditions will a function $f: X \to Y$ not be continuous at a point $p \in X$?

Solution:

   A function $f: X \to Y$ is continuous at $p \in X$ iff, for every open set $H \subset Y$ containing $f(p)$, $f^{-1}[H]$ is a superset of an open set containing p. Hence f is not continuous at $p \in X$ if there exists at least one open set $H \subset Y$ containing $f(p)$ such that $f^{-1}[H]$ does not contain an open set containing p.

   Equivalently, $f: X \to Y$ is not continuous at $p \in X$ iff ∃ a neighborhood N of $f(p)$ such that $f^{-1}[N]$ is not a neighborhood of p.

14. Consider the following topology defined on $X = \{a, b, c, d\}$:

$$\mathcal{T} = \{X, \varnothing, \{a\}, \{b\}, \{a, b\}, \{b, c, d\}\}$$

Let the function $f: X \to X$ be defined by the adjoining diagram.

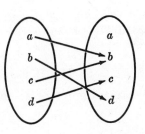

(i)   Show that f is not continuous at c.

(ii) Show that f is continuous at d.

Solution:

(i)   Observe that $\{a, b\}$ is an open set containing $f(c) = b$ and that $f^{-1}[\{a, b\}] = \{a, c\}$. Hence f is not continuous at c since there exists no open set containing c which is contained in $\{a, c\}$.

(ii) The only open sets containing $f(d) = c$ are $\{b, c, d\}$ and X. Note that $f^{-1}[\{b, c, d\}] = X$ and $f^{-1}[X] = X$. Hence f is continuous at d since the inverse of each open set containing $f(d)$ is an open set containing d.

15. Suppose a singleton set $\{p\}$ is an open subset of a topological space X. Show that for any topological space Y and any function $f: X \to Y$, f is continuous at $p \in X$.

Solution:

   Let $H \subset Y$ be an open set containing $f(p)$. But

$$f(p) \in H \quad \Rightarrow \quad p \in f^{-1}[H] \quad \Rightarrow \quad \{p\} \subset f^{-1}[H]$$

Hence f is continuous at p.

16. Prove: If $f: X \to Y$ is continuous at $p \in X$, then the restriction of f to a subset containing p is also continuous at p. More precisely, let A be a subset of a topological space $(X, \mathcal{T})$ such that $p \in A \subset X$, and let $f_A: A \to Y$ denote the restriction of $f: X \to Y$ to A. Then if f is $\mathcal{T}$-continuous at p, f_A will be $\mathcal{T}_A$-continuous at p where $\mathcal{T}_A$ is the relative topology on A.

Solution:

   Let $H \subset Y$ be an open set containing $f(p)$. Since f is continuous at p,

$$\exists G \in \mathcal{T} \quad \text{such that} \quad p \in G \subset f^{-1}[H]$$

and so

$$p \in A \cap G \subset A \cap f^{-1}[H] = f_A^{-1}[H]$$

But, by definition of the induced topology, $A \cap G \in \mathcal{T}_A$; hence f_A is $\mathcal{T}_A$-continuous at p.

17. Prove Theorem 7.5: Let X and Y be topological spaces. Then a function $f: X \to Y$ is continuous if and only if it is continuous at every point $p \in X$.

Solution:

   Assume f is continuous, and let $H \subset Y$ be an open set containing $f(p)$. But then $p \in f^{-1}[H]$, and $f^{-1}[H]$ is open. Hence f is continuous at p.

   Now suppose f is continuous at every point $p \in X$, and let $H \subset Y$ be open. For every $p \in f^{-1}[H]$, there exists an open set $G_p \subset X$ such that $p \in G_p \subset f^{-1}[H]$. Hence $f^{-1}[H] = \bigcup \{G_p : p \in f^{-1}[H]\}$ a union of open sets. Accordingly, $f^{-1}[H]$ is open and so f is continuous.

18. Prove Proposition 7.6: If a function $f : X \to Y$ is continuous at $p \in X$, then it is sequentially continuous at p, i.e. $a_n \to p \Rightarrow f(a_n) \to f(p)$.

Solution:

We need to show that any neighborhood N of $f(p)$ contains almost all the terms of the sequence $\langle f(a_1), f(a_2), \ldots \rangle$.

Let N be a neighborhood of $f(p)$. By hypothesis, f is continuous at p; hence $M = f^{-1}[N]$ is a neighborhood of p. If the sequence $\langle a_n \rangle$ converges to p, then M contains almost all the terms of the sequence $\langle a_1, a_2, \ldots \rangle$, i.e. $a_n \in M$ for almost all $n \in N$. But

$$a_n \in M \quad \Rightarrow \quad f(a_n) \in f[M] = f[f^{-1}[N]] = N$$

Hence $f(a_n) \in N$ for almost all $n \in N$, and so the sequence $\langle f(a_n) \rangle$ converges to $f(p)$. Accordingly, f is sequentially continuous at p.

OPEN AND CLOSED FUNCTIONS, HOMEOMORPHISMS

19. Give an example of a real function $f : \mathbf{R} \to \mathbf{R}$ such that f is continuous and closed, but not open.

Solution:

Let f be a constant function, say $f(x) = 1$ for all $x \in \mathbf{R}$. Then $f[A] = \{1\}$ for any $A \subset \mathbf{R}$. Hence f is a closed function and is not an open function. Furthermore, f is continuous.

20. Let the real function $f : \mathbf{R} \to \mathbf{R}$ be defined by $f(x) = x^2$. Show that f is not open.

Solution:

Let $A = (-1, 1)$, an open set. Note that $f[A] = [0, 1)$, which is not open; hence f is not an open function.

21. Let $\mathcal{B}$ be a base for a topological space X. Show that if $f : X \to Y$ has the property that $f[B]$ is open for every $B \in \mathcal{B}$, then f is an open function.

Solution:

We want to show that the image of every open subset of X is open in Y. Let $G \subset X$ be open. By definition of a base, $G = \cup_i B_i$ where $B_i \in \mathcal{B}$. Now $f[G] = f[\cup_i B_i] = \cup_i f[B_i]$. By hypothesis, each $f[B_i]$ is open in Y and so $f[G]$, a union of open sets, is also open in Y; hence f is an open function.

22. Show that the closed interval $A = [a, b]$ is homeomorphic to the closed unit interval $I = [0, 1]$.

Solution:

The linear function $f : I \to A$ defined by $f(x) = (b - a)x + a$ is one-one, onto and bicontinuous. Hence f is a homeomorphism.

23. Show that area is not a topological property.

Solution:

The open disc $D = \{\langle r, \theta \rangle : r < 1\}$ with radius 1 is homeomorphic to the open disc $D^* = \{\langle r, \theta \rangle : r < 2\}$ with radius 2. In fact, the function $f : D \to D^*$ defined by $f(\langle r, \theta \rangle) = \langle 2r, \theta \rangle$ is a homeomorphism. Here $\langle r, \theta \rangle$ denotes the polar coordinates of a point in the plane $\mathbf{R}^2$.

24. Let $f : (X, \mathcal{T}) \to (Y, \mathcal{T}^*)$ be one-one and open, let $A \subset X$, and let $f[A] = B$. Show that the function $f_A : (A, \mathcal{T}_A) \to (B, \mathcal{T}_B^*)$ is also one-one and open. Here f_A denotes the restriction of f to A, and $\mathcal{T}_A$ and $\mathcal{T}_B^*$ are the relative topologies.

Solution:

If f is one-one, then every restriction of f is also one-one; hence we need only show that f_A is open.

Let $H \subset A$ be $\mathcal{T}_A$-open. Then by definition of the relative topology, $H = A \cap G$ where $G \in \mathcal{T}$. Since f is one-one, $f[A \cap G] = f[A] \cap f[G]$, and so

$$f_A[H] \;=\; f[H] \;=\; f[A \cap G] \;=\; f[A] \cap f[G] \;=\; B \cap f[G]$$

Since f is open and $G \in \mathcal{T}$, $f[G] \in \mathcal{T}^*$. Thus $B \cap f[G] \in \mathcal{T}_B^*$ and so $f_A : (A, \mathcal{T}_A) \to (B, \mathcal{T}_B^*)$ is open.

25. Let $f : (X, \mathcal{T}) \to (Y, \mathcal{T}^*)$ be a homeomorphism and let $(A, \mathcal{T}_A)$ be any subspace of $(X, \mathcal{T})$. Show that $f_A : (A, \mathcal{T}_A) \to (B, \mathcal{T}_B^*)$ is also a homeomorphism where f_A is the restriction of f to A, $f[A] = B$, and $\mathcal{T}_B^*$ is the relative topology on B.

Solution:

Since f is one-one and onto, $f_A : A \to B$, where $B = f[A]$, is also one-one and onto. Hence we need only show that f_A is bicontinuous, i.e. open and continuous. By the preceding problem f_A is open. Furthermore, the restriction of any continuous function is also continuous; hence $f_A : (A, \mathcal{T}_A) \to (B, \mathcal{T}_B^*)$ is a homeomorphism.

26. Show that any interval $A = (a, b)$ is connected as a subspace of the real line $\mathbf{R}$. (See Example 4.3 for the definition of connectedness.)

Solution:

Suppose A is not connected. Then $\exists$ open sets $G, H \subset \mathbf{R}$ such that $A \cap G$ and $A \cap H$ are non-empty, disjoint and satisfy $(A \cap G) \cup (A \cap H) = A$. Define the function $f : A \to \mathbf{R}$ by

$$f(x) \;=\; \begin{cases} 1 & \text{if } x \in A \cap G \\ 0 & \text{if } x \in A \cap H \end{cases}$$

Then f is continuous, for the inverse of any open set is either $A \cap G$, $A \cap H$, $\emptyset$ or A and so is open. But then the intermediate value theorem applies, so $\exists x_0 \in A$ for which $f(x_0) = \frac{1}{2}$. But this is impossible, so A is connected.

27. Show that the following subsets of the plane $\mathbf{R}^2$ are not homeomorphic, where the topologies are the relativized usual topologies:

$$X \;=\; \{x : d(x, p_0) = 1 \text{ or } d(x, p_1) = 1;\; p_0 = \langle 0, -1 \rangle,\; p_1 = \langle 0, 1 \rangle\}$$

$$Y \;=\; \{x : d(x, p) = 1,\; p = \langle 0, 5 \rangle\}$$

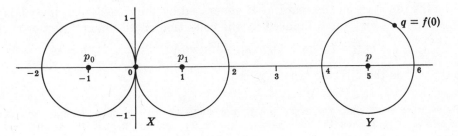

Solution:

Suppose there exists a homeomorphism $f : X \to Y$; let $q = f(0)$, $X^* = X \setminus \{0\}$, and $Y^* = Y \setminus \{q\}$. Then $f : X^* \to Y^*$ is also a homeomorphism with respect to the relative topologies (see Problem 25).

We show that Y^* is connected. For if $q = \langle 5 + \cos\theta_0, \sin\theta_0 \rangle$, then the function

$$g : (0, 2\pi) \to Y^* \quad \text{defined by} \quad g(\theta) = \langle 5 + \cos(\theta_0 + \theta),\, \sin(\theta_0 + \theta) \rangle$$

is a homeomorphism. But the interval $(0, 2\pi)$ is connected, so Y^* is also connected.

On the other hand, X^* is not connected; for the sets

$$G = \{\langle x, y \rangle : x > 0\} \quad \text{and} \quad H = \{\langle x, y \rangle : x < 0\}$$

are both open in $\mathbf{R}^2$, so $G^* = X^* \cap G$ and $H^* = X^* \cap H$ are open subsets of X^*. Furthermore, G^* and H^* are non-empty, disjoint and satisfy $G^* \cup H^* = X^*$. Since connectedness is a topological property, X^* is not homeomorphic to Y^* and therefore there can exist no such function f.

TOPOLOGIES INDUCED BY FUNCTIONS

28. Let $\{f_i : X \to (Y_i, \mathcal{T}_i)\}$ be a collection of constant functions from an arbitrary set X into the topological spaces $(Y_i, \mathcal{T}_i)$. Determine the coarsest topology on X with respect to which the functions f_i are continuous.

Solution:

Recall (see Problem 1) that a constant function $f : X \to Y$ is continuous with respect to every topology on X. Hence all the constant functions f_i are continuous with respect to the indiscrete topology $\{X, \emptyset\}$ on X. Since the indiscrete topology $\{X, \emptyset\}$ on X is the coarsest topology on X, it is also the coarsest topology on X with respect to which the constant functions are continuous.

29. Consider the following topology on $Y = \{a, b, c, d\}$:

$$\mathcal{T} = \{Y, \emptyset, \{c\}, \{a, b, c\}, \{c, d\}\}$$

Let $X = \{1, 2, 3, 4\}$ and let the functions $f : X \to (Y, \mathcal{T})$ and $g : X \to (Y, \mathcal{T})$ be defined by

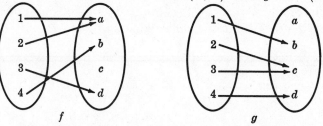

Find the defining subbase $\mathcal{S}$ for the topology $\mathcal{T}^*$ on X induced by f and g, i.e. the coarsest topology with respect to which f and g are continuous.

Solution:

Recall that
$$\mathcal{S} = \{f^{-1}[H] : H \in \mathcal{T}\} \cup \{g^{-1}[H] : H \in \mathcal{T}\}$$

that is, $\mathcal{S}$ consists of the inverses under f and g of the open subsets of Y. Hence

$$\mathcal{S} = \{X, \emptyset, \{1, 2, 4\}, \{3\}, \{2, 3\}, \{1, 2, 3\}, \{2, 3, 4\}\}$$

30. Let $\mathcal{T}$ be the topology on the real line $\mathbf{R}$ generated by the closed-open intervals $[a, b)$, and let $\mathcal{T}^*$ be the topology on $\mathbf{R}$ induced by the collection of all linear functions

$$f : \mathbf{R} \to (\mathbf{R}, \mathcal{T}) \quad \text{defined by} \quad f(x) = ax + b, \quad a, b \in \mathbf{R}$$

Show that $\mathcal{T}^*$ is the discrete topology on $\mathbf{R}$.

Solution:

We want to show that, for every $p \in \mathbf{R}$, the singleton set $\{p\}$ is a $\mathcal{T}^*$-open set. Consider the $\mathcal{T}$-open set $A = [1, 2)$ and the functions $f : \mathbf{R} \to (\mathbf{R}, \mathcal{T})$ and $g : \mathbf{R} \to (\mathbf{R}, \mathcal{T})$ defined by

$$f(x) = x - p + 1 \quad \text{and} \quad g(x) = -x - p + 1$$

and illustrated below.

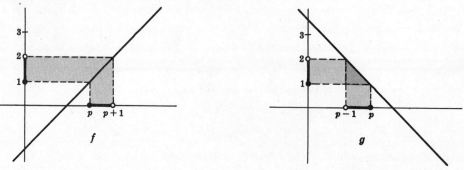

Now $A \in \mathcal{T}$ implies

$$f^{-1}[A] = [p, p+1) \quad \text{and} \quad g^{-1}[A] = (p-1, p]$$

belong to the defining subbase $\mathcal{S}$ for the topology $\mathcal{T}^*$. Hence the intersection

$$(p-1, p] \cap [p, p+1) = \{p\}$$

belongs to $\mathcal{T}^*$, and so $\mathcal{T}^*$ is the discrete topology on $\mathbf{R}$.

31. Prove Theorem 7.9: Let $\{f_i : X \to (Y_i, \mathcal{T}_i)\}$ be a collection of functions defined on an arbitrary non-empty set X, let

$$\mathcal{S} = \bigcup_i \{f_i^{-1}[H] : H \in \mathcal{T}_i\}$$

and let $\mathcal{T}$ be the topology on X generated by $\mathcal{S}$. Then:

(i)   All the functions f_i are continuous relative to $\mathcal{T}$.

(ii)   If $\mathcal{T}^*$ is the intersection of all topologies on X with respect to which the functions f_i are continuous, then $\mathcal{T} = \mathcal{T}^*$.

(iii) $\mathcal{T}$ is the coarsest topology on X with respect to which the functions f_i are continuous.

(iv) $\mathcal{S}$ is a subbase for $\mathcal{T}$.

Solution:

(i)   For any function $f_i : (X, \mathcal{T}) \to (Y_i, \mathcal{T}_i)$, if $H \in \mathcal{T}_i$ then $f_i^{-1}[H] \in \mathcal{S} \subset \mathcal{T}$. Hence all the f_i are continuous with respect to $\mathcal{T}$.

(ii)   By Problem 9, all the functions f_i are also continuous with respect to $\mathcal{T}^*$; hence $\mathcal{S} \subset \mathcal{T}^*$ and, since $\mathcal{T}$ is the topology generated by $\mathcal{S}$, $\mathcal{T} \subset \mathcal{T}^*$. On the other hand, $\mathcal{T}$ is one of the topologies with respect to which the f_i are continuous; hence $\mathcal{T}^* \subset \mathcal{T}$ and so $\mathcal{T} = \mathcal{T}^*$.

(iii)   Follows from (ii).

(iv)   Follows from the fact that any class of sets is a subbase of the topology it generates.

Supplementary Problems

CONTINUOUS FUNCTIONS

32. Prove that $f : X \to Y$ is continuous if and only if $f^{-1}[A^\circ] \subset (f^{-1}[A])^\circ$ for every $A \subset X$.

33. Let X and Y be topological spaces with $X = E \cup F$. Let $f : E \to Y$ and $g : F \to Y$, with $f = g$ on $E \cap F$, be continuous with respect to the relative topologies. Note that $h = f \cup g$ is a function from X into Y. (i) Show, by an example, that h need not be continuous. (ii) Prove: If E and F are both open, then h is continuous. (iii) Prove: If E and F are both closed, then h is continuous.

34. Let $f : X \to Y$ be continuous. Show that $f : X \to f[X]$ is also continuous where $f[X]$ has the relative topology.

35. Let X be a topological space and let $\chi_A : X \to \mathbf{R}$ be the characteristic function for some subset A of X. Show that χ_A is continuous at $p \in X$, if and only if p is not an element of the boundary of A. (Recall $\chi_A(x) = 1$ if $x \in A$, and $\chi_A(x) = 0$ if $x \in A^c$.)

36. Consider the real line $\mathbf{R}$ with the usual topology. Show that if every function $f : X \to \mathbf{R}$ is continuous, then X is a discrete space.

OPEN AND CLOSED FUNCTIONS

37. Let $f : (X, \mathcal{T}) \to (Y, \mathcal{T}^*)$. Prove the following:

(i) f is closed if and only if $\overline{f[A]} \subset f[\bar{A}]$ for every $A \subset X$;

(ii) f is open if and only if $f[A^\circ] \subset (f[A])^\circ$ for every $A \subset X$.

38. Show that the function $f : (0, \infty) \to [-1, 1]$ defined by $f(x) = \sin(1/x)$ is continuous, but neither open nor closed, where $(0, \infty)$ and $[-1, 1]$ have the relativized usual topologies.

39. Prove: Let $f : (X, \mathcal{T}) \to (Y, \mathcal{T}^*)$ be open and onto, and let $\mathcal{B}$ be a base for $\mathcal{T}$. Then $\{f[B] : B \in \mathcal{B}\}$ is a base for $\mathcal{T}^*$.

40. Give an example of a function $f : X \to Y$ and a subset $A \subset X$ such that f is open but f_A, the restriction of f to A, is not open.

HOMEOMORPHISMS, TOPOLOGICAL PROPERTIES

41. Let $f : X \to Y$ and $g : Y \to Z$ be continuous. Show that if $g \circ f : X \to Z$ is a homeomorphism, then g one-one (or f onto) implies that f and g are homeomorphisms.

42. Prove that each of the following is a topological property: (i) accumulation point, (ii) interior, (iii) boundary, (iv) density, and (v) neighborhood.

43. Prove: Let $f : X \to Y$ be a homeomorphism and let $A \subset X$ have the property that $A \cap A' = \emptyset$. Then $f[A] \cap (f[A])' = \emptyset$. (A subset $A \subset X$ having the property $A \cap A' = \emptyset$ is called *isolated*. The property of being isolated is thus a topological property.)

TOPOLOGIES INDUCED BY FUNCTIONS

44. Consider the following topology on $Y = \{a, b, c, d\}$: $\mathcal{T} = \{Y, \emptyset, \{a, b\}, \{c, d\}\}$. Let $X = \{1, 2, 3, 4, 5\}$ and let $f : X \to Y$ and $g : X \to Y$ be as follows:

$$f = \{\langle 1, a \rangle, \langle 2, a \rangle, \langle 3, b \rangle, \langle 4, b \rangle, \langle 5, d \rangle\}, \qquad g = \{\langle 1, c \rangle, \langle 2, b \rangle, \langle 3, d \rangle, \langle 4, a \rangle, \langle 5, c \rangle\}$$

Find the defining subbase for the topology on X induced by f and g.

45. Let $f : X \to (Y, \mathcal{T}^*)$. Show that if $\mathcal{S}$ is the defining subbase for the topology $\mathcal{T}$ induced by the one function f, then $\mathcal{S} = \mathcal{T}$.

46. Prove: Let $\{f_i : X \to (Y_i, \mathcal{T}_i)\}$ be a collection of functions defined on an arbitrary set X, and let $\mathcal{S}_i$ be a subbase for the topology $\mathcal{T}_i$ on Y_i. Then the class $\mathcal{S}^* = \bigcup_i \{f_i^{-1}[S] : S \in \mathcal{S}_i\}$ has the following properties: (i) $\mathcal{S}^*$ is a subclass of the defining subbase $\mathcal{S}$ of the topology $\mathcal{T}$ on X induced by the functions f_i; (ii) $\mathcal{S}^*$ is also a subbase for $\mathcal{T}$.

47. Show that the coarsest topology on the real line $\mathbf{R}$ with respect to which the linear functions

$$f : \mathbf{R} \to (\mathbf{R}, \mathcal{U}) \qquad \text{defined by} \qquad f(x) = ax + b, \quad a, b \in \mathbf{R}$$

are continuous is also the usual topology $\mathcal{U}$.

Answers to Supplementary Problems

33. (i) Let $X = (0, 2)$ and let $E = (0, 1)$ and $F = [1, 2)$. Then $f(x) = 1$ and $g(x) = 2$ are each continuous, but $h = f \cup g$ is not continuous.

44. $\{X, \emptyset, \{1, 2, 3, 4\}, \{5\}, \{2, 4\}, \{1, 3, 5\}\}$

45. *Hint.* Show that $\mathcal{S}$ is a topology.

Chapter 8

Metric and Normed Spaces

METRICS

Let X be a non-empty set. A real-valued function d defined on $X \times X$, i.e. ordered pairs of elements in X, is called a *metric* or *distance function* on X iff it satisfies, for every $a, b, c \in X$, the following axioms:

[M₁] $d(a, b) \geqq 0$ and $d(a, a) = 0$.

[M₂] (Symmetry) $d(a, b) = d(b, a)$.

[M₃] (Triangle Inequality) $d(a, c) \leqq d(a, b) + d(b, c)$.

[M₄] If $a \neq b$, then $d(a, b) > 0$.

The real number $d(a, b)$ is called the *distance* from a to b.

Observe that **[M₁]** states that the distance from any point to another is never negative, and that the distance from a point to itself is zero. The axiom **[M₂]** states that the distance from a point a to a point b is the same as the distance from b to a; hence we speak of the distance *between* a and b.

[M₃] is called the Triangle Inequality because if a, b and c are points in the plane $\mathbf{R}^2$ as illustrated on the right, then **[M₃]** states that the length $d(a, c)$ of one side of the triangle is less than or equal to the sum $d(a, b) + d(b, c)$ of the lengths of the other two sides of the triangle. The last axiom **[M₄]** states that the distance between two distinct points is positive.

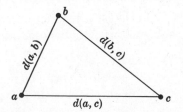

We now give some examples of metrics. That they actually satisfy the required axioms will be verified later.

Example 1.1: The function d defined by $d(a, b) = |a - b|$, where a and b are real numbers, is a metric and called the *usual* metric on the real line $\mathbf{R}$. Furthermore, the function d defined by

$$d(p, q) = \sqrt{(a_1 - b_1)^2 + (a_2 - b_2)^2}$$

where $p = \langle a_1, a_2 \rangle$ and $q = \langle b_1, b_2 \rangle$ are points in the plane $\mathbf{R}^2$, is a metric and called the *usual* metric on $\mathbf{R}^2$. We shall assume these metrics on $\mathbf{R}$ and $\mathbf{R}^2$, respectively, unless otherwise specified.

Example 1.2: Let X be any non-empty set and let d be the function defined by

$$d(a, b) = \begin{cases} 0 & \text{if } a = b \\ 1 & \text{if } a \neq b \end{cases}$$

Then d is a metric on X. This distance function d is usually called the *trivial* metric on X.

Example 1.3: Let $C[0, 1]$ denote the class of continuous functions on the closed unit interval $[0, 1]$. A metric is defined on the class $C[0, 1]$ as follows:

$$d(f, g) = \int_0^1 |f(x) - g(x)| \, dx$$

111

Here $d(f, g)$ is precisely the area of the region which lies between the functions as illustrated below.

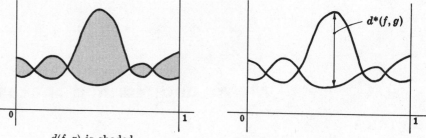

$d(f, g)$ is shaded

Example 1.4: Again let $C[0, 1]$ denote the collection of continuous functions on $[0, 1]$. Another metric is defined on $C[0, 1]$ as follows:
$$d^*(f, g) = \sup \{|f(x) - g(x)| : x \in [0, 1]\}$$
Here $d^*(f, g)$ is precisely the greatest vertical gap between the functions as illustrated above.

Example 1.5: Let $p = \langle a_1, a_2 \rangle$ and $q = \langle b_1, b_2 \rangle$ be arbitrary points in the plane $\mathbf{R}^2$, i.e. ordered pairs of real numbers. The functions d_1 and d_2 defined by
$$d_1(p, q) = \max(|a_1 - b_1|, |a_2 - b_2|), \qquad d_2(p, q) = |a_1 - b_1| + |a_2 - b_2|$$
are distinct metrics on $\mathbf{R}^2$.

A function ρ satisfying [M_1], [M_2] and [M_3], i.e. not necessarily [M_4], is called a *pseudometric*. Many of the results for metrics are also true for pseudometrics.

DISTANCE BETWEEN SETS, DIAMETERS

Let d be a metric on a set X. The *distance* between a point $p \in X$ and a non-empty subset A of X is denoted and defined by
$$d(p, A) = \inf \{d(p, a) : a \in A\}$$
i.e. the greatest lower bound of the distances from p to points of A. The *distance* between two non-empty subsets A and B of X is denoted and defined by
$$d(A, B) = \inf \{d(a, b) : a \in A, b \in B\}$$
i.e. the greatest lower bound of the distances from points in A to points in B.

The *diameter* of a non-empty subset A of X is denoted and defined by
$$d(A) = \sup \{d(a, a') : a, a' \in A\}$$
i.e. the least upper bound of the distances between points in A. If the diameter of A is finite, i.e. $d(A) < \infty$, then A is said to be *bounded*; if not, i.e. $d(A) = \infty$, then A is said to be *unbounded*.

Example 2.1: Let d be the trivial metric on a non-empty set X. Then for $p \in X$ and $A, B \subset X$,
$$d(p, A) = \begin{cases} 1 & \text{if } p \notin A \\ 0 & \text{if } p \in A \end{cases}, \qquad d(A, B) = \begin{cases} 1 & \text{if } A \cap B = \emptyset \\ 0 & \text{if } A \cap B \neq \emptyset \end{cases}$$

Example 2.2: Consider the following intervals on the real line $\mathbf{R}$: $A = [0, 1)$, $B = (1, 2]$.

If d denotes the usual metric on $\mathbf{R}$, then $d(A, B) = 0$. On the other hand, if d^* denotes the trivial metric on $\mathbf{R}$, then $d^*(A, B) = 1$ since A and B are disjoint.

The next proposition clearly follows from the above definitions:

Proposition 8.1: Let A and B be non-empty subsets of X and let $p \in X$. Then:
    (i) $d(p, A)$, $d(A, B)$ and $d(A)$ are non-negative real numbers.
    (ii) If $p \in A$, then $d(p, A) = 0$.

(iii) If $A \cap B$ is non-empty, then $d(A, B) = 0$.

(iv) If A is finite, then $d(A) < \infty$, i.e. A is bounded.

Observe that the converses of (ii), (iii) and (iv) are not true.

For the empty set $\emptyset$, the following conventions are adopted:

$$d(p, \emptyset) = \infty, \quad d(A, \emptyset) = d(\emptyset, A) = \infty, \quad d(\emptyset) = -\infty$$

OPEN SPHERES

Let d be a metric on a set X. For any point $p \in X$ and any real number $\delta > 0$, we shall let $S_d(p, \delta)$ or simply $S(p, \delta)$ denote the set of points within a distance of δ from p:

$$S(p, \delta) = \{x : d(p, x) < \delta\}$$

We call $S(p, \delta)$ the *open sphere*, or simply *sphere*, with center p and radius δ. It is also called a *spherical neighborhood* or *ball*.

Example 3.1: Consider the point $p = \langle 0, 0 \rangle$ in the plane $\mathbf{R}^2$, and the real number $\delta = 1$. If d is the usual metric on $\mathbf{R}^2$, then $S_d(p, \delta)$ is the open unit disc illustrated on the right. On the other hand, if d_1 and d_2 are the metrics on $\mathbf{R}^2$ which are defined in Example 1.5, then $S_{d_1}(p, \delta)$ and $S_{d_2}(p, \delta)$ are the subsets of $\mathbf{R}^2$ which are illustrated below.

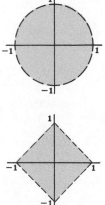

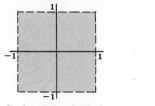

$S_{d_1}(p, \delta)$ is shaded $S_{d_2}(p, \delta)$ is shaded

Example 3.2: Let d denote the trivial metric on some set X, and let $p \in X$. Recall that the distance between p and every other point in X is exactly 1. Hence

$$S(p, \delta) = \begin{cases} X & \text{if } \delta > 1 \\ \{p\} & \text{if } \delta \leq 1 \end{cases}$$

Example 3.3: Let d be the usual metric on the real line $\mathbf{R}$, i.e. $d(a, b) = |a - b|$. Then the open sphere $S(p, \delta)$ is the open interval $(p - \delta, p + \delta)$.

Example 3.4: Let d be the metric on the collection $C[0, 1]$ of all continuous functions on $[0, 1]$ defined by

$$d(f, g) = \sup \{|f(x) - g(x)| : x \in [0, 1]\}$$

(see Example 1.4). Given $\delta > 0$ and a function $f_0 \in C[0, 1]$, then the open sphere $S(f_0, \delta)$ consists of all continuous functions g which lie in the area bounded by $f_0 - \delta$ and $f_0 + \delta$, as indicated in the diagram below:

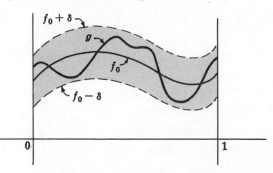

One important property of open spheres in metric spaces is given in the next lemma.

Lemma 8.2: Let S be an open sphere with center p and radius δ. Then for every point $q \in S$ there exists an open sphere T centered at q such that T is contained in S. (See the adjacent Venn diagram.)

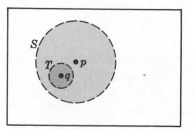

METRIC TOPOLOGIES, METRIC SPACES

In general, the intersection of two open spheres need not be an open sphere. However, we will show that every point in the intersection of two open spheres does belong to an open sphere contained in the intersection. Namely,

Lemma 8.3: Let S_1 and S_2 be open spheres and let $p \in S_1 \cap S_2$. Then there exists an open sphere S_p with center p such that $p \in S_p \subset S_1 \cap S_2$.

Hence by virtue of Theorem 6.1 we have

Theorem 8.4: The class of open spheres in a set X with metric d is a base for a topology on X.

Definition: Let d be a metric on a non-empty set X. The topology $\mathcal{T}$ on X generated by the class of open spheres in X is called the *metric topology* (or, the topology *induced* by the metric d). Furthermore, the set X together with the topology $\mathcal{T}$ induced by the metric d is called a *metric space* and is denoted by (X, d).

Thus a metric space is a topological space in which the topology is induced by a metric. Accordingly, all concepts defined for topological spaces are also defined for metric spaces. For example, we can speak about open sets, closed sets, neighborhoods, accumulation points, closure, etc., for metric spaces.

> **Example 4.1:** If d is the usual metric on the real line $\mathbf{R}$, i.e. $d(a, b) = |a - b|$, then the open spheres in $\mathbf{R}$ are precisely the finite open intervals. Hence the usual metric on $\mathbf{R}$ induces the usual topology on $\mathbf{R}$. Similarly, the usual metric on the plane $\mathbf{R}^2$ induces the usual topology on $\mathbf{R}^2$.

> **Example 4.2:** Let d be the trivial metric on some set X. Note that for any $p \in X$, $S(p, \frac{1}{2}) = \{p\}$. Hence every singleton set is open and so every set is open. In other words, the trivial metric on X induces the discrete topology on X.

> **Example 4.3:** Let (X, d) be a metric space and let Y be a non-empty subset of X. The restriction of the function d to the points in the subset Y, also denoted by d, is a metric on Y. We call (Y, d) a *metric subspace* of (X, d). In fact, (Y, d) is a subspace of (X, d), i.e. has the relative topology.

Frequently the same symbol, say X, is used to denote both a metric space and the underlying set on which the metric is defined.

PROPERTIES OF METRIC TOPOLOGIES

Since the topology of a metric space X is derived from a metric, one would correctly expect that the topological properties of X are related to the distance properties of X. For example,

Theorem 8.5: Let p be a point in a metric space X. Then the countable class of open spheres, $\{S(p, 1), S(p, \frac{1}{2}), S(p, \frac{1}{3}), \ldots\}$ is a local base at p.

Theorem 8.6: The closure $\bar{A}$ of a subset A of a metric space X is the set of points whose distance from A is zero, i.e. $\bar{A} = \{x : d(x, A) = 0\}$.

Observe that axiom [M_4] implies that the only point with zero distance from a singleton set $\{p\}$ is the point p itself, i.e.,

$$d(x, \{p\}) = 0 \quad \text{implies} \quad x = p$$

Hence by the preceding theorem, singleton sets $\{p\}$ in a metric space are closed. Accordingly, finite unions of singleton sets, i.e. finite sets, are also closed. We state this result formally:

Corollary 8.7: In a metric space X all finite sets are closed.

Thus we see that a metric space X possesses certain topological properties which do not hold for topological spaces in general.

Next follows an important "separation" property of metric spaces.

Theorem 8.8 (Separation Axiom): Let A and B be closed disjoint subsets of a metric space X. Then there exist disjoint open sets G and H such that $A \subset G$ and $B \subset H$. (See Venn diagram below.)

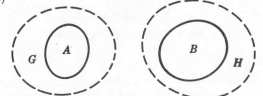

One might suspect from the above theorem that the distance between two disjoint closed sets is greater than zero. The next example shows that this is not true.

Example 5.1: Consider the following sets in the plane $\mathbf{R}^2$ which are illustrated below:

$$A = \{(x, y) : xy \geq -1,\ x < 0\}, \quad B = \{(x, y) : xy \geq 1,\ x > 0\}$$

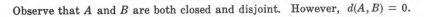

Observe that A and B are both closed and disjoint. However, $d(A, B) = 0$.

EQUIVALENT METRICS

Two metrics d and d^* on a set X are said to be equivalent iff they induce the same topology on X, i.e. iff the d-open spheres and the d^*-open spheres in X are bases for the same topology on X.

Example 6.1: The usual metric d and the metrics d_1 and d_2, defined in Example 1.5, all induce the usual topology on the plane $\mathbf{R}^2$, since the class of open spheres of each metric (illustrated below) is a base for the usual topology on $\mathbf{R}^2$.

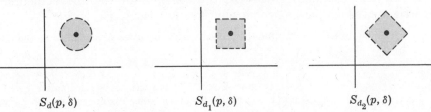

$$S_d(p, \delta) \qquad\qquad S_{d_1}(p, \delta) \qquad\qquad S_{d_2}(p, \delta)$$

Hence the metrics are equivalent.

Example 6.2: Consider the metric d on a non-empty set X defined by

$$d(a, b) \;=\; \begin{cases} 2 & \text{if } a \neq b \\ 0 & \text{if } a = b \end{cases}$$

Observe that $s_d(p, 1) = \{p\}$; so singleton sets are open and d induces the discrete topology on X. Accordingly, d is equivalent to the trivial metric on X which also induces the discrete topology.

The next proposition clearly follows from the above definition.

Proposition 8.9: The relation "d is equivalent to d^*" is an equivalence relation in any collection of metrics on a set X.

METRIZATION PROBLEM

Given any topological space $(X, \mathcal{T})$, it is natural to ask whether or not there exists a metric d on X which induces the topology $\mathcal{T}$. The topological space $(X, \mathcal{T})$ is said to be *metrizable* if such a metric exists.

Example 7.1: Every discrete space $(X, \mathcal{D})$ is metrizable since the trivial metric on X induces the discrete topology $\mathcal{D}$.

Example 7.2: Consider the topological space $(\mathbf{R}, \mathcal{U})$, the real line $\mathbf{R}$ with the usual topology $\mathcal{U}$. Observe that $(\mathbf{R}, \mathcal{U})$ is metrizable since the usual metric on $\mathbf{R}$ induces the usual topology on $\mathbf{R}$. Similarly, the plane $\mathbf{R}^2$ with the usual topology is metrizable.

Example 7.3: An indiscrete space $(X, \mathcal{J})$ where X consists of more than one point is not metrizable. For X and $\emptyset$ are the only closed sets in an indiscrete space $(X, \mathcal{J})$. But by Corollary 8.7 all finite sets in a metric space are closed. Hence X and $\emptyset$ cannot be the only closed sets in a topology on X induced by a metric. Accordingly, $(X, \mathcal{J})$ is not metrizable.

The *metrization problem* in topology consists of finding necessary and sufficient topological conditions for a topological space to be metrizable. An important partial solution to this problem was given in 1924 by Urysohn as a result of his celebrated Urysohn's Lemma. It was not until 1950 that a complete solution to this problem was given independently by a number of mathematicians. We will prove Urysohn's results later. The complete solution to the metrization problem is beyond the scope of this text and the reader is referred to the classical text of Kelley, *General Topology*.

ISOMETRIC METRIC SPACES

A metric space (X, d) is *isometric* to a metric space (Y, e) iff there exists a one-one, onto function $f : X \to Y$ which preserves distances, i.e. for all $p, q \in X$,

$$d(p, q) \;=\; e(f(p), f(q))$$

Observe that the relation "(X, d) is isometric to (Y, e)" is an equivalence relation in any collection of metric spaces. Furthermore,

Theorem 8.10: If the metric space (X, d) is isometric to (Y, e), then (X, d) is also homeomorphic to (Y, e).

The next example shows that the converse of the above theorem is not true, i.e. two metric spaces can be homeomorphic but not isometric.

Example 8.1: Let d be the trivial metric on a set X and let e be the metric on a set Y defined by

$$e(a, b) \;=\; \begin{cases} 2 & \text{if } a \neq b \\ 0 & \text{if } a = b \end{cases}$$

Assume that X and Y have the same cardinality greater than one. Then (X, d) and (Y, e) are not isometric since distances between points in each space are different. But both d and e induce the discrete topology, and two discrete spaces with the same cardinality are homeomorphic; so (X, d) and (Y, e) are homeomorphic.

EUCLIDEAN m-SPACE

Recall that $\mathbf{R}^m$ denotes the product set of m copies of the set $\mathbf{R}$ of real numbers, i.e. consists of all m-tuples $\langle a_1, a_2, \ldots, a_m \rangle$ of real numbers. The function d defined by

$$d(p, q) \;=\; \sqrt{(a_1 - b_1)^2 + \cdots + (a_m - b_m)^2} \;=\; \sqrt{\sum_{i=1}^{m} (a_i - b_i)^2} \;=\; \sqrt{\sum_{i=1}^{m} |a_i - b_i|^2}$$

where $p = \langle a_1, \ldots, a_m \rangle$ and $q = \langle b_1, \ldots, b_m \rangle$, is a metric, called the *Euclidean metric* on $\mathbf{R}^m$. We assume this metric on $\mathbf{R}^m$ unless otherwise specified. The metric space $\mathbf{R}^m$ with the Euclidean metric is called *Euclidean m-space* and will also be denoted by E^m.

Theorem 8.11:   Euclidean m-space is a metric space.

Observe that Euclidean 1-space is precisely the real line $\mathbf{R}$ with the usual metric, and Euclidean 2-space is the plane $\mathbf{R}^2$ with the usual metric.

HILBERT SPACE

The class of all infinite real sequences

$$\langle a_1, a_2, \ldots \rangle \quad \text{such that} \quad \sum_{n=1}^{\infty} a_n^2 < \infty$$

i.e. such that the series $a_1^2 + a_2^2 + \cdots$ converges, is denoted by $\mathbf{R}^\infty$.

> **Example 9.1:**   Consider the sequences
>
> $$p = \langle 1, 1, 1, \ldots \rangle \quad \text{and} \quad q = \langle 1, \tfrac{1}{2}, \tfrac{1}{4}, \tfrac{1}{8}, \ldots \rangle$$
>
> Since $1^2 + 1^2 + \cdots$ does not converge, p is not a point in $\mathbf{R}^\infty$. On the other hand, the series $1^2 + (\tfrac{1}{2})^2 + (\tfrac{1}{4})^2 + \cdots$ does converge; hence q is a point in $\mathbf{R}^\infty$.

Now let $p = \langle a_n \rangle$ and $q = \langle b_n \rangle$ belong to $\mathbf{R}^\infty$. The function d defined by

$$d(p, q) \;=\; \sqrt{\sum_{n=1}^{\infty} |a_n - b_n|^2}$$

is a metric and called the l_2-*metric* on $\mathbf{R}^\infty$. We assume this metric on $\mathbf{R}^\infty$ unless otherwise specified. The metric space consisting of $\mathbf{R}^\infty$ with the l_2-metric is called *Hilbert space* or l_2-*space* and will also be denoted by $\mathbf{H}$. We formally state:

Theorem 8.12:   Hilbert space (or l_2-space) is a metric space.

> **Example 9.2:**   Let $\mathbf{H}_m$ denote the subspace of Hilbert space $\mathbf{H}$ consisting of all sequences of the form
> $$\langle a_1, a_2, \ldots, a_{m-1}, a_m, 0, 0, 0, \ldots \rangle$$
> Observe that $\mathbf{H}_m$ is isometric and hence homeomorphic to Euclidean m-space by the natural identification
> $$\langle a_1, \ldots, a_m \rangle \;\leftrightarrow\; \langle a_1, \ldots, a_m, 0, 0, \ldots \rangle$$

Hilbert space exhibits two phenomena (not occurring in Euclidean m-space) described in the examples below:

> **Example 9.3:**   Consider the sequence $\langle p_n \rangle$ of points in Hilbert space where $p_k = \langle a_{1k}, a_{2k}, \ldots \rangle$ is defined by $a_{ik} = \delta_{ik}$; i.e. $a_{ik} = 1$ if $i = k$, and $a_{ik} = 0$ if $i \neq k$. Observe, as illustrated below, that the projection $\langle \pi_i(p_n) \rangle$ of $\langle p_n \rangle$ into each coordinate space converges to zero:
>
> $$\begin{aligned} p_1 &= \langle 1, 0, 0, 0, \ldots \rangle \\ p_2 &= \langle 0, 1, 0, 0, \ldots \rangle \\ p_3 &= \langle 0, 0, 1, 0, \ldots \rangle \\ p_4 &= \langle 0, 0, 0, 1, \ldots \rangle \end{aligned}$$
>
> $$\downarrow \; \downarrow \; \downarrow \; \downarrow$$
> $$\mathbf{0} = \langle 0, 0, 0, 0, \ldots \rangle$$
>
> But the sequence $\langle p_n \rangle$ does not converge to $\mathbf{0}$, since $d(p_k, \mathbf{0}) = 1$ for every $k \in \mathbf{N}$; in fact, $\langle p_n \rangle$ has no convergent subsequence.

Example 9.4:   Let $\mathbf{H}^*$ denote the proper subspace of $\mathbf{H}$ which consists of all points in $\mathbf{H}$ whose first coordinate is zero. Observe that the function $f : \mathbf{H} \to \mathbf{H}^*$ defined by $f(\langle a_1, a_2, \ldots \rangle) = \langle 0, a_1, a_2, \ldots \rangle$ is one-one, onto and preserves distances. Hence Hilbert space is isometric to a proper subspace of itself.

CONVERGENCE AND CONTINUITY IN METRIC SPACES

The following definitions of convergence and continuity in metric spaces are frequently used. Observe their similarity to the usual $\epsilon - \delta$ definitions.

Definition: The sequence $\langle a_1, a_2, \ldots \rangle$ of points in a metric space (X, d) converges to $b \in X$ if for every $\epsilon > 0$ there exists a positive integer n_0 such that

$$n > n_0 \quad \text{implies} \quad d(a_n, b) < \epsilon$$

Definition: Let (X, d) and (Y, d^*) be metric spaces. A function f from X into Y is continuous at $p \in X$ if for every $\epsilon > 0$ there exists a $\delta > 0$ such that

$$d(p, x) < \delta \quad \text{implies} \quad d^*(f(p), f(x)) < \epsilon$$

The above definitions are equivalent to the definitions of convergence and continuity (in the metric topology) which were given for topological spaces in general.

NORMED SPACES

Let $\mathbf{V}$ be a real linear vector space, that is, $\mathbf{V}$ under an operation of vector addition and of scalar multiplication by real numbers satisfies the axioms $[\mathbf{V_1}]$, $[\mathbf{V_2}]$ and $[\mathbf{V_3}]$ of Chapter 2, Page 22. A function which assigns to each vector $v \in \mathbf{V}$ the real number $||v||$ is a *norm* on $\mathbf{V}$ iff it satisfies, for all $v, w \in \mathbf{V}$ and $k \in \mathbf{R}$, the following axioms:

$[\mathbf{N_1}]$ $||v|| \geqq 0$ and $||v|| = 0$ iff $v = 0$.

$[\mathbf{N_2}]$ $||v + w|| \leq ||v|| + ||w||$.

$[\mathbf{N_3}]$ $||kv|| = |k| \, ||v||$.

A linear space $\mathbf{V}$ together with a norm is called a *normed linear vector space* or simply a *normed space*. The real number $||v||$ is called the *norm* of the vector v.

Theorem 8.13:   Let $\mathbf{V}$ be a normed space. The function d defined by

$$d(v, w) = ||v - w||$$

where $v, w \in \mathbf{V}$, is a metric, called the *induced metric* on $\mathbf{V}$.

Thus every normed space with the induced metric is a metric space and hence is also a topological space.

Example 10.1:   The product set $\mathbf{R}^m$ is a linear vector space with addition defined by

$$\langle a_1, \ldots, a_m \rangle + \langle b_1, \ldots, b_m \rangle = \langle a_1 + b_1, \ldots, a_m + b_m \rangle$$

and scalar multiplication defined by

$$k \langle a_1, \ldots, a_m \rangle = \langle k \, a_1, \ldots, k \, a_m \rangle$$

The function on $\mathbf{R}^m$ defined by

$$|| \langle a_1, \ldots, a_m \rangle || = \sqrt{a_1^2 + \cdots + a_m^2} = \sqrt{\sum a_i^2} = \sqrt{\sum |a_i|^2}$$

is a norm and called the *Euclidean norm* on $\mathbf{R}^m$. Note that the Euclidean norm on $\mathbf{R}^m$ induces the Euclidean metric on $\mathbf{R}^m$. If $p = \langle a_1, a_2, a_3 \rangle$ is a point in $\mathbf{R}^3$, then $||p||$ corresponds precisely to the "length" of the arrow (or vector) from the origin to the point p as illustrated below.

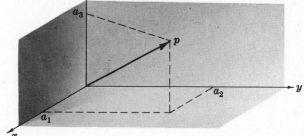

Example 10.2:   The following two functions are also norms on the linear space $\mathbf{R}^m$:

$$\| \langle a_1, \ldots, a_m \rangle \| \quad = \quad \max (|a_1|, |a_2|, \ldots, |a_m|)$$

$$\| \langle a_1, \ldots, a_m \rangle \| \quad = \quad |a_1| + |a_2| + \cdots + |a_m|$$

Let $\mathcal{F}(X, \mathbf{R})$ be the collection of all real-valued functions on a non-empty set X.   Recall (see Theorem 2.9) that $\mathcal{F}(X, \mathbf{R})$ is a linear space with vector addition and scalar multiplication defined as follows:

$$(f + g)(x) \equiv f(x) + g(x) \qquad \text{and} \qquad (kf)(x) \equiv k f(x)$$

We shall frequently want to study classes of functions with certain other properties such as boundedness, continuity, etc.   We shall use the following result from linear algebra:

Proposition 8.14:   Let $\mathcal{A}(X, \mathbf{R})$ be a non-empty subcollection of $\mathcal{F}(X, \mathbf{R})$ satisfying the following two properties:

   (i)   If $f, g \in \mathcal{A}(X, \mathbf{R})$, then the sum $f + g \in \mathcal{A}(X, \mathbf{R})$.

   (ii)   If $f \in \mathcal{A}(X, \mathbf{R})$ and $k \in \mathbf{R}$, then the scalar multiple $kf \in \mathcal{A}(X, \mathbf{R})$.

   Then $\mathcal{A}(X, \mathbf{R})$ is, itself, a linear vector space.

Example 10.3:   The class $C[0, 1]$ of all continuous real functions on the interval $I = [0, 1]$ is a linear space since the sum and scalar multiples of continuous functions are continuous.   The function on $C[0, 1]$ defined by

$$\|f\| \quad = \quad \int_0^1 |f(x)|\ dx$$

is a norm which induces the metric on $C[0, 1]$ defined in Example 1.3.

Example 10.4:   The function on the linear space $C[0, 1]$ defined by

$$\|f\| \quad := \quad \sup \{|f(x)| : x \in [0, 1]\}$$

is also a norm.   This norm induces the metric on $C[0, 1]$ defined in Example 1.4.

Example 10.5:   Let $\mathcal{B}(X, \mathbf{R})$ denote the subcollection of $\mathcal{F}(X, \mathbf{R})$ consisting of all bounded functions $f : X \to \mathbf{R}$.   Then $\mathcal{B}(X, \mathbf{R})$ is a linear space since the sum and scalar multiples of bounded functions are also bounded.   The function on $\mathcal{B}(X, \mathbf{R})$ defined by

$$\|f\| \quad = \quad \sup \{|f(x)| : x \in X\}$$

is a norm.

Example 10.6:   We show later that the class $\mathbf{R}^\infty$ of all real sequences $\langle a_n \rangle$ such that $\sum |a_n|^2 < \infty$ is a linear space.   The function on $\mathbf{R}^\infty$ defined by

$$\| \langle a_n \rangle \| \quad = \quad \sqrt{\sum |a_n|^2}$$

is a norm and called the l_2-norm on $\mathbf{R}^\infty$.   Observe that this norm induces the l_2-metric in Hilbert space.

Solved Problems

METRICS

1. Show that in the definition of a metric the axiom $[M_3]$ can be replaced by the following (weaker) axiom:

    $[M_3^*]$ If $a, b, c \in X$ are distinct then $d(a, c) \leqq d(a, b) + d(b, c)$.

    Solution:

    Suppose $a = b$. Then
    $$d(a, c) = d(b, c) = d(b, b) + d(b, c) \leqq d(a, b) + d(b, c)$$
    If $b = c$, the argument is similar. Lastly, suppose $a = c$; then
    $$d(a, c) = 0 \leqq d(a, b) + d(b, c)$$
    Thus the Triangle Inequality follows from $[M_1]$ if the points a, b and c are not all distinct.

2. Show that the trivial metric on a set X is a metric, i.e. that the function d defined by
    $$d(a, b) = \begin{cases} 1 & \text{if } a \neq b \\ 0 & \text{if } a = b \end{cases}$$
    satisfies $[M_1]$, $[M_2]$, $[M_3^*]$ and $[M_4]$.

    Solution:

    Let $a, b \in X$. Then $d(a, b) = 1$ or $d(a, b) = 0$. In either case, $d(a, b) \geqq 0$. Also, if $a = b$ then, by definition of d, $d(a, b) = 0$. Hence d satisfies $[M_1]$.

    Let $a, b \in X$. If $a \neq b$, then $b \neq a$. Hence $d(a, b) = 1$ and $d(b, a) = 1$. Accordingly, $d(a, b) = d(b, a)$. On the other hand, if $a = b$ then $b = a$ and therefore $d(a, b) = 0 = d(b, a)$. Hence d satisfies $[M_2]$.

    Now let $a, b, c \in X$ be distinct points. Then $d(a, c) = 1$, $d(a, b) = 1$ and $d(b, c) = 1$. Hence
    $$d(a, c) = 1 \leqq 1 + 1 = d(a, b) + d(b, c)$$
    and d satisfies $[M_3^*]$.

    Lastly, let $a, b \in X$ and $a \neq b$. Then $d(a, b) = 1$. Hence $d(a, b) \neq 0$, and d satisfies $[M_4]$.

3. Let d be a metric on a non-empty set X. Show that the function e defined by
    $$e(a, b) = \min(1, d(a, b))$$
    where $a, b \in X$, is also a metric on X.

    Solution:

    Let $a, b \in X$. Since d is a metric, $d(a, b)$ is non-negative. Hence $e(a, b)$, which is either 1 or $d(a, b)$, is also non-negative. Furthermore, if $a = b$ then
    $$e(a, b) = \min(1, d(a, b)) = \min(1, 0) = 0$$
    Hence e satisfies $[M_1]$.

    Now let $a, b \in X$. By definition $e(a, b) = d(a, b)$ or $e(a, b) = 1$. Suppose $e(a, b) = d(a, b)$; then $d(a, b) < 1$. Since d is a metric, $d(b, a) = d(a, b) < 1$. Consequently,
    $$e(b, a) = d(b, a) = d(a, b) = e(a, b)$$
    On the other hand, suppose $e(a, b) = 1$; then $d(a, b) \geqq 1$. Hence $d(b, a) = d(a, b) \geqq 1$. Consequently,
    $$e(b, a) = 1 = e(a, b)$$
    In either case e satisfies $[M_2]$.

    Now let $a, b, c \in X$. We want to prove the Triangle Inequality
    $$e(a, c) \leqq e(a, b) + e(b, c)$$
    Observe that $e(a, c) = \min(1, d(a, c)) \leqq 1$. Hence if $e(a, b) = 1$ or $e(b, c) = 1$, the Triangle Inequality holds. But if both $e(a, b) < 1$ and $e(b, c) < 1$, then $e(a, b) = d(a, b)$ and $e(b, c) = d(b, c)$. Accordingly,
    $$e(a, c) = \min(1, d(a, c)) \leqq d(a, c) \leqq d(a, b) + d(b, c) = e(a, b) + e(b, c)$$
    Thus in all cases the Triangle Inequality holds. Hence e satisfies $[M_3]$.

    Finally, let $a, b \in X$ and $a \neq b$. Then $d(a, b) \neq 0$. Hence $e(a, b) = \min(1, d(a, b))$ is also not zero. Thus e satisfies $[M_4]$.

4. Let d be a metric on a non-empty set X. Show that the function e defined by

$$e(a, b) \;=\; \frac{d(a, b)}{1 + d(a, b)}$$

where $a, b \in X$, is also a metric on X.

Solution:

Since d is a metric, e clearly satisfies **[M₁]**, **[M₂]** and **[M₄]**. Hence we only need to show that e satisfies **[M₃]**, the Triangle Inequality. Let $a, b, c \in X$; then

$$\frac{d(a, b)}{1 + d(a, b) + d(b, c)} \;\leq\; \frac{d(a, b)}{1 + d(a, b)} \;=\; e(a, b)$$

and

$$\frac{d(b, c)}{1 + d(a, b) + d(b, c)} \;\leq\; \frac{d(b, c)}{1 + d(b, c)} \;=\; e(b, c)$$

Since d is a metric, $d(a, c) \leq d(a, b) + d(b, c)$. Hence

$$e(a, c) \;=\; \frac{d(a, c)}{1 + d(a, c)} \;\leq\; \frac{d(a, b) + d(b, c)}{1 + d(a, b) + d(b, c)}$$

$$=\; \frac{d(a, b)}{1 + d(a, b) + d(b, c)} \;+\; \frac{d(b, c)}{1 + d(a, b) + d(b, c)} \;\leq\; e(a, b) + e(b, c)$$

Thus e is a metric.

OPEN SPHERES

5. Prove Lemma 8.2: Let S be an open sphere with center p and radius δ, i.e. $S = S(p, \delta)$. Then for every point $q \in S$ there exists an open sphere T centered at q such that T is contained in S.

Solution:

Now $d(q, p) < \delta$ since $q \in S = S(p, \delta)$. Hence

$$\epsilon \;=\; \delta - d(q, p) \;>\; 0$$

We claim that the open sphere $T = S(q, \epsilon)$, with center q and radius ϵ, is a subset of S.

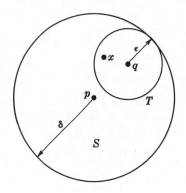

Let $x \in T = S(q, \epsilon)$. Then $d(x, q) < \epsilon = \delta - d(q, p)$. So, by the Triangle Inequality,

$$d(x, p) \;\leq\; d(x, q) + d(q, p) \;<\; [\delta - d(q, p)] + d(q, p) \;=\; \delta$$

Thus $x \in S = S(p, \delta)$ since its distance from p is less than δ. So $x \in T$ implies $x \in S$, i.e. T is a subset of S (as indicated in the adjacent Venn diagram).

6. Let δ_1 and δ_2 be real numbers such that $0 < \delta_1 \leq \delta_2$. Show that the open sphere $S(p, \delta_1)$ is a subset of the open sphere $S(p, \delta_2)$.

Solution:

Let $x \in S(p, \delta_1)$. Then $d(x, p) < \delta_1 \leq \delta_2$. Hence $x \in S(p, \delta_2)$ and thus $S(p, \delta_1) \subset S(p, \delta_2)$.

7. Show that if S and T are open spheres with the same center, then one of them is a subset of the other.

Solution:

Say $S = S(p, \delta_1)$ and $T = S(p, \delta_2)$, i.e. S and T have the same center p with radii δ_1 and δ_2 respectively. But either $\delta_1 \leq \delta_2$ or $\delta_2 \leq \delta_1$. Hence by the preceding problem either $S \subset T$ or $T \subset S$.

8. Prove Lemma 8.3: Let S_1 and S_2 be open spheres and let $p \in S_1 \cap S_2$. Then there exists an open sphere S_p with center p such that $p \in S_p \subset S_1 \cap S_2$.

Solution:

Since $p \in S_1$ and S_1 is an open sphere, there exists by Lemma 8.2 an open sphere S_1^* with center p such that $p \in S_1^* \subset S_1$. Similarly there exists an open sphere S_2^* with center p such that $p \in S_2^* \subset S_2$. Now S_1^* and S_2^* each has center p; so by Problem 7 one of them, say S_1^*, is contained in the other. Thus we have

$$p \in S_1^* \subset S_1 \quad \text{and} \quad p \in S_1^* \subset S_2^* \subset S_2$$

Accordingly, $p \in S_1^* \subset S_1 \cap S_2$. Hence we may take $S_p = S_1^*$. (See adjacent diagram.)

METRIC TOPOLOGIES

9. Prove: Let X be a metric space, and let $\mathcal{D}_p$ denote the class of open spheres with center $p \in X$. Then $\mathcal{D}_p$ is a local base at p.

Solution:

Let G be an open subset of X containing p. Since the open spheres in X form a base for the metric topology, ∃ an open sphere S such that $p \in S \subset G$. But by Lemma 8.2 ∃ an open sphere $S_p \in \mathcal{D}_p$, i.e. with center p, such that $p \in S_p \subset S \subset G$. Hence $\mathcal{D}_p$ is a local base at p.

10. Prove Theorem 8.5: Let X be a metric space. Then the countable class of open spheres

$$\mathcal{Z} \;=\; \{S(p, 1),\ S(p, \tfrac{1}{2}),\ S(p, \tfrac{1}{3}),\ \ldots\}$$

with center $p \in X$, is a local base at p.

Solution:

Let G be an open subset of X containing p. By the preceding problem, ∃ an open sphere $S(p, \delta)$ with center p such that $p \in S(p, \delta) \subset G$. Since $\delta > 0$,

$$\exists n_0 \in \mathbf{N} \quad \text{such that} \quad 1/n_0 < \delta$$

Accordingly, $p \in S(p, 1/n_0) \subset S(p, \delta) \subset G$ where $S(p, 1/n_0) \in \mathcal{Z}$. Hence $\mathcal{Z}$ is a local base at p.

11. Prove Theorem 8.6: The closure $\bar{A}$ of a subset A of a metric space X is the set of points whose distance from A is zero: $\bar{A} = \{x : d(x, A) = 0\}$.

Solution:

Suppose $d(p, A) = 0$. Then every open sphere with center p, and therefore every open set G containing p, also contains at least one point of A. Hence $p \in A$ or p is a limit point of A, and so $p \in \bar{A}$.

On the other hand, suppose $d(p, A) = \epsilon > 0$. Then the open sphere $S(p, \tfrac{1}{2}\epsilon)$ with center p contains no point of A. Hence p belongs to the exterior of A, and so $p \notin \bar{A}$. Accordingly, $\bar{A} = \{x : d(x, A) = 0\}$.

12. Show that a subset F of a metric space X is closed if and only if $\{x : d(x, F) = 0\} \subset F$.

Solution:

This follows directly from Problem 11 and the fact that a set is closed iff it is equal to its closure.

13. If F is a closed subset of a metric space X and $p \in X$ does not belong to F, i.e. $p \notin F$, then $d(p, F) \neq 0$.

Solution:

If $d(p, F) = 0$ and F is closed, then by Problem 12, $p \in F$. But by hypothesis $p \notin F$; so $d(p, F) \neq 0$.

14. **Prove Theorem 8.8:** Let A and B be closed disjoint subsets of a metric space X. Then there exist disjoint open sets G and H such that $A \subset G$ and $B \subset H$.

Solution:

If either A or B is empty, say $A = \emptyset$, then $\emptyset$ and X are open disjoint sets such that $A \subset \emptyset$ and $B \subset X$. Hence we may assume A and B are non-empty.

Let $a \in A$. Since A and B are disjoint, $a \notin B$. But B is closed; hence by the preceding problem, $d(a, B) = \delta_a > 0$. Similarly, if $b \in B$, then $d(b, A) = \delta_b > 0$. Set

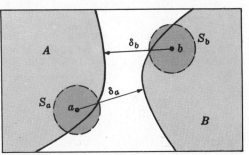

$$S_a = S(a, \tfrac{1}{3}\delta_a) \quad \text{and} \quad S_b = S(b, \tfrac{1}{3}\delta_b)$$

so $a \in S_a$ and $b \in S_b$. (See the adjacent Venn diagram.)

We claim that the sets

$$G = \bigcup\{S_a : a \in A\} \quad \text{and} \quad H = \bigcup\{S_b : b \in B\}$$

satisfy the required conditions of the theorem. Now G and H are open since they are each the union of open spheres. Furthermore, $a \in S_a$ implies $A \subset G$, and $b \in S_b$ implies $B \subset H$. We must show that $G \cap H = \emptyset$.

Suppose $G \cap H \neq \emptyset$, say $p \in G \cap H$. Then

$$\exists \; a_0 \in A, \; b_0 \in B \quad \text{such that} \quad p \in S_{a_0}, \; p \in S_{b_0}$$

Let $d(a_0, b_0) = \epsilon > 0$. Then $d(a_0, B) = \delta_{a_0} \leq \epsilon$ and $d(b_0, A) = \delta_{b_0} \leq \epsilon$. But $p \in S_{a_0}$ and $p \in S_{b_0}$, so

$$d(a_0, p) < \tfrac{1}{3}\delta_{a_0} \quad \text{and} \quad d(p, b_0) < \tfrac{1}{3}\delta_{b_0}$$

Therefore by the Triangle Inequality,

$$d(a_0, b_0) = \epsilon \leq d(a_0, p) + d(p, b_0) < \tfrac{1}{3}\delta_{a_0} + \tfrac{1}{3}\delta_{b_0} \leq \tfrac{1}{3}\epsilon + \tfrac{1}{3}\epsilon = \tfrac{2}{3}\epsilon$$

an impossibility. Hence G and H are disjoint and the theorem is true.

EQUIVALENT METRICS

15. Let d and e be metrics on a set X such that for each d-open sphere S_d with center $p \in X$ there exists an e-open sphere S_e with center p such that $S_e \subset S_d$. Show that the topology $\mathcal{T}_d$ induced by d is coarser (smaller) than the topology $\mathcal{T}_e$ induced by e, i.e. $\mathcal{T}_d \subset \mathcal{T}_e$.

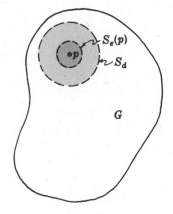

Solution:

Let $G \in \mathcal{T}_d$. We want to show that G is also an e-open set. Let $p \in G$. Since G is d-open there exists a d-open sphere S_d with center p such that $p \in S_d \subset G$. By hypothesis, there exists an e-open sphere $S_e(p)$ with center p such that $p \in S_e(p) \subset S_d \subset G$. Accordingly, $G = \bigcup\{S_e(p) : p \in G\}$. Thus G is the union of e-open spheres, and so it is e-open. Hence $\mathcal{T}_d \subset \mathcal{T}_e$.

16. Let d and e be metrics on a set X such that for each d-open sphere S_d with center $p \in X$ there exists an e-open sphere S_e with center p such that $S_e \subset S_d$, and for each e-open sphere S_e^* with center $p \in X$ there exists a d-open sphere S_d^* such that $S_d^* \subset S_e^*$. Show that d and e are equivalent metrics, i.e. that they induce the same topology on X.

Solution:

By Problem 15, the topology $\mathcal{T}_d$ induced by d is coarser than the topology $\mathcal{T}_e$ induced by e, i.e. $\mathcal{T}_d \subset \mathcal{T}_e$. Also by Problem 15, $\mathcal{T}_e \subset \mathcal{T}_d$. Therefore $\mathcal{T}_d = \mathcal{T}_e$.

17. Show that the usual metric d on the plane $\mathbf{R}^2$ is equivalent to the metrics d_1 and d_2 on $\mathbf{R}^2$ defined in Example 1.5.

Solution:

Observe that we can inscribe a square in any circle as shown in Fig. (a) below, and we can inscribe a circle in a square as shown in Fig. (b). Now the points inside a circle form a d-open sphere and the points inside a square form a d_1-open sphere, so the metrics d and d_1 are equivalent by Problem 16.

Furthermore, we can inscribe a "diamond" in any circle as shown in Fig. (c), and we can inscribe a circle in any diamond as shown in Fig. (d). Since the points inside a "diamond" form a d_2-open sphere, the metrics d and d_2 are equivalent by Problem 16.

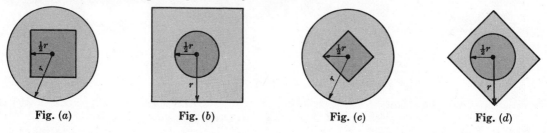

Fig. (a)   Fig. (b)   Fig. (c)   Fig. (d)

18. Let $C[0,1]$ denote the collection of all real continuous functions defined on $I = [0,1]$. Consider the metrics d and e on $C[0,1]$ defined by

$$d(f,g) = \sup\{|f(x) - g(x)| : x \in I\}, \qquad e(f,g) = \int_0^1 |f(x) - g(x)|\, dx$$

(see Example 1.3 and Example 1.4). Show that the topology $\mathcal{T}_d$ induced by d is not coarser than the topology $\mathcal{T}_e$ induced by e, i.e. $\mathcal{T}_d \not\subset \mathcal{T}_e$.

Solution:

Let p be the constant function $p(x) = 2$ and let $\epsilon = 1$. Then the sphere $S_d(p, \epsilon)$ consists of all functions g for which g lies between the functions $p-1$ and $p+1$, i.e. such that $1 < g(x) < 3$ for all $x \in I$.

It is sufficient to show that $S_d(p, \epsilon)$ contains no e-open sphere with center p; i.e. for every $\delta > 0$, $S_e(p, \delta) \not\subset S_d(p, \epsilon)$. Let $\delta > 0$. Consider the function q consisting of the line segments between the points $(0, 4)$ and $(\frac{1}{2}\delta, 2)$ and between $(\frac{1}{2}\delta, 2)$ and $(1, 2)$, i.e. defined by

$$g(x) = \begin{cases} (-4x/\delta) + 4 & \text{if } 0 \le x < \frac{1}{2}\delta \\ 2 & \text{if } \frac{1}{2}\delta \le x \le 1 \end{cases}$$

(see diagram above). Observe that the "area" between p and q is $\frac{1}{2}\delta$, i.e. $e(p, q) = \frac{1}{2}\delta$. Then $q \in S_e(p, \delta)$. But $d(p, q) = 2$; so $q \notin S_d(p, \epsilon)$. Thus $S_e(p, \delta) \not\subset S_d(p, \epsilon)$ for any $\delta > 0$. Hence $\mathcal{T}_d \not\subset \mathcal{T}_e$.

19. Let $C[a, b]$ denote the collection of all continuous functions on a closed interval $X = [a, b]$. Consider the metrics d and e on $C[a, b]$ defined by

$$d(f,g) = \sup\{|f(x) - g(x)| : x \in X\}, \qquad e(f,g) = \int_a^b |f(x) - g(x)|\, dx$$

Show that the topology $\mathcal{T}_e$ induced by e is coarser than the topology $\mathcal{T}_d$ induced by d, i.e. $\mathcal{T}_e \subset \mathcal{T}_d$.

Solution:

Let $S_e(p, \epsilon)$ be any e-open sphere in $C[a, b]$ with center $p \in C[a, b]$. Let $\delta = \epsilon/(b-a)$. In view of Problem 15 it is sufficient to show that $S_d(p, \delta)$, the d-open sphere with center p and radius δ, is a subset of $S_e(p, \epsilon)$, i.e. $S_d(p, \delta) \subset S_e(p, \epsilon)$.

Let $f \in S_d(p, \delta)$; then $\qquad \sup\{|p(x) - f(x)|\} < \delta = \epsilon/(b-a)$

Hence

$$e(p, f) = \int_a^b |p(x) - f(x)| \, dx \leq \int_a^b \sup\{|p(x) - f(x)|\} \, dx < \int_a^b \epsilon/(b-a) \, dx = \epsilon$$

So $f \in S_e(p, \epsilon)$ and therefore $S_d(p, \delta) \subset S_e(p, \epsilon)$.

NORMED SPACES

20. Prove Theorem 8.13:   The function d defined by $d(v, w) = \|v - w\|$, where v and w are vectors in a normed space **V**, is a metric on **V**.

Solution:

Note that by $[\mathbf{N_1}]$,

$$d(v, w) = \|v - w\| \geq 0 \quad\text{and}\quad d(v, v) = \|v - v\| = \|0\| = 0$$

Hence d satisfies $[\mathbf{M_1}]$. Also, by $[\mathbf{N_3}]$,

$$d(v, w) = \|v - w\| = \|(-1)(w - v)\| = |-1| \, \|w - v\| = \|w - v\| = d(w, v)$$

Hence d satisfies $[\mathbf{M_2}]$. By $[\mathbf{N_2}]$, $\|v + w\| \leq \|v\| + \|w\|$ for all $v, w \in \mathbf{V}$. Accordingly if $a, b, c \in \mathbf{V}$, then substituting $v = a - b$ and $w = b - c$ we have

$$\|a - c\| = \|(a - b) + (b - c)\| = \|v + w\| \leq \|v\| + \|w\| = \|a - b\| + \|b - c\|$$

that is, $d(a, c) \leq d(a, b) + d(b, c)$. Hence d satisfies $[\mathbf{M_3}]$.

Finally, if $v \neq w$ then $v - w \neq 0$; hence by $[\mathbf{N_1}]$, $d(v, w) = \|v - w\| > 0$. Thus d satisfies $[\mathbf{M_4}]$.

21. Prove the Cauchy-Schwarz Inequality: For any pair of points $p = \langle a_1, \ldots, a_m \rangle$ and $q = \langle b_1, \ldots, b_m \rangle$ in $\mathbf{R}^m$,

$$\sum_{i=1}^m |a_i b_i| \leq \|p\| \, \|q\| = \sqrt{\sum_{i=1}^m |a_i|^2} \sqrt{\sum_{i=1}^m |b_i|^2}$$

where $\|p\|$ is the Euclidean norm.

Solution:

If $p = 0$ or $q = 0$, then the inequality reduces to $0 \leq 0$ and is therefore true. So we need only consider the case in which $p \neq 0$ and $q \neq 0$, i.e. in which $\|p\| \neq 0$ and $\|q\| \neq 0$.

Now for any real numbers $x, y \in \mathbf{R}$, $0 \leq (x - y)^2 = x^2 - 2xy + y^2$ or, equivalently,

$$2xy \leq x^2 + y^2 \tag{1}$$

Since x and y are arbitrary real numbers, we can let $x = |a_i|/\|p\|$ and $y = |b_i|/\|q\|$ in (1). So, for any i,

$$2 \frac{|a_i|}{\|p\|} \frac{|b_i|}{\|q\|} \leq \frac{|a_i|^2}{\|p\|^2} + \frac{|b_i|^2}{\|q\|^2} \tag{2}$$

But by definition of the Euclidean norm, $\sum |a_i|^2 = \|p\|^2$ and $\sum |b_i|^2 = \|q\|^2$. So summing (2) with respect to i and using $|a_i b_i| = |a_i| \, |b_i|$, we have

$$2 \frac{\sum_{i=1}^m |a_i b_i|}{\|p\| \, \|q\|} \leq \frac{\sum_{i=1}^m |a_i|^2}{\|p\|^2} + \frac{\sum_{i=1}^m |b_i|^2}{\|q\|^2} = \frac{\|p\|^2}{\|p\|^2} + \frac{\|q\|^2}{\|q\|^2} = 2$$

that is,

$$\frac{\sum_{i=1}^m |a_i b_i|}{\|p\| \, \|q\|} \leq 1$$

Multiplying both sides by $\|p\| \, \|q\|$ gives us the required inequality.

22. Prove Minkowski's Inequality: For any pair of points $p = \langle a_1, \ldots, a_m \rangle$ and $q = \langle b_1, \ldots, b_m \rangle$ in $\mathbf{R}^m$,

$$\|p+q\| \leq \|p\| + \|q\| \quad \text{i.e.} \quad \sqrt{\sum |a_i + b_i|^2} \leq \sqrt{\sum |a_i|^2} + \sqrt{\sum |b_i|^2}$$

Solution:

If $\|p+q\| = 0$, the inequality clearly holds. Hence we need only consider the case in which $\|p+q\| \neq 0$.

Observe that, for real numbers $a_i, b_i \in \mathbf{R}$, we have $|a_i + b_i| \leq |a_i| + |b_i|$. Hence

$$\|p+q\|^2 = \sum |a_i + b_i|^2 = \sum |a_i + b_i| \, |a_i + b_i|$$
$$\leq \sum |a_i + b_i| \, (|a_i| + |b_i|)$$
$$= \sum |a_i + b_i| \, |a_i| + \sum |a_i + b_i| \, |b_i|$$

But by the Cauchy-Schwarz Inequality,

$$\sum |a_i + b_i| \, |a_i| \leq \|p+q\| \, \|p\| \quad \text{and} \quad \sum |a_i + b_i| \, |b_i| \leq \|p+q\| \, \|q\|$$

Then

$$\|p+q\|^2 \leq \|p+q\| \, \|p\| + \|p+q\| \, \|q\| = \|p+q\| \, (\|p\| + \|q\|)$$

Since we are considering the case $\|p+q\| \neq 0$, we can divide by $\|p+q\|$; this yields the required inequality.

23. Prove that the Euclidean norm,

$$\|p\| = \sqrt{\sum |a_i|^2} \quad \text{where} \quad p = \langle a_1, \ldots, a_m \rangle \in \mathbf{R}^m$$

satisfies the required axioms **[N₁]**, **[N₂]** and **[N₃]**.

Solution:

Now **[N₁]** follows from properties of the real numbers, and **[N₂]** is Minkowski's Inequality which was proven in the preceding problem. Hence we only need to show that **[N₃]** holds. But for any vector $p = \langle a_1, \ldots, a_m \rangle$ and any real number $k \in \mathbf{R}$,

$$\|kp\| = \|k\langle a_1, \ldots, a_m \rangle\| = \|\langle ka_1, \ldots, ka_m \rangle\|$$
$$= \sqrt{\sum |ka_i|^2} = \sqrt{\sum |k|^2 \, |a_i|^2} = \sqrt{|k|^2 \sum |a_i|^2}$$
$$= \sqrt{|k|^2} \, \sqrt{\sum |a_i|^2} = |k| \sqrt{\sum |a_i|^2} = |k| \, \|p\|$$

Hence **[N₃]** also holds.

24. Prove Theorem 8.11: Euclidean m-space is a metric space, i.e. the Euclidean metric on $\mathbf{R}^m$ satisfies the axioms **[M₁]** to **[M₄]**.

Solution:

Use Problem 23 and the fact that the Euclidean metric on $\mathbf{R}^m$ is induced by the Euclidean norm on $\mathbf{R}^m$.

25. Let $\langle a_1, a_2, \ldots \rangle$ be a convergent sequence of real numbers with the property that $a_n \leq b$ for all $n \in \mathbf{N}$. Show that $\lim a_n \leq b$.

Solution:

Suppose $\lim a_n = a > b$ and set $\epsilon = a - b > 0$. Since $a_n \to a$,

$$\exists n_0 \in \mathbf{N} \quad \text{such that} \quad a - a_{n_0} \leq |a - a_{n_0}| < \epsilon = a - b$$

Thus $-a_{n_0} < -b$ and therefore $b < a_{n_0}$, which contradicts the hypothesis. Accordingly, $\lim a_n \leq b$.

26. Prove Minkowski's inequality for infinite sums: If $\langle a_n \rangle, \langle b_n \rangle \in \mathbf{R}^\infty$, then

$$\|\langle a_n + b_n \rangle\| \leq \|\langle a_n \rangle\| + \|\langle b_n \rangle\| \quad \text{i.e.} \quad \sqrt{\sum_{n=1}^{\infty} |a_n + b_n|^2} \leq \sqrt{\sum_{n=1}^{\infty} |a_n|^2} + \sqrt{\sum_{n=1}^{\infty} |b_n|^2}$$

Solution:

By Minkowski's inequality for finite sums,

$$\sqrt{\sum_{n=1}^{m} |a_n + b_n|^2} \;\leq\; \sqrt{\sum_{n=1}^{m} |a_n|^2} + \sqrt{\sum_{n=1}^{m} |a_n|^2} \;\leq\; \sqrt{\sum_{n=1}^{\infty} |a_n|^2} + \sqrt{\sum_{n=1}^{\infty} |b_n|^2}$$

Since the above is true for every $m \in \mathbf{N}$, by the preceding problem it is also true in the limit.

27. Show that the l_2-norm on $\mathbf{R}^\infty$, i.e. $\|\langle a_n \rangle\| = \sqrt{\sum |a_n|^2}$, satisfies the required axioms $[\mathbf{N}_1]$, $[\mathbf{N}_2]$ and $[\mathbf{N}_3]$.

Solution:
This is similar to the proof in Problem 23 that the Euclidean norm satisfies the axioms $[\mathbf{N}_1]$, $[\mathbf{N}_2]$ and $[\mathbf{N}_3]$.

28. Prove Theorem 8.12: Hilbert space (or l_2-space) is a metric space.

Solution:
Use Problem 27 and the fact that the l_2-metric on $\mathbf{R}^\infty$ is induced by the l_2-norm.

29. Let a and b be real numbers with the property that $a \leq b + \epsilon$ for every $\epsilon > 0$. Show that $a \leq b$.

Solution:
Suppose $a > b$. Then $a = b + \delta$ where $\delta > 0$. Set $\epsilon = \frac{1}{2}\delta$. Now $a > b + \frac{1}{2}\delta = b + \epsilon$ where $\epsilon > 0$. But this contradicts the hypothesis; so $a \leq b$.

30. Let $I = [0, 1]$. Show that the following is a norm on $C[0,1]$: $\|f\| = \sup\{|f(x)|\}$.

Solution:
Recall that a real continuous function on a closed interval is bounded; so $\|f\|$ is well-defined. Since $|f(x)| \geq 0$ for every $x \in I$, $\|f\| \geq 0$; also $\|f\| = 0$ iff $|f(x)| = 0$ for every $x \in I$, i.e. iff $f = 0$. Thus $[\mathbf{N}_1]$ is satisfied.

Let $\epsilon > 0$. Then $\exists x_0 \in I$ such that

$$\begin{aligned}
\|f + g\| \;=\; \sup\{|f(x) + g(x)|\} \;&\leq\; |f(x_0) + g(x_0)| + \epsilon \\
&\leq\; |f(x_0)| + |g(x_0)| + \epsilon \\
&\leq\; \sup\{|f(x)|\} + \sup\{|g(x)|\} + \epsilon \\
&=\; \|f\| + \|g\| + \epsilon
\end{aligned}$$

Hence by Problem 29, $\|f + g\| \leq \|f\| + \|g\|$ and $[\mathbf{N}_2]$ is satisfied.

Now let $k \in \mathbf{R}$. Then

$$\begin{aligned}
\|kf\| \;=\; \sup\{|(kf)(x)|\} \;&=\; \sup\{|k\,f(x)|\} \;=\; \sup\{|k|\,|f(x)|\} \\
&=\; |k|\sup\{|f(x)|\} \;=\; |k|\,\|f\|
\end{aligned}$$

and $[\mathbf{N}_3]$ is satisfied.

Supplementary Problems

METRICS

31. Let $\mathcal{B}(X, Y)$ be the collection of all bounded functions from an arbitrary set X into a metric space (Y, d). Show that the function e is a metric on $\mathcal{B}(X, Y)$:

$$e(f, g) \;=\; \sup\{d(f(x), g(x)) : x \in X\}$$

32. Let $d_1, \ldots, d_m$ be metrics on $X_1, \ldots, X_m$ respectively. Show that the following functions are metrics on the product set $X = \prod_i X_i$:

$$d(p, q) = \max\{d_1(a_1, b_1), \ldots, d_m(a_m, b_m)\}, \qquad e(p, q) = d_1(a_1, b_1) + \cdots + d_m(a_m, b_m)$$

Here, $p = \langle a_1, \ldots, a_m \rangle, \, q = \langle b_1, \ldots, b_m \rangle \in X = \prod_i X_i$.

33. Let $\mathbf{R}^* = \mathbf{R} \cup \{\infty, -\infty\}$ be the extended real line and let $f : \mathbf{R}^* \to [-1, 1]$ be defined by $f(x) = x/(1 + |x|)$ if $x \in \mathbf{R}$, $f(\infty) = 1$ and $f(-\infty) = -1$. Show that the following function is a metric on $\mathbf{R}^*$: $d(x, y) = |f(x) - f(y)|$.

34. Let $\mathbf{R}^+$ denote the non-negative real numbers, and let $f : \mathbf{R}^+ \to \mathbf{R}^+$ be a continuous function such that (i) $f(0) = 0$, (ii) $f(x + y) \leq f(x) + f(y)$, and (iii) $x < y$ implies $f(x) < f(y)$. Show that if d is a metric on any set X then the composition function $f \circ d$ is also a metric on X.

35. Let ρ be a pseudometric on some set X. Let $\sim$ be the relation in X defined by
$$a \sim b \quad \text{iff} \quad \rho(a, b) = 0$$
    (i)   Show that $\sim$ is an equivalence relation in X.
    (ii) Show that the following function is a metric on the quotient set $X/\sim \;=\; \{[a] : a \in X\}$: $d([a], [b]) = \rho(a, b)$. Here $[a]$ denotes the equivalence class of $a \in X$.

36. Let $\mathcal{R}[0, 1]$ denote the collection of (Riemann) integrable functions on $[0, 1]$. Show that the following function is a pseudometric on $\mathcal{R}[0, 1]$:
$$\rho(f, g) \;=\; \int_0^1 |f(x) - g(x)| \, dx$$
Also show by a counterexample that ρ is not a metric.

37. Show that a function d is a metric on a set X iff it satisfies the following two conditions:
    (i) $d(a, b) = 0$ iff $a = b$;   (ii) $d(a, c) \leq d(a, b) + d(c, b)$.

DISTANCES BETWEEN SETS, DIAMETERS

38. Give an example of two closed subsets A and B of the real line $\mathbf{R}$ such that
$$d(A, B) = 0 \quad \text{but} \quad A \cap B = \emptyset$$

39. Let d be a metric on X. Show that for any subsets $A, B \subset X$:
    (i) $d(A \cup B) \leq d(A) + d(B) + d(A, B)$ and   (ii) $d(\bar{A}) = d(A)$.

40. Let d be a metric on X and let A be any arbitrary subset of X. Show that the function $f : X \to \mathbf{R}$ defined by $f(x) = d(x, A)$ is continuous.

41. Consider the function $d : \mathbf{R}^2 \to \mathbf{R}$ defined by $d(\langle a, b \rangle) = |a - b|$ (i.e. the usual metric on $\mathbf{R}$). Show that d is continuous with respect to the usual topologies on the line $\mathbf{R}$ and the plane $\mathbf{R}^2$.

42. Let A be any subset of a metric space X. Show that $d(A) = d(\bar{A})$.

METRIC TOPOLOGIES

43. Let (A, d) be a metric subspace of (X, d). Show that (A, d) is also a topological subspace of (X, d), i.e. the restriction of d to A induces the relative topology on A.

44. Prove: If the topological space $(X, \mathcal{T})$ is homeomorphic to a metric space (Y, d), then $(X, \mathcal{T})$ is metrizable.

45. Prove Theorem 8.10: If (X, d) is isometric to (Y, e), then (X, d) is also homeomorphic to (Y, e).

46. Give an example to show that the closure of an open sphere
$$S(p, \delta) \;=\; \{x : d(p, x) < \delta\}$$
need not be the "closed sphere"
$$\bar{S}(p, \delta) \;=\; \{x : d(p, x) \leq \delta\}$$

47. Show that a closed sphere $\bar{S}(p, \delta) = \{x : d(p, x) \leq \delta\}$ is closed.

48. Prove: The sequence $\langle a_1, a_2, \ldots \rangle$ converges to the point p in a metric space X if and only if the sequence of real numbers $\langle d(a_1, p), d(a_2, p), \ldots \rangle$ converges to $0 \in \mathbf{R}$, i.e. $\lim a_n = p$ iff $\lim d(a_n, p) = 0$.

49. Prove: If $\lim a_n = p$ and $\lim b_n = q$ in a metric space X, then the sequence of real numbers $\langle d(a_1, b_1), d(a_2, b_2), \ldots \rangle$ converges to $d(p, q) \in \mathbf{R}$, i.e. $\lim d(a_n, b_n) = d(\lim a_n, \lim b_n)$.

EQUIVALENT METRICS

50. Let d be a metric on X. Show that the following metric is equivalent to d: $e(a, b) = \min\{1, d(a, b)\}$.

51. Let d be a metric on X. Show that the following metric is equivalent to d: $e(a, b) = \dfrac{d(a, b)}{1 + d(a, b)}$.

52. Let d and e be metrics on X. Suppose $\exists k, k' \in \mathbf{R}$ such that, for every $a, b \in X$,
$$d(a, b) \leq k\, e(a, b) \qquad \text{and} \qquad e(a, b) \leq k'\, d(a, b)$$
Show that d and e are equivalent metrics.

EUCLIDEAN m-SPACE, HILBERT SPACE

53. Let $p_1 = \langle a_{11}, a_{12}, \ldots, a_{1m} \rangle$, $p_2 = \langle a_{21}, a_{22}, \ldots, a_{2m} \rangle$, $\ldots$ be points in Euclidean m-space. Show that $p_n \to q = \langle b_1, b_2, \ldots, b_m \rangle$ if and only if, for $k = 1, \ldots, m$, $\langle a_{1k}, a_{2k}, a_{3k}, \ldots \rangle$ converges to b_k; i.e. the projection $\langle \pi_k(p_n) \rangle$ converges to $\pi_k(q)$ in each coordinate space.

54. Show that if G is an open subset of Hilbert Space $\mathbf{H}$, then $\exists p = \langle a_n \rangle \in G$ such that $a_1 \neq 0$.

55. Let H^* denote the proper subspace of Hilbert Space $\mathbf{H}$ which consists of all points in $\mathbf{H}$ whose first coordinate is zero. (i) Show that H^* is closed. (ii) Show that H^* is nowhere dense in $\mathbf{H}$, i.e. $\text{int}(\overline{H^*}) = \emptyset$.

56. Let $p_1 = \langle a_{11}, a_{12}, \ldots \rangle$, $p_2 = \langle a_{21}, a_{22}, \ldots \rangle$, $\ldots$ be points in $\mathbf{R}^\infty$ and suppose that the sequence of real numbers $\langle \pi_k(p_n) \rangle = \langle a_{1k}, a_{2k}, a_{3k}, \ldots \rangle$ converge to $b_k \in \mathbf{R}$ for every $k \in \mathbf{N}$.

(i)   Show that $q = \langle b_1, b_2, \ldots \rangle$ belongs to $\mathbf{R}^\infty$.

(ii)   Show that the sequence $\langle p_1, p_2, \ldots \rangle$ converges to q.

HILBERT CUBE

57. The set $\mathbf{I}$ of all real sequences $\langle a_1, a_2, \ldots \rangle$ such that $0 \leq a_n \leq \dfrac{1}{n}$, for every $n \in \mathbf{N}$, is called the *Hilbert cube*.

(i)   Show that $\mathbf{I}$ is a subset of $\mathbf{R}^\infty$.

(ii)   Show that $\mathbf{I}$ is a closed and bounded subset of $\mathbf{R}^\infty$.

NORMED SPACES

58. Let $\mathcal{B}(X, \mathbf{R})$ denote the class of all real bounded functions $f : X \to \mathbf{R}$ defined on some non-empty set X. Show that the following is a norm on $\mathcal{B}(X, \mathbf{R})$: $\|f\| = \sup\{|f(x)| : x \in X\}$.

59. Two norms, $\|\cdots\|_1$ and $\|\cdots\|_2$, on a linear space X are equivalent iff they induce equivalent metrics on X, i.e. iff they determine the same topology on X. Show that $\|\cdots\|_1$ is equivalent to $\|\cdots\|_2$ if and only if $\exists a_1, a_2, b_1, b_2 \in \mathbf{R}$ such that, for all $x \in X$,
$$a_1 \|x\|_1 < \|x\|_2 < b_1 \|x\|_1 \qquad \text{and} \qquad a_2 \|x\|_2 < \|x\|_1 < b_2 \|x\|_2$$

60. Let $||\cdots||$ be the Euclidean norm and let d be the induced Euclidean metric on the plane $\mathbf{R}^2$. Consider the function e defined by

$$e(p,q) \;=\; \begin{cases} ||p|| + ||q|| & \text{if } ||p|| \neq ||q|| \\ d(p,q) & \text{if } ||p|| = ||q|| \end{cases}$$

    (i)   Show that e is a metric on $\mathbf{R}^2$.

    (ii)   Describe an open sphere in the metric space $(\mathbf{R}^2, e)$.

61. Show that the following is a norm on $C[0,1]$: $||f|| = \displaystyle\int_0^1 |f(x)|\, dx$.

62. Let X be a normed space. Show that the function $f : X \to R$ defined by $f(x) = ||x||$ is continuous.

Answers to Supplementary Problems

36. The function $f : [0,1] \to \mathbf{R}$ defined by

$$f(x) \;=\; \begin{cases} 1 & \text{if } x = 0 \\ 0 & \text{if } 0 < x \leq 1 \end{cases}$$

is (Riemann) integrable, i.e. belongs to $\mathcal{R}[0,1]$. The zero function $g : [0,1] \to \mathbf{R}$, i.e. $g(x) = 0$ for all $x \in [0,1]$, also belongs to $\mathcal{R}[0,1]$. But $\rho(f,g) = 0$ and $f \neq g$. Hence ρ is not a metric as it does not satisfy $[\mathbf{M}_4]$.

38. Let $A = \{2, 3, 4, 5, \ldots\}$ and $B = \{2\tfrac{1}{2}, 3\tfrac{1}{3}, 4\tfrac{1}{4}, \ldots\}$.

46. Let d be the trivial metric on a set X containing more than one point. Then, for any $p \in X$,

$$S(p,1) \;=\; \{x : d(p,x) < 1\} \;=\; \{p\}$$
$$\overline{S}(p,1) \;=\; \{x : d(p,x) \leq 1\} \;=\; X$$

But d induces the discrete topology on X, and so every subset of X is both open and closed. Thus

$$\overline{S(p,1)} \;=\; \overline{\{p\}} \;=\; \{p\} \;\neq\; \overline{S}(p,1)$$

58. *Hint.* Proof is similar to that of Problem 30.

60. (ii) If $||p|| \geq \delta$, then $S(p,\delta)$ is an arc of the circle $\{x : ||x|| = ||p||\}$. If $||p|| < \delta$, then $S(p,\delta)$ consists of the points interior to the circle $\{x : ||x|| = \delta - ||p||\}$ and the points on an arc of the circle $\{x : ||x|| = ||p||\}$.

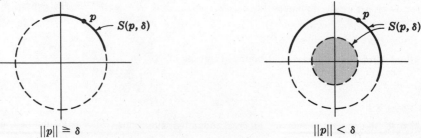

$$||p|| \geq \delta \qquad\qquad\qquad\qquad ||p|| < \delta$$

61. $||f + g|| \;=\; \displaystyle\int_0^1 |f(x) + g(x)|\, dx \;\leq\; \int_0^1 (|f(x)| + |g(x)|)\, dx$

           $= \displaystyle\int_0^1 |f(x)|\, dx \;+\; \int_0^1 |g(x)|\, dx \;=\; ||f|| + ||g||$

Chapter 9

Countability

FIRST COUNTABLE SPACES

A topological space X is called a *first countable space* if it satisfies the following axiom, called the *first axiom of countability*.

[C₁] For each point $p \in X$ there exists a countable class $\mathcal{B}_p$ of open sets containing p such that every open set G containing p also contains a member of $\mathcal{B}_p$.

In other words, a topological space X is a first countable space iff there exists a countable local base at every point $p \in X$. Observe that **[C₁]** is a *local* property of a topological space X, i.e. it depends only upon the properties of arbitrary neighborhoods of the point $p \in X$.

> **Example 1.1:** Let X be a metric space and let $p \in X$. Recall that the countable class of open spheres $\{S(p,1), S(p,\frac{1}{2}), S(p,\frac{1}{3}), \ldots\}$ with center p is a local base at p. Hence every metric space satisfies the first axiom of countability.

> **Example 1.2:** Let X be any discrete space. Now the singleton set $\{p\}$ is open and is contained in every open set G containing $p \in X$. Hence every discrete space satisfies **[C₁]**.

First countable spaces possess the following property which was proven for the special case of the real line **R**.

Theorem 9.1: A function defined on a first countable space X is continuous at $p \in X$ if and only if it is sequentially continuous at p.

In other words, if X satisfies **[C₁]**, then $f: X \to Y$ is continuous at $p \in X$ iff for every sequence $\langle a_n \rangle$ converging to p in X, the sequence $\langle f(a_n) \rangle$ converges to $f(p)$ in Y, i.e.,

$$a_n \to p \quad \text{implies} \quad f(a_n) \to f(p)$$

Remark: Let $\mathcal{B}_p$ be a countable local base at the point $p \in X$. Then we can index the members of $\mathcal{B}_p$ by **N**, i.e. we can write $\mathcal{B}_p = \{B_1, B_2, \ldots\}$. (We permit repetitions in the case that $\mathcal{B}_p$ is finite.) If, in addition, $B_1 \supset B_2 \supset B_3 \supset \ldots$, then we call $\mathcal{B}_p$ a *nested local base* at p. We show, as a solved problem, that we can always construct a nested local base from a countable local base.

SECOND COUNTABLE SPACES

A topological space $(X, \mathcal{T})$ is called a *second countable space* if it satisfies the following axiom, called the *second axiom of countability*.

[C₂] There exists a countable base $\mathcal{B}$ for the topology $\mathcal{T}$.

Observe that second countability is a global rather than a local property of a topological space.

> **Example 2.1:** The class $\mathcal{B}$ of open intervals (a, b) with rational endpoints, i.e. $a, b \in \mathbf{Q}$, is countable and is a base for the usual topology on the real line **R**. Thus **R** is a second countable space, i.e. **R** satisfies **[C₂]**.

Example 2.2: Consider the discrete topology $\mathcal{D}$ on the real line **R**. Recall that a class $\mathcal{B}$ is a base for a discrete topology if and only if it contains all singleton sets. But **R**, and hence the class of singleton subsets $\{p\}$ of **R**, are non-countable. Accordingly, $(\mathbf{R}, \mathcal{D})$ does not satisfy the second axiom of countability.

Now if $\mathcal{B}$ is a countable base for a space X, and if $\mathcal{B}_p$ consists of the members of $\mathcal{B}$ which contain the point $p \in X$, then $\mathcal{B}_p$ is a countable local base at p. In other words,

Proposition 9.2: A second countable space is also first countable.

On the other hand, the real line **R** with the discrete topology does not satisfy [C$_2$] by Example 2.2 but does satisfy [C$_1$] by Example 1.2. Thus we see that the converse of Proposition 9.2 is not true.

LINDELÖF'S THEOREMS

It is convenient to introduce some terminology. Let $A \subset X$ and let $\mathcal{A}$ be a class of subsets of X such that

$$A \subset \bigcup \{E : E \in \mathcal{A}\}$$

Then $\mathcal{A}$ is called a *cover* (or, *covering*) of A, or $\mathcal{A}$ is said to *cover* A. If each member of $\mathcal{A}$ is an open subset of X, then $\mathcal{A}$ is called an *open cover* of A. Furthermore, if $\mathcal{A}$ contains a countable (finite) subclass which also is a cover of A, then $\mathcal{A}$ is said to be *reducible to a countable (finite) cover*, or $\mathcal{A}$ is said to contain a countable (finite) *subcover*.

The central facts about second countable spaces are contained in the next two theorems, due to Lindelöf.

Theorem 9.3: Let A be any subset of a second countable space X. Then every open cover of A is reducible to a countable cover.

Theorem 9.4: Let X be a second countable space. Then every base $\mathcal{B}$ for X is reducible to a countable base for X.

The preceding theorem motivates the definition of a Lindelöf space. A topological space X is called a *Lindelöf space* if every open cover of X is reducible to a countable cover. Hence every second countable space is a Lindelöf space.

SEPARABLE SPACES

A topological space X is said to be *separable* if it satisfies the following axiom.

[S] X contains a countable dense subset.

In other words, X is separable iff there exists a finite or a denumerable subset A of X such that the closure of A is the entire space, i.e. $\bar{A} = X$.

Example 3.1: The real line **R** with the usual topology is a separable space since the set **Q** of rational numbers is denumerable and is dense in **R**, i.e. $\bar{\mathbf{Q}} = \mathbf{R}$.

Example 3.2: Consider the real line **R** with the discrete topology $\mathcal{D}$. Recall that every subset of **R** is both $\mathcal{D}$-open and $\mathcal{D}$-closed; so the only $\mathcal{D}$-dense subset of **R** is **R** itself. But **R** is not a countable set; hence $(\mathbf{R}, \mathcal{D})$ is not a separable space.

We will show that every second countable space is also separable. Namely,

Proposition 9.5: If X satisfies the second axiom of countability, then X is separable.

The real line **R** with the topology generated by the closed-open intervals $[a, b)$ is a classical example of a separable space which does not satisfy the second axiom of countability. So the converse of the previous proposition is not true in general. We do, though, have the following special case.

Theorem 9.6:   Every separable metric space is second countable.

> **Example 3.3:**   Let $C[0, 1]$ denote the linear space of all continuous functions on the closed interval $[0, 1]$ with the norm defined by
>
> $$\|f\| = \sup \{|f(x)| : 0 \leq x \leq 1\}$$
>
> By the Weierstrass Approximation Theorem, for any function $f \in C[0, 1]$ and any $\epsilon > 0$, there exists a polynomial p with rational coefficients such that
>
> $$\|f - p\| < \epsilon \quad \text{i.e.} \quad |f(x) - p(x)| < \epsilon \quad \text{for all} \quad x \in [0, 1]$$
>
> Hence the collection $\mathcal{P}$ of all such polynomials is dense in $C[0, 1]$. But $\mathcal{P}$ is a countable set; so $C[0, 1]$ is separable and, by Theorem 9.6, second countable.

In our last example we show that a metric space need not be separable.

> **Example 3.4:**   Consider the metric e on the plane **R**^{2} defined by
>
> $$e(p, q) = \begin{cases} \|p\| + \|q\| & \text{if } \|p\| \neq \|q\| \\ d(p, q) & \text{if } \|p\| = \|q\| \end{cases}$$
>
> where $\|\cdots\|$ is the Euclidean norm on **R**^{2} and d is the induced usual metric (see Problem 60, Chapter 8).
>
> Recall that if $p \neq \langle 0, 0 \rangle$ and $\delta < \|p\|$, then the e-open sphere $S(p, \delta)$ consists only of points on the circle
>
> $$P = \{x : \|x\| = \|p\|\}$$

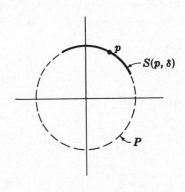

> and so p cannot be an accumulation point of any $A \subset$ **R**^{2} unless A contains points of the circle P. But there are an uncountable number of circles with center $\langle 0, 0 \rangle$, so $A \subset$ **R**^{2} cannot be dense in **R**^{2} unless A is uncountable. Thus the metric space (**R**^{2}, e) is not separable.

HEREDITARY PROPERTIES

A property P of a topological space X is said to be *hereditary* iff every subspace of X also possesses property P. We will show that every subspace of a second countable space is second countable and every subspace of a first countable space is first countable. In other words, the properties [**C**$_1$] and [**C**$_2$] are both hereditary. On the other hand, we will show by a counterexample that a subspace of a separable space need not be separable, i.e. separability is not hereditary.

We conclude with the following diagram which gives the only relationship between the three axioms in this chapter:

$$\text{separable} \quad \leftarrow \quad \text{second countable} \quad \rightarrow \quad \text{first countable}$$

Here an arrow denotes implication as stated in Propositions 9.2 and 9.5.

Solved Problems

FIRST COUNTABLE SPACES

1. Show that any subspace $(Y, \mathcal{T}_Y)$ of a first countable space $(X, \mathcal{T})$ is also first countable.

Solution:

Let $p \in Y$. Since $Y \subset X$, $p \in X$. By hypothesis, $(X, \mathcal{T})$ is a first countable space, so $\exists$ a countable $\mathcal{T}$-local base $\mathcal{B}_p = \{B_n : n \in \mathbf{N}\}$ at p. By a previous problem, $\mathcal{B}_p^* = \{Y \cap B_n : n \in \mathbf{N}\}$ is a $\mathcal{T}_Y$-local base at p. Since $\mathcal{B}_p^*$ is countable, $(Y, \mathcal{T}_y)$ satisfies $[\mathbf{C_1}]$.

2. Let $\mathcal{B}_p = \{G_1, G_2, \ldots\}$ be a countable local base at $p \in X$. Show that:

(i) There exists a nested local base at p.

(ii) If X satisfies $[\mathbf{C_1}]$ then there exists a nested local base at every $p \in X$.

Solution:

(i) Set
$$B_1 = G_1, \quad B_2 = G_1 \cap G_2, \quad \ldots, \quad B_n = G_1 \cap \cdots \cap G_n, \quad \ldots$$

Then $B_1 \supset B_2 \supset \cdots$ and each B_k is open and contains p. Furthermore, if G is an open set containing p, then
$$\exists n_0 \in \mathbf{N} \quad \text{such that} \quad B_{n_0} \subset G_{n_0} \subset G$$

Accordingly, $\{B_1, B_2, \ldots\}$ is a nested local base at p.

(ii) If X satisfies $[\mathbf{C_1}]$ and $p \in X$, then $\exists$ a countable local base at p by $[\mathbf{C_1}]$ and, by (i), there exists a nested local base at p.

3. Let $\mathcal{B}_p = \{B_1, B_2, \ldots\}$ be a nested local base at $p \in X$ and let $\langle a_1, a_2, \ldots \rangle$ be a sequence such that $a_1 \in B_1$, $a_2 \in B_2$, $\ldots$. Show that $\langle a_n \rangle$ converges to p.

Solution:

Let G be an open set containing p. Since $\mathcal{B}_p$ is a local base at p,
$$\exists n_0 \in \mathbf{N} \quad \text{such that} \quad B_{n_0} \subset G$$

But $\mathcal{B}_p$ is nested; hence $n > n_0$ implies $a_n \in B_{n_0} \subset G$, and so $a_n \to p$.

4. Let $\mathcal{T}$ be the cofinite topology on the real line $\mathbf{R}$, i.e. $\mathcal{T}$ contains $\emptyset$ and the complements of finite sets. Show that $(\mathbf{R}, \mathcal{T})$ does not satisfy the first axiom of countability.

Solution:

Suppose that $(\mathbf{R}, \mathcal{T})$ does satisfy $[\mathbf{C_1}]$. Then $1 \in \mathbf{R}$ possesses a countable open local base $\mathcal{B}_1 = \{B_n : n \in \mathbf{N}\}$. Since each B_n is $\mathcal{T}$-open, its complement B_n^c is $\mathcal{T}$-closed and hence finite. Accordingly, $A = \bigcup\{B_n^c : n \in \mathbf{N}\}$ is the countable union of finite sets and is therefore countable. But $\mathbf{R}$ is not countable; hence there exists a point $p \in \mathbf{R}$ different from 1 which does not belong to A, i.e. $p \in A^c$.

Now, by DeMorgan's Law we have
$$p \in A^c = (\bigcup\{B_n^c : n \in \mathbf{N}\})^c = \bigcap\{B_n^{cc} : n \in \mathbf{N}\} = \bigcap\{B_n : n \in \mathbf{N}\}$$

Hence $p \in B_n$ for every $n \in \mathbf{N}$. On the other hand, $\{p\}^c$ is a $\mathcal{T}$-open set since it is the complement of a finite set, and $\{p\}^c$ contains 1 since p is different from 1. Since $\mathcal{B}_1$ is a local base, there exists a member $B_{n_0} \in \mathcal{B}_1$ such that $B_{n_0} \subset \{p\}^c$. Hence $p \notin B_{n_0}$. But this contradicts the statement that $p \in B_n$ for every $n \in \mathbf{N}$. Consequently, the original assumption that $(\mathbf{R}, \mathcal{T})$ satisfies the first axiom of countability is false.

5. Prove Theorem 9.1: Let X satisfy the first axiom of countability. Then $f : X \to Y$ is continuous at $p \in X$ if and only if it is sequentially continuous at p.

Solution:

It suffices to show that if f is sequentially continuous at p then f is continuous at p, since the converse has been proven for an arbitrary topological space. We shall in fact prove the contrapositive statement: if f is not continuous at p then f is not sequentially continuous at p.

Let $\mathcal{B}_p = \{B_1, B_2, \ldots\}$ be a nested local base at p and suppose f is not continuous at p. Then there exists an open subset H of Y such that

$$f(p) \in H \quad \text{but} \quad B_n \not\subset f^{-1}[H] \quad \text{for every} \quad n \in \mathbf{N}$$

Hence, for every $n \in \mathbf{N}$,

$$\exists a_n \in B_n \quad \text{such that} \quad a_n \notin f^{-1}[H] \quad \text{which implies} \quad f(a_n) \notin H$$

Now by a previous problem the sequence $\langle a_n \rangle$ converges to p; but the sequence $\langle f(a_n) \rangle$ does not converge to $f(p)$, since the open set H containing $f(p)$ does not contain any of the terms of the sequence. Accordingly, f is not sequentially continuous at p.

SECOND COUNTABLE SPACES

6. Show that the plane $\mathbf{R}^2$ with the usual topology satisfies the second axiom of countability.

    Solution:
    Let $\mathcal{B}$ be the class of open discs in $\mathbf{R}^2$ with rational radii and centers whose coordinates are rational. Then $\mathcal{B}$ is a countable set and, furthermore, is a base for the usual topology on $\mathbf{R}^2$. Hence $\mathbf{R}^2$ is a second countable space.

7. Show that every subspace of a second countable space is second countable.

    Solution:
    Let $\mathcal{B} = \{B_n : n \in \mathbf{N}\}$ be a countable base for the second countable space X, and let Y be a subspace of X. By a previous problem, $\mathcal{B}_y = \{Y \cap B_n : n \in \mathbf{N}\}$ is a base for Y. Since $\mathcal{B}_Y$ is countable, Y satisfies $[\mathbf{C}_2]$.

8. Prove Theorem (Lindelöf) 9.3: Let A be any subset of a second countable space X. If $\mathcal{G}$ is an open cover of A, then $\mathcal{G}$ is reducible to a countable cover.

    Solution:
    Let $\mathcal{B}$ be a countable base for X. Since $A \subset \bigcup\{G : G \in \mathcal{G}\}$, for every $p \in A$, $\exists G_p \in \mathcal{G}$ such that $p \in G_p$. Since $\mathcal{B}$ is a base for X, for every $p \in A$,

    $$\exists B_p \in \mathcal{B} \quad \text{such that} \quad p \in B_p \subset G_p$$

    Hence $A \subset \bigcup\{B_p : p \in A\}$. But $\{B_p : p \in A\} \subset \mathcal{B}$, so it is countable; hence

    $$\{B_p : p \in A\} \quad = \quad \{B_n : n \in N\}$$

    where N is a countable index set. For each $n \in N$ choose one set $G_n \in \mathcal{G}$ such that $B_n \subset G_n$. Then

    $$A \subset \bigcup\{B_n : n \in N\} \subset \bigcup\{G_n : n \in N\}$$

    and so $\{G_n : n \in N\}$ is a countable subcover of $\mathcal{G}$.

9. Prove Theorem (Lindelöf) 9.4: Let $\mathcal{G}$ be a base for a second countable space X. Then $\mathcal{G}$ is reducible to a countable base for X.

    Solution:
    Since X is second countable, X has a countable base $\mathcal{B} = \{B_n : n \in \mathbf{N}\}$. Since $\mathcal{G}$ is also a base for X, for each $n \in \mathbf{N}$,
    $$B_n = \bigcup\{G : G \in \mathcal{G}_n\} \quad \text{with} \quad \mathcal{G}_n \subset \mathcal{G}$$

    So $\mathcal{G}_n$ is an open cover of B_n and, by the preceding theorem, is reducible to a countable cover $\mathcal{G}_n^*$, i.e., for each $n \in \mathbf{N}$, $\quad B_n = \bigcup\{G : G \in \mathcal{G}_n^*\} \quad \text{with} \quad \mathcal{G}_n^* \subset \mathcal{G} \text{ and } \mathcal{G}_n^* \text{ countable}$

    But $$\mathcal{G}^* = \{G : G \in \mathcal{G}_n^*, n \in \mathbf{N}\}$$

    is a base for X since $\mathcal{B}$ is. Furthermore, $\mathcal{G}^* \subset \mathcal{G}$ and $\mathcal{G}^*$ is countable.

SEPARABILITY

10. Let $\mathcal{T}$ be the cofinite topology on any set X. Show that $(X, \mathcal{T})$ is separable, i.e. contains a countable dense subset.

   Solution:

   If X is itself countable, then clearly X is a countable dense subset of $(X, \mathcal{T})$. On the other hand, suppose X is not countable. Then X contains a denumerable, i.e. non-finite countable, subset A. Recall that the only $\mathcal{T}$-closed sets are the finite sets and X; hence the closure of the non-finite set A is the entire space X, i.e. $\bar{A} = X$.   But A is countable; hence $(X, \mathcal{T})$ is separable.

11. Show that a discrete space X is separable if and only if X is countable.

   Solution:

   Recall that every subset of a discrete space X is both open and closed. Hence the only dense subset of X is X itself. Hence X contains a countable dense subset iff X is countable, i.e. X is separable iff X is countable.

12. Prove Proposition 9.5:   If X satisfies the second axiom of countability, then X is separable.

   Solution:

   Since X satisfies $[\mathbf{C}_2]$, X has a countable base $\mathcal{B} = \{B_n : n \in \mathbf{N}\}$.   For each $n \in \mathbf{N}$, choose a point $a_n \in B_n$. Then the set $A = \{a_n : n \in \mathbf{N}\}$ is also countable. We show that $\bar{A} = X$ or, equivalently, that each point $p \in A^c$, the complement of A, is an accumulation point of A.

   Let G be an open set containing p. Then G contains at least one set $B_{n_0} \in \mathcal{B}$. Hence $a_{n_0} \in B_{n_0} \subset G$. Now a_{n_0} is different from p since $p \in A^c$ but $a_{n_0} \in A$. Accordingly, p is an accumulation point of A since every open set G containing p also contains a point of A different from p.

13. Let $\mathcal{T}$ be the topology on the plane $\mathbf{R}^2$ generated by the half-open rectangles.

$$[a, b) \times [c, d)   =   \{\langle x, y \rangle : a \leqq x < b,\ c \leqq y < d\}$$

   Show that $(\mathbf{R}^2, \mathcal{T})$ is separable.

   Solution:

   Now there are always rational numbers x_0 and y_0 such that $a < x_0 < b$ and $c < y_0 < d$, so the above open rectangle contains the point $p = \langle x_0, y_0 \rangle$ with rational coordinates. Hence the set $A = Q \times Q$ consisting of all points in $\mathbf{R}^2$ with rational coordinates is dense in $\mathbf{R}^2$. But A is a countable set; thus $(\mathbf{R}^2, \mathcal{T})$ is separable.

14. Show by a counterexample that a subspace of a separable space need not be separable, i.e. separability is not a hereditary property.

   Solution:

   Consider the separable topological space $(\mathbf{R}^2, \mathcal{T})$ of the preceding problem. Recall (see Problem 25 of Chapter 6) that the relative topology $\mathcal{T}_y$ on the line $Y = \{\langle x, y \rangle : x + y = 0\}$ is the discrete topology since each singleton subset $\{p\}$ of Y is $\mathcal{T}_y$-open. But an uncountable discrete space is not separable. Thus the separability of $(\mathbf{R}^2, \mathcal{T})$ is not inherited by the subspace $(Y, \mathcal{T}_y)$.

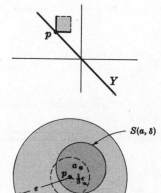

15. Let $S(p, \epsilon)$ be an open sphere in a metric space X, and let $d(p, a) < \frac{1}{3}\epsilon$. Show that if $\frac{1}{3}\epsilon < \delta < \frac{2}{3}\epsilon$, then

$$p \in S(a, \delta) \subset S(p, \epsilon)$$

   Solution:

   Now $d(p, a) < \frac{1}{3}\epsilon < \delta$, so $p \in S(a, \delta)$. Accordingly, we need only show that $S(a, \delta) \subset S(p, \epsilon)$.

Let $x \in S(a, \delta)$. Then $d(a, x) < \delta$ and, by the Triangle Inequality,

$$d(p, x) \;\leq\; d(p, a) + d(a, x) \;<\; \tfrac{1}{3}\epsilon + \delta \;<\; \tfrac{1}{3}\epsilon + \tfrac{2}{3}\epsilon \;=\; \epsilon$$

Hence $x \in S(p, \epsilon)$, or $S(a, \delta) \subset S(p, \epsilon)$.

16. Prove Theorem 9.6: Let X be a separable metric space. Then X satisfies $[\mathbf{C_2}]$, i.e. X contains a countable base.

Solution:

Since X is separable, X contains a countable dense subset A. Let $\mathcal{B}$ be the class of all open spheres with centers in A and with rational radii, i.e.,

$$\mathcal{B} \;=\; \{S(a, \delta) : a \in A, \; \delta \in Q\}$$

Note that $\mathcal{B}$ is a countable set. We claim that $\mathcal{B}$ is a base for the topology on X, i.e. for every open set $G \subset X$ and every $p \in G$,

$$\exists \; S(a, \delta) \in \mathcal{B} \qquad \text{such that} \qquad p \in S(a, \delta) \subset G$$

Since $p \in G$, $\exists$ an open sphere $S(p, \epsilon)$ with center p such that $p \in S(p, \epsilon) \subset G$. Since A is dense in X,

$$\exists \; a_0 \in A \qquad \text{such that} \qquad d(p, a_0) < \tfrac{1}{3}\epsilon$$

Let δ_0 be a rational number such that $\tfrac{1}{3}\epsilon < \delta_0 < \tfrac{2}{3}\epsilon$. Then, by the preceding problem,

$$p \in S(a_0, \delta_0) \subset S(p, \epsilon) \subset G$$

But $S(a_0, \delta_0) \in \mathcal{B}$, and so $\mathcal{B}$ is a countable base for the topology on X.

Supplementary Problems

FIRST COUNTABLE SPACES

17. Show that the property of being a first countable space is a topological property.

18. Let $\mathcal{B}_p = \{B_1, B_2, \ldots\}$ be a nested local base at $p \in X$. Show that any subsequence $\{B_{i_1}, B_{i_2}, \ldots\}$ of $\mathcal{B}_p$ is also a nested local base at p.

19. Let $\mathcal{T}$ be the topology on the real line $\mathbf{R}$ generated by the closed-open intervals $[a, b)$. Show that $(\mathbf{R}, \mathcal{T})$ satisfies $[\mathbf{C_1}]$ by exhibiting a countable local base at any point $p \in \mathbf{R}$.

20. Let $\mathcal{T}$ be the topology on the plane $\mathbf{R}^2$ generated by the half-open rectangles

$$[a, b) \times [c, d) \;=\; \{\langle x, y \rangle : a \leq x < b, \; c \leq y < d\}$$

Show that $(\mathbf{R}^2, \mathcal{T})$ satisfies $[\mathbf{C_1}]$ by exhibiting a countable local base at any point $p \in \mathbf{R}^2$.

21. Let $\mathcal{T}$ and $\mathcal{T}^*$ be topologies on X with $\mathcal{T}$ coarser than $\mathcal{T}^*$, i.e. $\mathcal{T} \subset \mathcal{T}^*$.
    (i)   Show that if $(X, \mathcal{T}^*)$ can be first countable, but $(X, \mathcal{T})$ not.
    (ii) Show that $(X, \mathcal{T})$ can be first countable, but $(X, \mathcal{T}^*)$ not.

SECOND COUNTABLE SPACES

22. Show that the property of being a second countable space is a topological property.

23. Show that if X has a countable subbase then X satisfies $[\mathbf{C_2}]$.

24. Exhibit a countable base for Euclidean m-space.

25. Let $\mathcal{A}$ be any collection of disjoint open subsets of a second countable space X. Show that $\mathcal{A}$ is a countable collection.

26. Let A be an uncountable subset of a second countable space X. Show that A has at least one point of accumulation.

27. Let $\mathcal{T}$ be the topology on the real line $\mathbf{R}$ generated by the closed-open intervals $[a, b)$. Show that $(\mathbf{R}, \mathcal{T})$ does not satisfy $[\mathbf{C_2}]$.

28. Show that l_2-space (Hilbert Space) is second countable.

SEPARABLE SPACES

29. Show that the property of being a separable space is a topological property.

30. Show that Euclidean m-space is separable.

31. Show that l_2-space (Hilbert Space) is separable.

32. Let $\mathcal{T}$ be the topology on the real line $\mathbf{R}$ generated by the closed-open intervals $[a, b)$. Show that $(\mathbf{R}, \mathcal{T})$ is separable.

33. Let $\mathcal{T}$ and $\mathcal{T}^*$ be topologies on X with $\mathcal{T}$ coarser than $\mathcal{T}^*$, i.e. $\mathcal{T} \subset \mathcal{T}^*$.
    (i)   Show that if $(X, \mathcal{T}^*)$ is separable, then $(X, \mathcal{T})$ is also separable.
    (ii) Show by a counterexample that the converse of (i) is not true.

34. Let $C[0, 1]$ denote the class of continuous functions on $[0, 1]$ with norm

$$||f|| \quad = \quad \int_0^1 |f(x)| \; dx$$

Show that $C[0, 1]$ is separable and therefore second countable.

LINDELÖF SPACES

35. Show that a continuous image of a Lindelöf space is also a Lindelöf space.

36. Let A be a closed subset of a Lindelöf space X. Show that A, with the relative topology, is also a Lindelöf space.

37. Show that a discrete space X is Lindelöf if and only if X is a countable set.

38. Let $\mathcal{T}$ be the topology on the plane $\mathbf{R}^2$ generated by the half-open rectangles

$$[a, b) \times [c, d) \quad = \quad \{\langle x, y \rangle : a \leqq x < b, \; c \leqq y < d\}$$

Recall (see Problem 14) that $\mathcal{T}$ induces the discrete topology on the line $Y = \{\langle x, y \rangle : x + y = 1\}$. Show that $(\mathbf{R}^2, \mathcal{T})$ is not Lindelöf and thus $(\mathbf{R}^2, \mathcal{T})$ is a separable first countable space which does not satisfy the second axiom of countability.

Chapter 10

Separation Axioms

INTRODUCTION

Many properties of a topological space X depend upon the distribution of the open sets in the space. Roughly speaking, a space is more likely to be separable, or first or second countable, if there are "few" open sets; on the other hand, an arbitrary function on X to some topological space is more likely to be continuous, or a sequence to have a unique limit, if the space has "many" open sets.

The *separation axioms* of Alexandroff and Hopf, discussed in this chapter, postulate the existence of "enough" open sets.

T_1-SPACES

A topological space X is a T_1-*space* iff it satisfies the following axiom:

[$\mathbf{T_1}$] Given any pair of distinct points $a, b \in X$, each belongs to an open set which does not contain the other.

In other words, there exist open sets G and H such that

$$a \in G, \ b \notin G \quad \text{and} \quad b \in H, a \notin H$$

The open sets G and H are not necessarily disjoint.

Our next theorem gives a very simple characterization of T_1-spaces.

Theorem 10.1: A topological space X is a T_1-space if and only if every singleton subset $\{p\}$ of X is closed.

Since finite unions of closed sets are closed, the above theorem implies:

Corollary 10.2: $(X, \mathcal{T})$ is a T_1-space if and only if $\mathcal{T}$ contains the cofinite topology on X.

> **Example 1.1:** Every metric space X is a T_1-space, since we proved that finite subsets of X are closed.
>
> **Example 1.2:** Consider the topology $\mathcal{T} = \{X, \emptyset, \{a\}\}$ on the set $X = \{a, b\}$. Observe that X is the only open set containing b, but it also contains a. Hence $(X, \mathcal{T})$ does not satisfy [$\mathbf{T_1}$], i.e. $(X, \mathcal{T})$ is not a T_1-space. Note that the singleton set $\{a\}$ is not closed since its complement $\{a\}^c = \{b\}$ is not open.
>
> **Example 1.3:** The cofinite topology on X is the coarsest topology on X for which $(X, \mathcal{T})$ is a T_1-space (Corollary 10.2). Hence the cofinite topology is also called the T_1-*topology*.

HAUSDORFF SPACES

A topological space X is a *Hausdorff space* or T_2-*space* iff it satisfies the following axiom:

[$\mathbf{T_2}$] Each pair of distinct points $a, b \in X$ belong respectively to disjoint open sets.

In other words, there exist open sets G and H such that

$$a \in G, \quad b \in H \quad \text{and} \quad G \cap H = \emptyset$$

Observe that a Hausdorff space is always a T_1-space.

> **Example 2.1:** We show that every metric space X is Hausdorff.
>
> Let $a, b \in X$ be distinct points; hence by $[\mathbf{M_4}]$ $d(a, b) = \epsilon > 0$. Consider the open spheres $G = S(a, \frac{1}{3}\epsilon)$ and $H = S(b, \frac{1}{3}\epsilon)$, centered at a and b respectively. We claim that G and H are disjoint. For if $p \in G \cap H$, then $d(a, p) < \frac{1}{3}\epsilon$ and $d(p, b) < \frac{1}{3}\epsilon$; hence by the Triangle Inequality,
>
> $$d(a, b) \;\le\; d(a, p) + d(p, b) \;<\; \tfrac{1}{3}\epsilon + \tfrac{1}{3}\epsilon \;=\; \tfrac{2}{3}\epsilon$$
>
> But this contradicts the fact that $d(a, b) = \epsilon$. Hence G and H are disjoint, i.e. a and b belong respectively to the disjoint open spheres G and H. Accordingly, X is Hausdorff.

We formally state the result in the preceding example, namely:

Theorem 10.3:   Every metric space is a Hausdorff space.

> **Example 2.2:** Let $\mathcal{T}$ be the cofinite topology, i.e. T_1-topology, on the real line $\mathbf{R}$. We show that $(\mathbf{R}, \mathcal{T})$ is not Hausdorff. Let G and H be any non-empty $\mathcal{T}$-open sets. Now G and H are infinite since they are complements of finite sets. If $G \cap H = \emptyset$, then G, an infinite set, would be contained in the finite complement of H; hence G and H are not disjoint. Accordingly, no pair of distinct points in $\mathbf{R}$ belongs, respectively, to disjoint $\mathcal{T}$-open sets. Thus T_1-spaces need not be Hausdorff.

As noted previously, a sequence $\langle a_1, a_2, \ldots \rangle$ of points in a topological space X could, in general, converge to more than one point in X. This cannot happen if X is Hausdorff:

Theorem 10.4:   If X is a Hausdorff space, then every convergent sequence in X has a unique limit.

The converse of the above theorem is not true unless we add additional conditions.

Theorem 10.5:   Let X be first countable. Then X is Hausdorff if and only if every convergent sequence has a unique limit.

Remark: The notion of a sequence has been generalized to that of a *net* (Moore-Smith sequence) and to that of a *filter* with the following results:

> **Theorem 10.4A:** X is a Hausdorff space if and only if every convergent net in X has a unique limit.

> **Theorem 10.4B:** X is a Hausdorff space if and only if every convergent filter in X has a unique limit.

The definitions of net and filter and the proofs of the above theorems lie beyond the scope of this text.

REGULAR SPACES

A topological space X is *regular* iff it satisfies the following axiom:

[R]   If F is a closed subset of X and $p \in X$ does not belong to F, then there exist disjoint open sets G and H such that $F \subset G$ and $p \in H$.

A regular space need not be a T_1-space, as seen by the next example.

> **Example 3.1:** Consider the topology $\mathcal{T} = \{X, \emptyset, \{a\}, \{b, c\}\}$ on the set $X = \{a, b, c\}$. Observe that the closed subsets of X are also $X, \emptyset, \{a\}$ and $\{b, c\}$ and that $(X, \mathcal{T})$ does satisfy $[\mathbf{R}]$. On the other hand, $(X, \mathcal{T})$ is not a T_1-space since there are finite sets, e.g. $\{b\}$, which are not closed.

A regular space X which also satisfies the separation axiom $[\mathbf{T_1}]$, i.e. a regular T_1-space, is called a T_3-*space*.

> **Example 3.2:** Let X be a T_3-space. Then X is also a Hausdorff space, i.e. a T_2-space. For let $a, b \in X$ be distinct points. Since X is a T_1-space, $\{a\}$ is a closed set; and since a and b are distinct, $b \notin \{a\}$. Accordingly, by $[\mathbf{R}]$, there exist disjoint open sets G and H such that $\{a\} \subset G$ and $b \in H$. Hence a and b belong respectively to disjoint open sets G and H.

NORMAL SPACES

A topological space X is normal iff X satisfies the following axiom:

$[\mathbf{N}]$ If F_1 and F_2 are disjoint closed subsets of X, then there exist disjoint open sets G and H such that $F_1 \subset G$ and $F_2 \subset H$.

A normal space can also be characterized as follows:

Theorem 10.6: A topological space X is normal if and only if for every closed set F and open set H containing F there exists an open set G such that $F \subset G \subset \bar{G} \subset H$.

> **Example 4.1:** Every metric space is normal by virtue of the Separation Theorem 8.8.

> **Example 4.2:** Consider the topology $\mathcal{T} = \{X, \emptyset, \{a\}, \{b\}, \{a, b\}\}$ on the set $X = \{a, b, c\}$. Observe that the closed sets are $X, \emptyset, \{b, c\}, \{a, c\}$ and $\{c\}$. If F_1 and F_2 are disjoint closed subsets of $(X, \mathcal{T})$, then one of them, say F_1, must be the empty set $\emptyset$. Hence $\emptyset$ and X are disjoint open sets and $F_1 \subset \emptyset$ and $F_2 \subset X$. In other words, $(X, \mathcal{T})$ is a normal space. On the other hand, $(X, \mathcal{T})$ is not a T_1-space since the singleton set $\{a\}$ is not closed. Furthermore, $(X, \mathcal{T})$ is not a regular space since $a \notin \{c\}$, and the only open superset of the closed set $\{c\}$ is X which also contains a.

A normal space X which also satisfies the separation axiom $[\mathbf{T_1}]$, i.e. a normal T_1-space, is called a T_4-*space*.

> **Example 4.3:** Let X be a T_4-space. Then X is also a regular T_1-space, i.e. T_3-space. For suppose F is a closed subset of X and $p \in X$ does not belong to F. By $[\mathbf{T_1}]$, $\{p\}$ is closed; and since F and $\{p\}$ are disjoint, by $[\mathbf{N}]$, there exist disjoint open sets G and H such that $F \subset G$ and $p \in \{p\} \subset H$.

Now a metric space is both a normal space and a T_1-space, i.e. a T_4-space. The following diagram illustrates the relationship between the spaces discussed in this chapter.

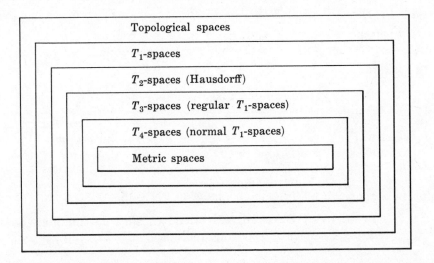

URYSOHN'S LEMMA AND METRIZATION THEOREM

Next comes the classical result of Urysohn.

Theorem (Urysohn's Lemma) 10.7:   Let F_1 and F_2 be disjoint closed subsets of a normal space X. Then there exists a continuous function $f : X \to [0, 1]$ such that

$$f[F_1] = \{0\} \quad \text{and} \quad f[F_2] = \{1\}$$

One important consequence of Urysohn's Lemma gives a partial solution to the metrization problem as discussed in Chapter 8. Namely,

Urysohn's Metrization Theorem 10.8:   Every second countable normal T_1-space is metrizable.

In fact, we will prove that every second countable normal T_1-space is homeomorphic to a subset of the Hilbert cube in $\mathbf{R}^\infty$.

FUNCTIONS THAT SEPARATE POINTS

Let $\mathcal{A} = \{f_i : i \in I\}$ be a class of functions from a set X into a set Y. The class $\mathcal{A}$ of functions is said to *separate points* iff for any pair of distinct points $a, b \in X$ there exists a function f in $\mathcal{A}$ such that $f(a) \neq f(b)$.

>   **Example 5.1:**   Consider the class of real-valued functions
> $$\mathcal{A} = \{f_1(x) = \sin x, \; f_2(x) = \sin 2x, \; f_3(x) = \sin 3x, \; \ldots\}$$
>   defined on $\mathbf{R}$. Observe that for every function $f_n \in \mathcal{A}$, $f_n(0) = f_n(\pi) = 0$. Hence the class $\mathcal{A}$ does not separate points.

>   **Example 5.2:**   Let $C(X, \mathbf{R})$ denote the class of all real-valued continuous functions on a topological space X. We show that if $C(X, \mathbf{R})$ separates points, then X is a Hausdorff space. Let $a, b \in X$ be distinct points. By hypothesis, there exists a continuous function $f : X \to \mathbf{R}$ such that $f(a) \neq f(b)$. But $\mathbf{R}$ is a Hausdorff space; hence there exist disjoint open subsets G and H of $\mathbf{R}$ containing $f(a)$ and $f(b)$ respectively. Accordingly, the inverses $f^{-1}[G]$ and $f^{-1}[H]$ are disjoint, open and contain a and b respectively. In other words, X is a Hausdorff space.

We formally state the result in the preceding example.

Proposition 10.9:   If the class $C(X, \mathbf{R})$ of all real-valued continuous functions on a topological space X separates points, then X is a Hausdorff space.

COMPLETELY REGULAR SPACES

A topological space X is *completely regular* iff it satisfies the following axiom:

[CR]   If F is a closed subset of X and $p \in X$ does not belong to F, then there exists a continuous function $f : X \to [0, 1]$ such that $f(p) = 0$ and $f[F] = \{1\}$.

We show later that

Proposition 10.10:   A completely regular space is also regular.

A completely regular space X which also satisfies **[T₁]**, i.e. a completely regular T_1-space, is called a *Tychonoff space*. By virtue of Urysohn's Lemma, a T_4-space is a Tychonoff space and, by Proposition 10.10, a Tychonoff space is a T_3-space. Hence a Tychonoff space, i.e. a completely regular T_1-space, is sometimes called a $T_{3\frac{1}{2}}$-space.

One important property of Tychonoff spaces is the following:

Theorem 10.11:   The class $C(X, \mathbf{R})$ of all real-valued continuous functions on a completely regular T_1-space X separates points.

Solved Problems

T_1-SPACES

1. Prove Theorem 10.1: A topological space X is a T_1-space if and only if every singleton subset of X is closed.

   Solution:

   Suppose X is a T_1-space and $p \in X$. We show that $\{p\}^c$ is open. Let $x \in \{p\}^c$. Then $x \neq p$, and so by **[T₁]**

   $$\exists \text{ an open set } G_x \quad \text{such that} \quad x \in G_x \text{ but } p \notin G_x$$

   Hence $x \in G_x \subset \{p\}^c$, and hence $\{p\}^c = \bigcup\{G_x : x \in \{p\}^c\}$. Accordingly $\{p\}^c$, a union of open sets, is open and $\{p\}$ is closed.

   Conversely, suppose $\{p\}$ is closed for every $p \in X$. Let $a, b \in X$ with $a \neq b$. Now $a \neq b \Rightarrow b \in \{a\}^c$; hence $\{a\}^c$ is an open set containing b but not containing a. Similarly $\{b\}^c$ is an open set containing a but not containing b. Accordingly, X is a T_1-space.

2. Show that the property of being a T_1-space is hereditary, i.e. every subspace of a T_1-space is also a T_1-space.

   Solution:

   Let $(X, \mathcal{T})$ be a T_1-space and let $(Y, \mathcal{T}_Y)$ be a subspace of $(X, \mathcal{T})$. We show that every singleton subset $\{p\}$ of Y is a $\mathcal{T}_Y$-closed set or, equivalently, that $Y \setminus \{p\}$ is $\mathcal{T}_Y$-open. Since $(X, \mathcal{T})$ is a T_1-space, $X \setminus \{p\}$ is $\mathcal{T}$-open. But

   $$p \in Y \subset X \quad \Rightarrow \quad Y \cap (X \setminus \{p\}) = Y \setminus \{p\}$$

   Hence by definition of subspace, $Y \setminus \{p\}$ is a $\mathcal{T}_Y$-open set. Thus $(Y, \mathcal{T}_Y)$ is also a T_1-space.

3. Show that a finite subset of a T_1-space X has no accumulation points.

   Solution:

   Suppose $A \subset X$ has n elements, say $A = \{a_1, \ldots, a_n\}$. Since A is finite it is closed and therefore contains all of its accumulation points. But $\{a_2, \ldots, a_n\}$ is also finite and hence closed. Accordingly, the complement $\{a_2, \ldots, a_n\}^c$ of $\{a_2, \ldots, a_n\}$ is open, contains a_1, and contains no points of A different from a_1. Hence a_1 is not an accumulation point of A. Similarly, no other point of A is an accumulation point of A and so A has no accumulation points.

4. Show that every finite T_1-space X is a discrete space.

   Solution:

   Every subset of X is finite and therefore closed. Hence every subset of X is also open, i.e. X is a discrete space.

5. Prove: Let X be a T_1-space. Then the following are equivalent:

   (i) $p \in X$ is an accumulation point of A.

   (ii) Every open set containing p contains an infinite number of points of A.

   Solution:

   By definition of an accumulation point of a set, (ii) $\Rightarrow$ (i); hence we only need to prove that (i) $\Rightarrow$ (ii).

   Suppose G is an open set containing p and only containing a finite number of points of A different from p; say

   $$B = (G \setminus \{p\}) \cap A = \{a_1, a_2, \ldots, a_n\}$$

   Now B, a finite subset of a T_1-space, is closed and so B^c is open. Set $H = G \cap B^c$. Then H is open, $p \in H$ and H contains no points of A different from p. Hence p is not an accumulation point of A and so (i) $\Rightarrow$ (ii).

6. Let X be a T_1-space and let $\mathcal{B}_p$ be a local base at $p \in X$. Show that if $q \in X$ is distinct from p, then some member of $\mathcal{B}_p$ does not contain q.

Solution:

Since $p \neq q$ and X satisfies $[\mathbf{T_1}]$, $\exists$ an open set $G \subset X$ containing p but not containing q. Now $\mathcal{B}_p$ is a local base at p, so G is a superset of some $B \in \mathcal{B}_p$ and B also does not contain q.

7. Let X be a T_1-space which satisfies the first axiom of countability. Show that if $p \in X$ is an accumulation point of $A \subset X$, then there exists a sequence of distinct terms in A converging to p.

Solution:

Let $\mathcal{B} = \{B_n\}$ be a nested local base at p. Set $B_{i_1} = B_1$. Since p is a limit point of A, B_{i_1} contains a point $a_1 \in A$ different from p. By the preceding problem,

$$\exists\ B_{i_2} \in \mathcal{B} \quad \text{such that} \quad a_1 \notin B_{i_2}$$

Similarly B_{i_2} contains a point $a_2 \in A$ different from p and, since $a_1 \notin B_{i_2}$, different from a_1. Again by the preceding problem,

$$\exists\ B_{i_3} \in \mathcal{B} \quad \text{such that} \quad a_2 \notin B_{i_3}$$

Furthermore,

$$a_2 \in B_{i_2},\ a_2 \notin B_{i_3} \quad \Rightarrow \quad B_{i_2} \supset B_{i_3}$$

Continuing in this manner we obtain a subsequence $\{B_{i_1}, B_{i_2}, \ldots\}$ of $\mathcal{B}$ and a sequence $\langle a_1, a_2, \ldots \rangle$ of distinct terms in A with $a_1 \in B_{i_1}$, $a_2 \in B_{i_2}$, $\ldots$. But $\{B_{i_n}\}$ is also a nested local base at p; hence $\langle a_n \rangle$ converges to p.

HAUSDORFF SPACES

8. Show that the property of being a Hausdorff space is hereditary, i.e. every subspace of a Hausdorff space is also Hausdorff.

Solution:

Let $(X, \mathcal{T})$ be a Hausdorff space and let $(Y, \mathcal{T}_Y)$ be a subspace of $(X, \mathcal{T})$. Furthermore, let $a, b \in Y \subset X$ with $a \neq b$. By hypothesis, $(X, \mathcal{T})$ is Hausdorff; hence

$$\exists\ G, H \in \mathcal{T} \quad \text{such that} \quad a \in G,\ b \in H \text{ and } G \cap H = \emptyset$$

By definition of a subspace, $Y \cap G$ and $Y \cap H$ are $\mathcal{T}_y$-open sets. Furthermore,

$$a \in G,\ a \in Y \quad \Rightarrow \quad a \in Y \cap G$$

$$b \in H,\ b \in Y \quad \Rightarrow \quad b \in Y \cap H$$

$$G \cap H = \emptyset \quad \Rightarrow \quad (Y \cap G) \cap (Y \cap H) = Y \cap (G \cap H) = Y \cap \emptyset = \emptyset$$

(as indicated in the diagram below). Accordingly $(Y, \mathcal{T}_y)$ is also a Hausdorff space.

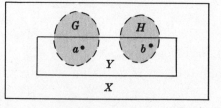

9. Let $\mathcal{T}$ be the topology on the real line $\mathbf{R}$ generated by the open-closed intervals $(a, b]$. Show that $(\mathbf{R}, \mathcal{T})$ is Hausdorff.

Solution:

Let $a, b \in \mathbf{R}$ with $a \neq b$, say $a < b$. Choose $G = (a - 1, a]$ and $H = (a, b]$. Then

$$G, H \in \mathcal{T},\quad a \in G,\quad b \in H \text{ and } G \cap H = \emptyset$$

Hence $(X, \mathcal{T})$ is Hausdorff.

10. Prove Theorem 10.4: Let X be a Hausdorff space. Then every convergent sequence in X has a unique limit.

Solution:

Suppose $\langle a_1, a_2, \ldots \rangle$ converges to a and b, and suppose $a \neq b$. Since X is Hausdorff, $\exists$ open sets G and H such that

$$a \in G, \quad b \in H \quad \text{and} \quad G \cap H = \emptyset$$

By hypothesis, $\langle a_n \rangle$ converges to a; hence

$$\exists \, n_0 \in \mathbf{N} \quad \text{such that} \quad n > n_0 \quad \text{implies} \quad a_n \in G$$

i.e. G contains all except a finite number of the terms of the sequence. But G and H are disjoint; hence H can only contain those terms of the sequence which do not belong to G and there are only a finite number of these. Accordingly, $\langle a_n \rangle$ cannot converge to b. But this violates the hypothesis; hence $a = b$.

11. Prove Theorem 10.5: Let X be a first countable space. Then the following are equivalent: (i) X is Hausdorff. (ii) Every convergent sequence has a unique limit.

Solution:

By the preceding problem, (i) $\Rightarrow$ (ii); hence we need only show that (ii) $\Rightarrow$ (i). Suppose X is not Hausdorff. Then $\exists \, a, b \in X$, $a \neq b$, with the property that every open set containing a has a non-empty intersection with every open set containing b.

Now let $\{G_n\}$ and $\{H_n\}$ be nested local bases at a and b respectively. Then $G_n \cap H_n \neq \emptyset$ for every $n \in \mathbf{N}$, and so

$$\exists \, \langle a_1, a_2, \ldots \rangle \quad \text{such that} \quad a_1 \in G_1 \cap H_1, \; a_2 \in G_2 \cap H_2, \; \ldots$$

Accordingly, $\langle a_n \rangle$ converges to both a and b. In other words, (ii) $\Rightarrow$ (i).

NORMAL SPACES AND URYSOHN'S LEMMA

12. Prove Theorem 10.6: Let X be a topological space. Then the following conditions are equivalent: (i) X is normal. (ii) If H is an open superset of a closed set F, then there exists an open set G such that $F \subset G \subset \bar{G} \subset H$.

Solution:

(i) $\Rightarrow$ (ii). Let $F \subset H$, with F closed and H open. Then H^c is closed, and $F \cap H^c = \emptyset$. But X is normal; hence

$$\exists \text{ open sets } G, G^* \quad \text{such that} \quad F \subset G, \; H^c \subset G^* \quad \text{and} \quad G \cap G^* = \emptyset$$

But

$$G \cap G^* = \emptyset \; \Rightarrow \; G \subset G^{*c} \quad \text{and} \quad H^c \subset G^* \; \Rightarrow \; G^{*c} \subset H$$

Furthermore, G^{*c} is closed; hence $F \subset G \subset \bar{G} \subset G^{*c} \subset H$.

(ii) $\Rightarrow$ (i). Let F_1 and F_2 be disjoint closed sets. Then $F_1 \subset F_2^c$, and F_2^c is open. By (ii),

$$\exists \text{ an open set } G \quad \text{such that} \quad F_1 \subset G \subset \bar{G} \subset F_2^c$$

But

$$\bar{G} \subset F_2^c \; \Rightarrow \; F_2 \subset \bar{G}^c \quad \text{and} \quad G \subset \bar{G} \; \Rightarrow \; G \cap \bar{G}^c = \emptyset$$

Furthermore, $\bar{G}^c$ is open. Thus $F_1 \subset G$ and $F_2 \subset \bar{G}^c$ with $G, \bar{G}^c$ disjoint open sets; hence X is normal.

13. Let $\mathcal{B}$ be a base for a normal T_1-space X. Show that for each $G_i \in \mathcal{B}$ and any point $p \in G_i$, there exists a member $G_j \in \mathcal{B}$ such that $p \in \bar{G}_j \subset G_i$.

Solution:

Since X is a T_1-space, $\{p\}$ is closed; hence G_i is an open superset of the closed set $\{p\}$. By Theorem 10.6,

$\exists$ an open set G such that $\{p\} \subset G \subset \bar{G} \subset G_i$

Since $p \in G$, there is a member G_j of the base $\mathcal{B}$ such that $p \in G_j \subset G$; so $p \in \bar{G}_j \subset \bar{G}$. But $\bar{G} \subset G_i$; hence $p \in \bar{G}_j \subset G_i$.

14. Let D be the set of dyadic fractions (fractions whose denominators are powers of 2) in the unit interval $[0, 1]$, i.e.,

$$D = \{\tfrac{1}{2}, \tfrac{1}{4}, \tfrac{3}{4}, \tfrac{1}{8}, \tfrac{3}{8}, \tfrac{5}{8}, \tfrac{7}{8}, \tfrac{1}{16}, \ldots, \tfrac{15}{16}, \ldots\}$$

Show that D is dense in $[0, 1]$.

Solution:

To show that $\bar{D} = [0, 1]$, it is sufficient to show that any open interval $(a - \delta, \, a + \delta)$ centered at any point $a \in [0, 1]$ contains a point of D. Observe that $\lim_{n \to \infty} \frac{1}{2^n} = 0$; hence there exists a power $q = 2^{n_0}$ such that $0 < 1/q < \delta$. Consider the intervals

$$\left[0, \tfrac{1}{q}\right], \quad \left[\tfrac{1}{q}, \tfrac{2}{q}\right], \quad \left[\tfrac{2}{q}, \tfrac{3}{q}\right], \quad \ldots, \quad \left[\tfrac{q-2}{q}, \tfrac{q-1}{q}\right], \quad \left[\tfrac{q-1}{q}, 1\right]$$

Since $[0, 1]$ is the union of the above intervals, one of them, say $\left[\tfrac{m}{q}, \tfrac{m+1}{q}\right]$ contains a, i.e. $\tfrac{m}{q} \le a \le \tfrac{m+1}{q}$. But $\tfrac{1}{q} < \delta$; hence

$$a - \delta \, < \, \frac{m}{q} \, \le \, a \, < \, a + \delta$$

In other words, the open interval $(a - \delta, \, a + \delta)$ contains the point m/q which belongs to D. Thus D is dense in $[0, 1]$.

15. Prove Theorem (Urysohn's Lemma) 10.7: Let F_1 and F_2 be disjoint closed subsets of a normal space X. Then there exists a continuous function $f : X \to [0, 1]$ such that $f[F_1] = \{0\}$ and $f[F_2] = \{1\}$.

Solution:

By hypothesis, $F_1 \cap F_2 = \emptyset$; hence $F_1 \subset F_2^c$. In particular, since F_2 is a closed set, F_2^c is an *open* superset of the closed set F_1. By Theorem 10.4, there exists an open set $G_{1/2}$ such that

$$F_1 \subset G_{1/2} \subset \bar{G}_{1/2} \subset F_2^c$$

Observe that $G_{1/2}$ is an open superset of the closed set F_1, and F_2^c is an open superset of the closed set $\bar{G}_{1/2}$. Hence, by Theorem 10.4, there exist open sets $G_{1/4}$ and $G_{3/4}$ such that

$$F_1 \subset G_{1/4} \subset \bar{G}_{1/4} \subset G_{1/2} \subset \bar{G}_{1/2} \subset G_{3/4} \subset \bar{G}_{3/4} \subset F_2^c$$

We continue in this manner and obtain for each $t \in D$, where D is the set of dyadic fractions in $[0, 1]$, an open set G_t with the property that if $t_1, t_2 \in D$ and $t_1 < t_2$ then $\bar{G}_{t_1} \subset G_{t_2}$.

Define the function f on X as follows:

$$f(x) = \begin{cases} \inf \{t : x \in G_t\} & \text{if } x \notin F_2 \\ 1 & \text{if } x \in F_2 \end{cases}$$

Observe that, for every $x \in X$, $0 \le f(x) \le 1$, i.e. f maps X into $[0, 1]$. Observe also that $F_1 \subset G_t$ for all $t \in D$; hence $f[F_1] = \{0\}$. Moreover, by definition, $f[F_2] = \{1\}$. Consequently, the only thing left for us to prove is that f is continuous.

Now f is continuous if the inverses of the sets $[0, a)$ and $(b, 1]$ are open subsets of X (see Problem 7, Chapter 7). We claim that

$$f^{-1}[[0, a)] = \bigcup \{G_t : t < a\} \tag{1}$$

$$f^{-1}[(b, 1]] = \bigcup \{\bar{G}_t^c : t > b\} \tag{2}$$

Then each is the union of open sets and is therefore open.

We first prove (1). Let $x \in f^{-1}[[0, a)]$. Then $f(x) \in [0, a)$, i.e. $0 \leqq f(x) < a$. Since D is dense in $[0, 1]$, there exists $t_x \in D$ such that $f(x) < t_x < a$. In other words,

$$f(x) = \inf\{t : x \in G_t\} < t_x < a$$

Accordingly $x \in G_{t_x}$ where $t_x < a$. Hence $x \in \bigcup\{G_t : t < a\}$. We have just shown that every element in $f^{-1}[[0, a)]$ also belongs to $\bigcup\{G_t : t < a\}$, i.e.,

$$f^{-1}[[0, a)] \subset \bigcup\{G_t : t < a\}$$

On the other hand, suppose $y \in \bigcup\{G_t : t < a\}$. Then $\exists\, t_y \in D$ such that $t_y < a$ and $y \in G_{t_y}$. Therefore

$$f(y) = \inf\{t : y \in G_t\} \leqq t_y < a$$

Hence y also belongs to $f^{-1}[[0, a)]$. In other words,

$$\bigcup\{G_t : t < a\} \subset f^{-1}[[0, a)]$$

The above two results imply (1).

We now prove (2). Let $x \in f^{-1}[(b, 1]]$. Then $f(x) \in (b, 1]$, i.e. $b < f(x) \leqq 1$. Since D is dense in $[0, 1]$, there exist $t_1, t_2 \in D$ such that $b < t_1 < t_2 < f(x)$. In other words,

$$f(x) = \inf\{t : x \in G_t\} > t_2$$

Hence $x \notin G_{t_2}$. Observe that $t_1 < t_2$ implies $\bar{G}_{t_1} \subset G_{t_2}$. Hence x does not belong to $\bar{G}_{t_1}$ either. Accordingly, $x \in \bar{G}_{t_1}^c$ where $t_1 > b$; hence $x \in \bigcup\{\bar{G}_t^c : t > b\}$. Consequently,

$$f^{-1}[(b, 1]] \subset \bigcup\{\bar{G}_t^c : t > b\}$$

On the other hand, let $y \in \bigcup\{\bar{G}_t^c : t > b\}$. Then there exists $t_y \in D$ such that $t_y > b$ and $y \in \bar{G}_{t_y}^c$; hence y does not belong to $\bar{G}_{t_y}$. But $t < t_y$ implies $G_t \subset G_{t_y} \subset \bar{G}_{t_y}$; hence $y \notin G_t$ for every t less than t_y. Consequently,

$$f(y) = \inf\{t : y \in G_t\} \geqq t_y > b$$

Hence $y \in f^{-1}((b, 1])$. In other words,

$$\bigcup\{\bar{G}_t^c : t > b\} \subset f^{-1}[(b, 1]]$$

The above two results imply (2). Hence f is continuous and Urysohn's Lemma is proven.

---

16. Prove Urysohn's Metrization Theorem 10.8: Every second countable normal T_1-space X is metrizable. (In fact, X is homeomorphic to a subset of the Hilbert cube $\mathbf{I}$ of $\mathbf{R}^\infty$.)

Solution:

    If X is finite, then X is a discrete space and hence X is homeomorphic to any subset of H with an equivalent number of points. If X is infinite, then X contains a denumerable base $\mathcal{B} = \{G_1, G_2, G_3, \ldots\}$ where none of the members of $\mathcal{B}$ is X itself.

    By a previous problem, for each G_i in $\mathcal{B}$ there exists some G_j in $\mathcal{B}$ such that $\bar{G}_j \subset G_i$. The class of *all* such pairs $\langle G_j, G_i \rangle$, where $\bar{G}_j \subset G_i$, is denumerable; hence we can denote them by $P_1, P_2, \ldots$ where $P_n = \langle G_{j_n}, G_{i_n} \rangle$. Observe that $\bar{G}_{j_n} \subset G_{i_n}$ implies that $\bar{G}_{j_n}$ and $G_{i_n}^c$ are disjoint closed subsets of X. Hence by Urysohn's Lemma there exists a function $f_n : X \to [0, 1]$ such that $f_n[\bar{G}_{j_n}] = \{0\}$ and $f_n[G_{i_n}^c] = \{1\}$.

    Now define a function $f : X \to \mathbf{I}$ as follows:

$$f(x) = \left\langle \frac{f_1(x)}{2}, \frac{f_2(x)}{2^2}, \frac{f_3(x)}{2^3}, \ldots \right\rangle$$

Observe that, for all $n \in N$, $0 \leqq f_n(x) \leqq 1$ implies $\left| \dfrac{f_n(x)}{2^n} \right| \leqq \dfrac{1}{n}$; hence $f(x)$ is a point in the Hilbert cube $\mathbf{I}$. (Recall that $\mathbf{I} = \{\langle a_n \rangle : a_n \in \mathbf{R},\, n \in \mathbf{N},\, 0 \leqq a_n \leqq 1/n\}$, see Page 129.)

    We now show that f is one-to-one. Let x and y be distinct points in X. Since X is a T_1-space, there exists a member G_i of the base $\mathcal{B}$ such that $x \in G_i$ but $y \notin G_i$. By a previous problem, there exists a pair $P_m = \langle G_j, G_i \rangle$ such that $x \in \bar{G}_j \subset G_i$. By definition, $f_m(x) = 0$ since $x \in \bar{G}_j$, and $f_m(y) = 1$ since $y \notin G_i$, i.e. $y \in G_i^c$. Hence $f(x) \neq f(y)$ since they differ in the mth coordinate. Thus f is one-to-one.

We now prove that f is continuous. Let $\epsilon > 0$. Observe that f is continuous at $p \in X$ if there exists an open neighborhood G of p such that $x \in G$ implies $\|f(x) - f(p)\| < \epsilon$ or, equivalently, $\|f(x) - f(p)\|^2 < \epsilon^2$. Recall that

$$\|f(x) - f(p)\|^2 = \sum_{n=1}^{\infty} \frac{|f_n(x) - f_n(p)|^2}{2^{2n}}$$

Furthermore, since the values of f_n lie in $[0,1]$, $(|f_n(x) - f_n(p)|^2)/2^{2n} \le 1/2^{2n}$. Note that $\sum_n 1/2^{2n}$ converges; hence there exists an $n_0 = n_0(\epsilon)$, which is independent of x and p, such that

$$\|f(x) - f(p)\|^2 = \sum_{n=1}^{n_0} \frac{|f_n(x) - f_n(p)|^2}{2^{2n}} + \frac{\epsilon^2}{2}$$

Now each function $f_n : X \to [0,1]$ is continuous; hence there exists an open neighborhood G_n of p such that $x \in G_n$ implies $|f_n(x) - f_n(p)|^2 < \epsilon^2/2n_0$. Let $G = G_1 \cap \cdots \cap G_{n_0}$. Since G is a finite intersection of open neighborhoods of p, G is also an open neighborhood of p. Furthermore, if $x \in G$ then

$$\|f(x) - f(p)\|^2 = \sum_{n=1}^{\infty} \frac{|f_n(x) - f_n(p)|^2}{2^{2n}} < n_0 \left(\frac{\epsilon^2}{2n_0} \right) + \frac{\epsilon^2}{2} = \epsilon^2$$

Hence f is continuous.

Now let Y denote the range of f, i.e. $Y = f[X] \subset \mathbf{I}$. We want to prove that $f^{-1} : Y \to X$ is also continuous. Observe that continuity in Y is equivalent to sequential continuity; hence f^{-1} is continuous at $f(p) \in Y$ if for every sequence $\langle f(y_n) \rangle$ converging to $f(p)$, the sequence $\langle y_n \rangle$ converges to p.

Suppose f^{-1} is not continuous, i.e. suppose $\langle y_n \rangle$ does not converge to p. Then there exists an open neighborhood G of p such that G does not contain an infinite number of the terms of $\langle y_n \rangle$. Hence we can choose a subsequence $\langle x_n \rangle$ of $\langle y_n \rangle$ such that all the terms of $\langle x_n \rangle$ lie outside of G. Since $p \in G$, there exists a member G_i in the base $\mathcal{B}$ such that $p \in G_i \subset G$. Furthermore, by a previous problem, there exists a pair $P_m = \langle G_j, G_i \rangle$ such that $p \in \bar{G}_j \subset G_i \subset G$. Observe that, for all $n \in \mathbf{N}$, $x_n \notin G$; hence $x_n \in G_i^c$. Accordingly, $f_m(p) = 0$ and $f_m(x_n) = 1$. Then $|f_m(x_n) - f_m(p)|^2 = 1$ and

$$\|f(x_n) - f(p)\|^2 = \sum_{k=1}^{\infty} \frac{|f_k(x_n) - f_k(p)|^2}{2^{2k}} \ge \frac{1}{2^{2m}}$$

In other words, for every $n \in \mathbf{N}$, $\|f(x_n) - f(p)\| > 1/2^m$. Therefore the sequence $\langle f(x_n) \rangle$ does not converge to $f(p)$. But this contradicts the fact that every subsequence of $\langle f(y_n) \rangle$ should also converge to $f(p)$. Hence f^{-1} is continuous. Hence f is a homeomorphism and X is homeomorphic to a subset of the Hilbert cube. Accordingly, X is metrizable.

REGULAR AND COMPLETELY REGULAR SPACES

17. Prove Proposition 10.10: A completely regular space X is also regular.

Solution:

Let F be a closed subset of X and suppose $p \in X$ does not belong to F. By hypothesis, X is completely regular; hence there exists a continuous function $f : X \to [0,1]$ such that $f(p) = 0$ and $f[F] = \{1\}$. But $\mathbf{R}$ and its subspace $[0,1]$ are Hausdorff spaces; hence there are disjoint open sets G and H containing 0 and 1 respectively. Accordingly, their inverses $f^{-1}[G]$ and $f^{-1}[H]$ are disjoint, open and contain p and F respectively. In other words, X is also regular.

18. Prove Theorem 10.11: The class $C(X, \mathbf{R})$ of all real-valued continuous functions on a completely regular T_1-space X separates points.

Solution:

Let a and b be distinct points in X. Since X is a T_1-space, $\{b\}$ is a closed set. Also, since a and b are distinct, $a \notin \{b\}$. By hypothesis, X is completely regular; hence there exists a real-valued continuous function f on X such that $f(a) = 0$ and $f[\{b\}] = \{1\}$. Accordingly, $f(a) \ne f(b)$.

19. Let $(Y, \mathcal{T}_Y)$ be a subspace of $(X, \mathcal{T})$ and let $p \in Y$ and $A \subset Y \subset X$. Show that if p does not belong to the $\mathcal{T}_Y$-closure of A, then $p \notin \bar{A}$, the $\mathcal{T}$-closure of A.

Solution:

Now, by a property of subspaces (see Problem 89, Chapter 5),

$$\mathcal{T}_Y\text{-closure of } A = Y \cap \bar{A}$$

But $p \in Y$ and $p \notin \mathcal{T}_Y$-closure of A; hence $p \notin \bar{A}$. (Observe that, in particular, if F is a $\mathcal{T}_Y$-closed subset of Y and $p \notin F$, then $p \notin \bar{F}$.)

20. Show that the property of being a regular space is hereditary, i.e. every subspace of a regular space is regular.

Solution:

Let $(X, \mathcal{T})$ be a regular space and let $(Y, \mathcal{T}_Y)$ be a subspace of $(X, \mathcal{T})$. Furthermore, let $p \in Y$ and let F be a $\mathcal{T}_Y$-closed subset of Y such that $p \notin F$. Now by Problem 19, $p \notin \bar{F}$, the $\mathcal{T}$-closure of F. By hypothesis, $(X, \mathcal{T})$ is regular; hence

$$\exists \ G, H \in \mathcal{T} \quad \text{such that} \quad \bar{F} \subset G, \ p \in H \ \text{ and } \ G \cap H = \emptyset$$

But $Y \cap G$ and $Y \cap H$ are $\mathcal{T}_Y$-open subsets of Y, and

$$F \subset Y, \ F \subset \bar{F} \subset G \quad \Rightarrow \quad F \subset Y \cap G$$

$$p \in Y, \ p \in H \quad \Rightarrow \quad p \in Y \cap H$$

$$G \cap H = \emptyset \quad \Rightarrow \quad (Y \cap G) \cap (Y \cap H) = \emptyset$$

Accordingly, $(Y, \mathcal{T}_Y)$ is also regular.

Supplementary Problems

T_1-SPACES

21. Show that the property of being a T_1-space is topological.

22. Show, by a counterexample, that the image of a T_1-space under a continuous map need not be T_1.

23. Let $(X, \mathcal{T})$ be a T_1-space and let $\mathcal{T} \precsim \mathcal{T}^*$. Show that $(X, \mathcal{T}^*)$ is also a T_1-space.

24. Prove: X is a T_1-space if and only if every $p \in X$ is the intersection of all open sets containing it, i.e. $\{p\} = \bigcap \{G : G \text{ open}, p \in G\}$.

25. A topological space X is called a T_0-space if it satisfies the following axiom:

[$\mathbf{T_0}$] For any pair of distinct points in X, there exists an open set containing one of the points but not the other.

(i) Give an example of a T_0-space which is not a T_1-space.

(ii) Show that every T_1-space is also a T_0-space.

26. Let X be a T_1-space containing at least two points. Show that if $\mathcal{B}$ is a base for X then $\mathcal{B} \setminus \{X\}$ is also a base for X.

HAUSDORFF SPACES

27. Show that the property of being a Hausdorff space is topological.

28. Let $(X, \mathcal{T})$ be a Hausdorff space and let $\mathcal{T} \precsim \mathcal{T}^*$. Show that $(X, \mathcal{T}^*)$ is also a Hausdorff space.

29. Show that if $a_1, \ldots, a_m$ are distinct points in a Hausdorff space X, then there exists a disjoint class $\{G_1, \ldots, G_m\}$ of open subsets of X such that $a_1 \in G_1, \ldots, a_m \in G_m$.

30. Prove: Let X be an infinite Hausdorff space. Then there exists an infinite disjoint class of open subsets of X.

31. Prove: Let $f : X \to Y$ and $g : X \to Y$ be continuous functions from a topological space X into a Hausdorff space Y. Then $A = \{x : f(x) = g(x)\}$ is a closed subset of X.

NORMAL SPACES

32. Show that the property of being a normal space is topological.

33. Let $\mathcal{T}$ be the topology on the real line $\mathbf{R}$ generated by the closed-open intervals $[a, b)$. Show that $(\mathbf{R}, \mathcal{T})$ is a normal space.

34. Let $\mathcal{T}$ be the topology on the plane $\mathbf{R}^2$ generated by the half-open rectangles,

$$[a, b) \times [c, d) \quad = \quad \{\langle x, y \rangle : a \le x < b, \ c \le y < d\}$$

Furthermore, let A consist of the points on the line $Y = \{\langle x, y \rangle : x + y = 1\} \subset \mathbf{R}^2$ whose coordinates are rational and let $B = Y \setminus A$.
   (i)   Show that A and B are closed subsets of $(\mathbf{R}^2, \mathcal{T})$.
   (ii) Show that there exist no disjoint $\mathcal{T}$-open subsets G and H of $\mathbf{R}^2$ such that $A \subset G$ and $B \subset H$; and so $(\mathbf{R}^2, \mathcal{T})$ is not normal.

35. Let A be a closed subset of a normal T_1-space. Show that A with the relative topology is also a normal T_1-space.

36. Let X be an ordered set and let $\mathcal{T}$ be the order topology on X, i.e. $\mathcal{T}$ is generated by the subsets of X of the form $\{x : x < a\}$ and $\{x : x > a\}$. Show that $(X, \mathcal{T})$ is a normal space.

37. Prove: Let X be a normal space. Then X is regular if and only if X is completely regular.

URYSOHN'S LEMMA

38. Prove: If for every two disjoint closed subsets F_1 and F_2 of a topological space X, there exists a continuous function $f : X \to [0, 1]$ such that $f[F_1] = \{0\}$ and $f[F_2] = \{1\}$, then X is a normal space. (Note that this is the converse of Urysohn's Lemma.)

39. Prove the following generalization of Urysohn's Lemma: Let F_1 and F_2 be disjoint closed subsets of a normal space X. Then there exists a continuous function $f : X \to [a, b]$ such that $f[F_1] = \{a\}$ and $f[F_2] = \{b\}$.

40. Prove the Tietze Extension Theorem: Let F be a closed subset of a normal space X and let $f : F \to [a, b]$ be a real continuous function. Then f has a continuous extension $f^* : X \to [a, b]$.

41. Prove Urysohn's Lemma using the Tietze Extension Theorem.

REGULAR AND COMPLETELY REGULAR SPACES

42. Show that the property of being a regular space is topological.

43. Show that the property of being completely regular is topological.

44. Show that the property of being a completely regular space is hereditary, that is, every subspace of a completely regular space is also completely regular.

45. Prove: Let X be a regular Lindelöf space. Then X is normal.

Answers to Supplementary Problems

25. (i) Let $X = \{a, b\}$ and $\mathcal{T} = \{X, \{a\}, \emptyset\}$.

Chapter 11

Compactness

COVERS

Let $\mathcal{A} = \{G_i\}$ be a class of subsets of X such that $A \subset \cup_i G_i$ for some $A \subset X$. Recall that $\mathcal{A}$ is then called a *cover* of A, and an *open cover* if each G_i is open. Furthermore, if a finite subclass of $\mathcal{A}$ is also a cover of A, i.e. if

$$\exists\ G_{i_1}, \ldots, G_{i_m} \in \mathcal{A} \qquad \text{such that} \qquad A \subset G_{i_1} \cup \cdots \cup G_{i_m}$$

then $\mathcal{A}$ is said to be *reducible to a finite cover*, or contains a *finite subcover*.

Example 1.1: Consider the class $\mathcal{A} = \{D_p : p \in \mathbf{Z} \times \mathbf{Z}\}$, where D_p is the open disc in the plane $\mathbf{R}^2$ with radius 1 and center $p = \langle m, n \rangle$, m and n integers. Then $\mathcal{A}$ is a cover of $\mathbf{R}^2$, i.e. every point in $\mathbf{R}^2$ belongs to at least one member of $\mathcal{A}$. On the other hand, the class of open discs $\mathcal{B} = \{D_p^* : p \in \mathbf{Z} \times \mathbf{Z}\}$, where D_p has center p and radius $\frac{1}{2}$, is not a cover of $\mathbf{R}^2$. For example, the point $\langle \frac{1}{2}, \frac{1}{2} \rangle \in \mathbf{R}^2$ does not belong to any member of $\mathcal{B}$, as shown in the figure.

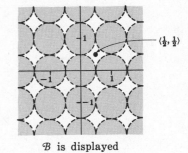

$\mathcal{B}$ is displayed

Example 1.2: Consider the classical

Heine-Borel Theorem: Let $A = [a, b]$ be a closed and bounded interval and let $\{G_i\}$ be a class of open sets such that $A \subset \cup_i G_i$. Then one can select a finite number of the open sets, say $G_{i_1}, \ldots, G_{i_m}$, so that $A \subset G_{i_1} \cup \cdots \cup G_{i_m}$.

By virtue of the above terminology, the Heine-Borel Theorem can be restated as follows:

Heine-Borel Theorem: Every open cover of a closed and bounded interval $A = [a, b]$ is reducible to a finite cover.

COMPACT SETS

The concept of *compactness* is no doubt motivated by the property of a closed and bounded interval as stated in the classical Heine-Borel Theorem. Namely,

Definition: A subset A of a topological space X is *compact* if every open cover of A is reducible to a finite cover.

In other words, if A is compact and $A \subset \cup_i G_i$, where the G_i are open sets, then one can select a finite number of the open sets, say $G_{i_1}, \ldots, G_{i_m}$, so that $A \subset G_{i_1} \cup \cdots \cup G_{i_m}$.

Example 2.1: By the Heine-Borel Theorem, every closed and bounded interval $[a, b]$ on the real line $\mathbf{R}$ is compact.

Example 2.2: Let A be any finite subset of a topological space X, say $A = \{a_1, \ldots, a_m\}$. Then A is necessarily compact. For if $\mathcal{G} = \{G_i\}$ is an open cover of A, then each point in A belongs to one of the members of $\mathcal{G}$, say $a_1 \in G_{i_1}, \ldots, a_m \in G_{i_m}$. Accordingly, $A \subset G_{i_1} \cup G_{i_2} \cup \cdots \cup G_{i_m}$.

Since a set A is compact iff every open cover of A contains a finite subcover, we only have to exhibit one open cover of A with no finite subcover to prove that A is not compact.

Example 2.3: The open interval $A = (0, 1)$ on the real line **R** with the usual topology is not compact. Consider, for example, the class of open intervals

$$G = \{(\tfrac{1}{3}, 1), (\tfrac{1}{4}, \tfrac{1}{2}), (\tfrac{1}{5}, \tfrac{1}{3}), (\tfrac{1}{6}, \tfrac{1}{4}), \ldots\}$$

Observe that $A = \cup_{n=1}^{\infty} G_n$, where $G_n = \left(\dfrac{1}{n+2}, \dfrac{1}{n}\right)$; hence G is an open cover of A.

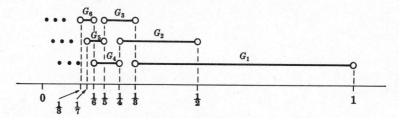

But G contains no finite subcover. For let

$$G^* = \{(a_1, b_1), (a_2, b_2), \ldots, (a_m, b_m)\}$$

be any finite subclass of G. If $\epsilon = \min(a_1, \ldots, a_m)$ then $\epsilon > 0$ and

$$(a_1, b_1) \cup \cdots \cup (a_m, b_m) \subset (\epsilon, 1)$$

But $(0, \epsilon]$ and $(\epsilon, 1)$ are disjoint; hence G^* is not a cover of A, and so A is not compact.

Example 2.4: We show that a continuous image of a compact set is also compact, i.e. if the function $f : X \to Y$ is continuous and A is a compact subset of X, then its image $f[A]$ is a compact subset of Y. For suppose $G = \{G_i\}$ is an open cover of $f[A]$, i.e. $f[A] \subset \cup_i G_i$. Then

$$A \subset f^{-1}[f[A]] \subset f^{-1}[\cup_i G_i] = \cup_i f^{-1}[G_i]$$

Hence $\mathcal{H} = \{f^{-1}[G_i]\}$ is a cover of A. Now f is continuous and each G_i is an open set, so each $f^{-1}[G_i]$ is also open. In other words, $\mathcal{H}$ is an open cover of A. But A is compact, so $\mathcal{H}$ is reducible to a finite cover, say

$$A \subset f^{-1}[G_{i_1}] \cup \cdots \cup f^{-1}[G_{i_m}]$$

Accordingly,

$$f[A] \subset f[f^{-1}[G_{i_1}] \cup \cdots \cup f^{-1}[G_{i_m}]] \subset G_{i_1} \cup \cdots \cup G_{i_m}$$

Thus $f[A]$ is compact.

We formally state the result in Example 2.4:

Theorem 11.1: Continuous images of compact sets are compact.

Compactness is an *absolute property* of a set. Namely,

Theorem 11.2: Let A be a subset of a topological space $(X, \mathcal{T})$. Then A is compact with respect to $\mathcal{T}$ if and only if A is compact with respect to the relative topology $\mathcal{T}_A$ on A.

Accordingly, we can frequently limit our investigation of compactness to those topological spaces which are themselves compact, i.e. to *compact spaces*.

SUBSETS OF COMPACT SPACES

A subset of a compact space need not be compact. For example, the closed unit interval $[0, 1]$ is compact by the Heine-Borel Theorem, but the open interval $(0, 1)$ is a subset of $[0, 1]$ which, by Example 2.3 above, is not compact. We do, however, have the following

Theorem 11.3: Let F be a closed subset of a compact space X. Then F is also compact.

Proof: Let $\mathcal{G} = \{G_i\}$ be an open cover of F, i.e. $F \subset \cup_i G_i$. Then $X = (\cup_i G_i) \cup F^c$, that is, $\mathcal{G}^* = \{G_i\} \cup \{F^c\}$ is a cover of X. But F^c is open since F is closed, so $\mathcal{G}^*$ is an open cover of X. By hypothesis, X is compact; hence $\mathcal{G}^*$ is reducible to a finite cover of X, say

$$X = G_{i_1} \cup \cdots \cup G_{i_m} \cup F^c, \qquad G_{i_k} \in \mathcal{G}$$

But F and F^c are disjoint; hence

$$F \subset G_{i_1} \cup \cdots \cup G_{i_m}, \qquad G_{i_k} \in \mathcal{G}$$

We have just shown that any open cover $\mathcal{G} = \{G_i\}$ of F contains a finite subcover, i.e. F is compact.

FINITE INTERSECTION PROPERTY

A class $\{A_i\}$ of sets is said to have the *finite intersection property* if every finite subclass $\{A_{i_1}, \ldots, A_{i_m}\}$ has a non-empty intersection, i.e. $A_{i_1} \cap \cdots \cap A_{i_m} \neq \emptyset$.

Example 3.1:   Consider the following class of open intervals:

$$\mathcal{A} = \{(0,1), (0,\tfrac{1}{2}), (0,\tfrac{1}{3}), (0,\tfrac{1}{4}), \ldots\}$$

Now $\mathcal{A}$ has the finite intersection property, for

$$(0, a_1) \cap (0, a_2) \cap \cdots \cap (0, a_m) = (0, b)$$

where $b = \min(a_1, \ldots, a_m) > 0$. Observe that $\mathcal{A}$ itself has an empty intersection.

Example 3.2:   Consider the following class of closed infinite intervals:

$$\mathcal{B} = \{\ldots, (-\infty, -2], (-\infty, -1], (-\infty, 0], (-\infty, 1], (-\infty, 2], \ldots\}$$

Note that $\mathcal{B}$ has an empty intersection, i.e. $\cap \{B_n : n \in \mathbf{Z}\} = \emptyset$ where $B_n = (-\infty, n]$. But any finite subclass of $\mathcal{B}$ has a non-empty intersection. In other words, $\mathcal{B}$ satisfies the finite intersection property.

With the above terminology, we can now state the notion of compactness in terms of the closed subsets of a topological space.

Theorem 11.4:   A topological space X is compact if and only if every class $\{F_i\}$ of closed subsets of X which satisfies the finite intersection property has, itself, a non-empty intersection.

COMPACTNESS AND HAUSDORFF SPACES

Here we relate the concept of compactness to the separation property of Hausdorff spaces.

Theorem 11.5:   Every compact subset of a Hausdorff space is closed.

The above theorem is not true in general; for example, finite sets are always compact and yet there exist topological spaces whose finite subsets are not all closed.

Theorem 11.6:   Let A and B be disjoint compact subsets of a Hausdorff space X. Then there exist disjoint open sets G and H such that $A \subset G$ and $B \subset H$.

In particular, suppose X is both Hausdorff and compact and F_1 and F_2 are disjoint closed subsets of X. By Theorem 11.3, F_1 and F_2 are compact and, by Theorem 11.6, F_1 and F_2 are subsets, respectively, of disjoint open sets. In other words,

Corollary 11.7:   Every compact Hausdorff space is normal.

Thus metric spaces and compact Hausdorff spaces are both contained in the class of T_4-spaces, i.e. normal T_1-spaces.

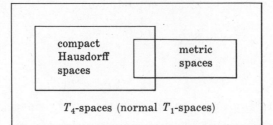

T_4-spaces (normal T_1-spaces)

The following theorem plays a very important role in geometry.

Theorem 11.8:   Let f be a one-one continuous function from a compact space X into a Hausdorff space Y.   Then X and $f[X]$ are homeomorphic.

The next example shows that the above theorem is not true in general.

Example 4.1:   Let f be the function from the half-open interval $X = [0, 1)$ into the plane $\mathbf{R}^2$ defined by $f(t) = \langle \cos 2\pi t, \sin 2\pi t \rangle$.   Observe that f maps X onto the unit circle and that f is one-one and continuous.

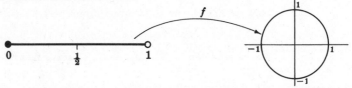

But the half-open interval $[0, 1)$ is not homeomorphic to the circle.   For example, if we delete the point $t = \frac{1}{2}$ from X, X will not be connected; but if we delete any point from a circle, the circle is still connected.   The reason that Theorem 11.8 does not apply in this case is that X is not compact.

Example 4.2:   Let f be a one-one continuous function from the closed unit interval $I = [0, 1]$ into Euclidean n-space $\mathbf{R}^n$.   Observe that I is compact by the Heine-Borel Theorem and that $\mathbf{R}^n$ is a metric space and therefore Hausdorff.   By virtue of Theorem 11.8, I and $f[I]$ are homeomorphic.

SEQUENTIALLY COMPACT SETS

A subset A of a topological space X is *sequentially compact* iff every sequence in A contains a subsequence which converges to a point in A.

Example 5.1:   Let A be a finite subset of a topological space X.   Then A is necessarily sequentially compact.   For if $\langle s_1, s_2, \ldots \rangle$ is a sequence in A, then at least one of the elements in A, say a_0, must appear an infinite number of times in the sequence.   Hence $\langle a_0, a_0, a_0, \ldots \rangle$ is a subsequence of $\langle s_n \rangle$, it converges, and furthermore it converges to the point a_0 belonging to A.

Example 5.2:   The open interval $A = (0, 1)$ on the real line $\mathbf{R}$ with the usual topology is not sequentially compact.   Consider, for example, the sequence $\langle s_n \rangle = \langle \frac{1}{2}, \frac{1}{3}, \frac{1}{4}, \ldots \rangle$ in A.   Observe that $\langle s_n \rangle$ converges to 0 and therefore every subsequence also converges to 0.   But 0 does not belong to A.   In other words, the sequence $\langle s_n \rangle$ in A does not contain a subsequence which converges to a point in A, i.e. A is not sequentially compact.

In general, there exist compact sets which are not sequentially compact and vice versa, although in metric spaces, as we show later, they are equivalent.

Remark: Historically, the term *bicompact* was used to denote a compact set, and the term compact was used to denote a sequentially compact set.

COUNTABLY COMPACT SETS

A subset A of a topological space X is *countably compact* iff every infinite subset B of A has an accumulation point in A. This definition is no doubt motivated by the classical

Bolzano-Weierstrass Theorem: Every bounded infinite set of real numbers has an accumulation point.

Example 6.1: Every bounded closed interval $A = [a, b]$ is countably compact. For if B is an infinite subset of A, then B is also bounded and, by the Bolzano-Weierstrass Theorem, B has an accumulation point p. Furthermore, since A is closed, the accumulation point p of B belongs to A, i.e. A is countably compact.

Example 6.2: The open interval $A = (0, 1)$ is not countably compact. For consider the infinite subset $B = \{\frac{1}{2}, \frac{1}{3}, \frac{1}{4}, \ldots\}$ of $A = (0, 1)$. Observe that B has exactly one limit point which is 0 and that 0 does not belong to A. Hence A is not countably compact.

The general relationship between compact, sequentially compact and countably compact sets is given in the following diagram and theorem.

$$\text{compact} \;\rightarrow\; \text{countably compact} \;\leftarrow\; \text{sequentially compact}$$

Theorem 11.9: Let A be a subset of a topological space X. If A is compact or sequentially compact, then A is also countably compact.

The next example shows that neither arrow in the above diagram can be reversed.

Example 6.3: Let $\mathcal{T}$ be the topology on $\mathbf{N}$, the set of positive integers, generated by the following sets:
$$\{1, 2\}, \; \{3, 4\}, \; \{5, 6\}, \; \ldots$$

Let A be a non-empty subset of $\mathbf{N}$, say $n_0 \in A$. If n_0 is odd, then $n_0 + 1$ is a limit point of A; and if n_0 is even, then $n_0 - 1$ is a limit point of A. In either case, A has an accumulation point. Accordingly, $(\mathbf{N}, \mathcal{T})$ is countably compact.

On the other hand, $(\mathbf{N}, \mathcal{T})$ is not compact since
$$\mathcal{A} \;=\; \{\{1, 2\}, \; \{3, 4\}, \; \{5, 6\}, \; \ldots\}$$

is an open cover of $\mathbf{N}$ with no finite subcover. Furthermore, $(\mathbf{N}, \mathcal{T})$ is not sequentially compact, since the sequence $\langle 1, 2, 3, \ldots \rangle$ contains no convergent subsequence.

LOCALLY COMPACT SPACES

A topological space X is *locally compact* iff every point in X has a compact neighborhood.

Example 7.1: Consider the real line $\mathbf{R}$ with the usual topology. Observe that each point $p \in \mathbf{R}$ is interior to a closed interval, e.g. $[p - \delta, p + \delta]$, and that the closed interval is compact by the Heine-Borel Theorem. Hence $\mathbf{R}$ is a locally compact space. On the other hand, $\mathbf{R}$ is not a compact space; for example, the class
$$\mathcal{A} \;=\; \{\ldots, (-3, -1), (-2, 0), (-1, 1), (0, 2), (1, 3), \ldots\}$$

is an open cover of $\mathbf{R}$ but contains no finite subcover.

Thus we see, by the above example, that a locally compact space need not be compact. On the other hand, since a topological space is always a neighborhood of each of its points, the converse is true. That is,

Proposition 11.10: Every compact space is locally compact.

COMPACTIFICATION

A topological space X is said to be *embedded* in a topological space Y if X is homeomorphic to a subspace of Y. Furthermore, if Y is a compact space, then Y is called a *compactification* of X. Frequently, the compactification of a space X is accomplished by adjoining one or more points to X and then defining an appropriate topology on the enlarged set so that the enlarged space is compact and contains X as a subspace.

> **Example 8.1:** Consider the real line $\mathbf{R}$ with the usual topology $\mathcal{U}$. We adjoin two new points, denoted by ∞ and $-\infty$, to $\mathbf{R}$ and call the enlarged set $\mathbf{R}^* = \mathbf{R} \cup \{-\infty, \infty\}$ the *extended real line*. The order relation in $\mathbf{R}$ can be extended to $\mathbf{R}^*$ by defining $-\infty < a < \infty$ for any $a \in \mathbf{R}$. The class of subsets of $\mathbf{R}^*$ of the form
>
> $$(a, b) = \{x : a < x < b\}, \quad (a, \infty] = \{x : a < x\} \quad \text{and} \quad [-\infty, a) = \{x : x < a\}$$
>
> is a base for a topology $\mathcal{U}^*$ on $\mathbf{R}^*$. Furthermore, $(\mathbf{R}^*, \mathcal{U}^*)$ is a compact space and contains $(\mathbf{R}, \mathcal{U})$ as a subspace, and so it is a compactification of $(\mathbf{R}, \mathcal{U})$.

Recall that the real line $\mathbf{R}$ with the usual topology is homeomorphic to any open interval (a, b) of real numbers. The above space $(\mathbf{R}^*, \mathcal{U}^*)$ can, in fact, be shown to be homeomorphic to any closed interval $[a, b]$ which is compact by the classical Heine-Borel Theorem.

> **Example 8.2:** Let C denote the $\langle x, y \rangle$-plane in Euclidian 3-space $\mathbf{R}^3$, and let S denote the sphere with center $\langle 0, 0, 1 \rangle$ on the z-axis and radius 1. The line passing through the "north pole" $\infty = \langle 0, 0, 2 \rangle \in S$ and any point $p \in C$ intersects the sphere S in exactly one point p' distinct from ∞, as shown in the figure.
>
>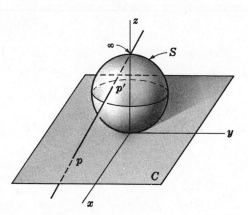
>
> Let $f : C \to S$ be defined by $f(p) = p'$. Then f is, in fact, a homeomorphism from the plane C, which is not compact, onto the subset $S \setminus \{\infty\}$ of the sphere S, and S is compact. Hence S is a compactification of C.

Now let $(X, \mathcal{T})$ be any topological space. We shall define the *Alexandrov* or *one-point compactification* of $(X, \mathcal{T})$ which we denote by $(X_\infty, \mathcal{T}_\infty)$. Here:

(1) $X_\infty = X \cup \{\infty\}$, where ∞, called the *point at infinity*, is distinct from every other point in X.

(2) $\mathcal{T}_\infty$ consists of the following sets:

    (i) each member of the topology $\mathcal{T}$ on X,

    (ii) the complement in X_∞ of any closed and compact subset of X.

We formally state:

Proposition 11.11: The above class $\mathcal{T}_\infty$ is a topology on X_∞, and $(X_\infty, \mathcal{T}_\infty)$ is a compactification of $(X, \mathcal{T})$.

In general, the space $(X_\infty, \mathcal{T}_\infty)$ may not possess properties similar to those of the original space. There does exist one important relationship between the two spaces; namely,

Theorem 11.12: If $(X, \mathcal{T})$ is a locally compact Hausdorff space, then $(X_\infty, \mathcal{T}_\infty)$ is a compact Hausdorff space.

Using Urysohn's lemma we obtain an important result used in measure and integration theory:

Corollary 11.13: Let E be a compact subset of a locally compact Hausdorff space X, and let E be a subset of an open set $G \neq X$. Then there exists a continuous function $f : X \to [0, 1]$ such that $f[E] = \{0\}$ and $f[G^c] = \{1\}$.

COMPACTNESS IN METRIC SPACES

Compactness in metric spaces can be summarized by the following

Theorem 11.14:   Let A be a subset of a metric space X. Then the following statements are equivalent: (i) A is compact, (ii) A is countably compact, and (iii) A is sequentially compact.

Historically, metric spaces were investigated before topological spaces; hence the above theorem gives the main reason that the terms compact and sequentially compact are sometimes used synonymously.

The proof of the above theorem requires the introduction of two auxiliary metric concepts which are interesting in their own right: that of a *totally bounded set* and that of a *Lebesgue number* for a cover.

TOTALLY BOUNDED SETS

Let A be a subset of a metric space X and let $\epsilon > 0$. A finite set of points $N = \{e_1, e_2, \ldots, e_m\}$ is called an ϵ-*net* for A if for every point $p \in A$ there exists an $e_{i_0} \in N$ with $d(p, e_{i_0}) < \epsilon$.

> **Example 9.1:**   Let $A = \{\langle x, y\rangle : x^2 + y^2 < 4\}$, i.e. A is the open disc centered at the origin and of radius 2. If $\epsilon = 3/2$, then the set
>
> $$N = \{\langle 1, -1\rangle,\ \langle 1, 0\rangle,\ \langle 1, 1\rangle,\ \langle 0, -1\rangle,\ \langle 0, 0\rangle,\ \langle 0, 1\rangle,\ \langle -1, -1\rangle,\ \langle -1, 0\rangle,\ \langle -1, 1\rangle\}$$
>
> is an ϵ-net for A. On the other hand, if $\epsilon = \frac{1}{2}$, then N is not an ϵ-net for A. For example, $p = \langle \frac{1}{2}, \frac{1}{2}\rangle$ belongs to A but the distance between p and any point in N is greater than $\frac{1}{2}$.

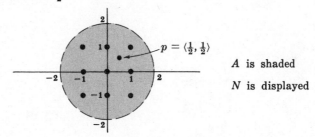

A is shaded

N is displayed

Recall that the *diameter* of A, $d(A)$, is defined by $d(A) = \sup \{d(a, a') : a, a' \in A\}$ and that A is bounded if $d(A) < \infty$.

> **Definition:**   A subset A of a metric space X is *totally bounded* if A possesses an ϵ-net for every $\epsilon > 0$.

A totally bounded set can also be described as follows:

Proposition 11.15:   A set A is totally bounded if and only if for every $\epsilon > 0$ there exists a decomposition of A into a finite number of sets, each with diameter less than ϵ.

We first show that a bounded set need not be totally bounded.

> **Example 9.2:**   Let A be the subset of Hilbert Space, i.e. of l_2-space, consisting of the following points:
>
> $$e_1 = \langle 1, 0, 0, \ldots \rangle$$
> $$e_2 = \langle 0, 1, 0, \ldots \rangle$$
> $$e_3 = \langle 0, 0, 1, \ldots \rangle$$
> $$\vdots \qquad \vdots$$

Observe that $d(e_i, e_j) = \sqrt{2}$ if $i \neq j$. Hence A is bounded; in fact,

$$d(A) = \sup \{d(e_i, e_j) : e_i, e_j \in A\} = \sqrt{2}$$

On the other hand, A is not totally bounded. For if $\epsilon = \frac{1}{2}$, the only non-empty subsets of A with diameter less than ϵ are the singleton sets, i.e. sets with one point. Accordingly, the infinite set A cannot be decomposed into a finite number of disjoint subsets each with diameter less than $\frac{1}{2}$.

The converse of the previous statement is true. Namely,

Proposition 11.16:   Totally bounded sets are bounded.

One relationship between compactness and total boundedness is as follows:

Lemma 11.17:   Sequentially compact sets are totally bounded.

LEBESGUE NUMBERS FOR COVERS

Let $\mathcal{A} = \{G_i\}$ be a cover for a subset A of a metric space X. A real number $\delta > 0$ is called a *Lebesgue number* for the cover if for each subset of A with diameter less than δ there is a member of the cover which contains A.

One relationship between compactness and Lebesgue number for a cover is as follows:

Lemma (Lebesgue) 11.18:   Every open cover of a sequentially compact subset of a metric space has a (positive) Lebesgue number.

Solved Problems

COMPACT SPACES

1. Let $\mathcal{T}$ be the cofinite topology on any set X. Show that $(X, \mathcal{T})$ is a compact space.

    Solution:

    Let $\mathcal{G} = \{G_i\}$ be an open cover of X. Choose $G_0 \in \mathcal{G}$. Since $\mathcal{T}$ is the cofinite topology, G_0^c is a finite set, say $G_0^c = \{a_1, \ldots, a_m\}$. Since $\mathcal{G}$ is a cover of X,

    $$\text{for each} \quad a_k \in G_0^c \quad \exists \ G_{i_k} \in \mathcal{G} \quad \text{such that} \quad a_k \in G_{i_k}$$

    Hence $G_0^c \subset G_{i_1} \cup \cdots \cup G_{i_m}$ and $X = G_0 \cup G_0^c = G_0 \cup G_{i_1} \cup \cdots \cup G_{i_m}$. Thus X is compact.

2. Show that any infinite subset A of a discrete topological space X is not compact.

    Solution:

    Recall that A is not compact if we can exhibit an open cover of A with no finite subcover. Consider the class $\mathcal{A} = \{\{a\} : a \in A\}$ of singleton subsets of A. Observe that: (i) $\mathcal{A}$ is a cover of A; in fact $A = \bigcup \{\{a\} : a \in A\}$. (ii) $\mathcal{A}$ is an open cover of A since all subsets of a discrete space are open. (iii) No proper subclass of $\mathcal{A}$ is a cover of A. (iv) $\mathcal{A}$ is infinite since A is infinite. Accordingly, the open cover $\mathcal{A}$ of A contains no finite subcover, so A is not compact.

    Since finite sets are always compact, we have also proven that a subset of a discrete space is compact if and only if it is finite.

3. Prove Theorem 11.2: Let A be a subset of a topological space $(X, \mathcal{T})$. Then the following are equivalent:

(i) A is compact with respect to $\mathcal{T}$.

(ii) A is compact with respect to the relative topology $\mathcal{T}_A$ on A.

Solution:

(i) $\Rightarrow$ (ii): Let $\{G_i\}$ be a $\mathcal{T}_A$-open cover of A. By definition of the relative topology,

$$\exists\ H_i \in \mathcal{T} \quad \text{such that} \quad G_i = A \cap H_i \subset H_i$$

Hence

$$A \subset \cup_i G_i \subset \cup_i H_i$$

and therefore $\{H_i\}$ is a $\mathcal{T}$-open cover of A. By (i), A is $\mathcal{T}$-compact, so $\{H_i\}$ contains a finite subcover, say

$$A \subset H_{i_1} \cup \cdots \cup H_{i_m}, \quad H_{i_k} \in \{H_i\}$$

But then

$$A \subset A \cap (H_{i_1} \cup \cdots \cup H_{i_m}) = (A \cap H_{i_1}) \cup \cdots \cup (A \cap H_{i_m}) = G_{i_1} \cup \cdots \cup G_{i_m}$$

Thus $\{G_i\}$ contains a finite subcover $\{G_{i_1}, \ldots, G_{i_m}\}$ and $(A, \mathcal{T}_A)$ is compact.

(ii) $\Rightarrow$ (i): Let $\{H_i\}$ be a $\mathcal{T}$-open cover of A. Set $G_i = A \cap H_i$; then

$$A \subset \cup_i H_i \quad \Rightarrow \quad A \subset A \cap (\cup_i H_i) = \cup_i (A \cap H_i) = \cup_i G_i$$

But $G_i \in \mathcal{T}_A$, so $\{G_i\}$ is a $\mathcal{T}_A$-open cover of A. By hypothesis, A is $\mathcal{T}_A$-compact; thus $\{G_i\}$ contains a finite subcover $\{G_{i_1}, \ldots, G_{i_m}\}$. Accordingly,

$$A \subset G_{i_1} \cup \cdots \cup G_{i_m} = (A \cap H_{i_1}) \cup \cdots \cup (A \cap H_{i_m}) = A \cap (H_{i_1} \cup \cdots \cup H_{i_m}) \subset H_{i_1} \cup \cdots \cup H_{i_m}$$

Thus $\{H_i\}$ is reducible to a finite cover $\{H_{i_1}, \ldots, H_{i_m}\}$ and therefore A is compact with respect to $\mathcal{T}$.

4. Let $(Y, \mathcal{T}^*)$ be a subspace of $(X, \mathcal{T})$ and let $A \subset Y \subset X$. Show that A is $\mathcal{T}$-compact if and only if A is $\mathcal{T}^*$-compact.

Solution:

Let $\mathcal{T}_A$ and $\mathcal{T}_A^*$ be the relative topologies on A. Then, by the preceding problem, A is $\mathcal{T}$- or $\mathcal{T}^*$-compact if and only if A is $\mathcal{T}_A$- or $\mathcal{T}_A^*$-compact; but $\mathcal{T}_A = \mathcal{T}_A^*$.

5. Prove that the following statements are equivalent:

(i) X is compact.

(ii) For every class $\{F_i\}$ of closed subsets of X, $\cap_i F_i = \emptyset$ implies $\{F_i\}$ contains a finite subclass $\{F_{i_1}, \ldots, F_{i_m}\}$ with $F_{i_1} \cap \cdots \cap F_{i_m} = \emptyset$.

Solution:

(i) $\Rightarrow$ (ii): Suppose $\cap_i F_i = \emptyset$. Then, by DeMorgan's Law,

$$X = \emptyset^c = (\cap_i F_i)^c = \cup_i F_i^c$$

so $\{F_i^c\}$ is an open cover of X, since each F_i is closed. But by hypothesis, X is compact; hence

$$\exists\ F_{i_1}^c, \ldots, F_{i_m}^c \in \{F_i^c\} \quad \text{such that} \quad X = F_{i_1}^c \cup \cdots \cup F_{i_m}^c$$

Thus by DeMorgan's Law,

$$\emptyset = X^c = (F_{i_1}^c \cup \cdots \cup F_{i_m}^c)^c = F_{i_1}^{cc} \cap \cdots \cap F_{i_m}^{cc} = F_{i_1} \cap \cdots \cap F_{i_m}$$

and we have shown that (i) $\Rightarrow$ (ii).

(ii) $\Rightarrow$ (i): Let $\{G_i\}$ be an open cover of X, i.e. $X = \cup_i G_i$. By DeMorgan's Law,

$$\emptyset = X^c = (\cup_i G_i)^c = \cap_i G_i^c$$

Since each G_i is open, $\{G_i^c\}$ is a class of closed sets and, by above, has an empty intersection. Hence by hypothesis,

$$\exists \; G_{i_1}^c, \ldots, G_{i_m}^c \in \{G_i^c\} \quad \text{such that} \quad G_{i_1}^c \cap \cdots \cap G_{i_m}^c = \emptyset$$

Thus by DeMorgan's Law,

$$X = \emptyset^c = (G_{i_1}^c \cap \cdots \cap G_{i_m}^c)^c = G_{i_1}^{cc} \cup \cdots \cup G_{i_m}^{cc} = G_{i_1} \cup \cdots \cup G_{i_m}$$

Accordingly, X is compact and so (ii) $\Rightarrow$ (i).

6. **Prove Theorem 11.4:** A topological space X is compact if and only if every class $\{F_i\}$ of closed subsets of X which satisfies the finite intersection property has, itself, a non-empty intersection.

Solution:

Utilizing the preceding problem, it suffices to show that the following statements are equivalent, where $\{F_i\}$ is any class of closed subsets of X:

(i) $\qquad\qquad F_{i_1} \cap \cdots \cap F_{i_m} \neq \emptyset \;\; \forall \, i_1, \ldots, i_m \;\; \Rightarrow \;\; \cap_i F_i \neq \emptyset$

(ii) $\qquad\qquad \cap_i F_i = \emptyset \;\; \Rightarrow \;\; \exists \, i_1, \ldots, i_m \;\; \text{s.t.} \;\; F_{i_1} \cap \cdots \cap F_{i_m} = \emptyset$

But these statements are contrapositives.

COMPACTNESS AND HAUSDORFF SPACES

7. **Prove:** Let A be a compact subset of a Hausdorff space X and suppose $p \in X \setminus A$. Then

$$\exists \; \text{open sets } G, H \quad \text{such that} \quad p \in G, \; A \subset H, \; G \cap H = \emptyset$$

Solution:

Let $a \in A$. Since $p \notin A$, $p \neq a$. By hypothesis, X is Hausdorff; hence

$$\exists \; \text{open sets } G_a, H_a \quad \text{such that} \quad p \in G_a, \; a \in H_a, \; G_a \cap H_a = \emptyset$$

Hence $A \subset \bigcup \{H_a : a \in A\}$, i.e. $\{H_a : a \in A\}$ is an open cover of A. But A is compact, so

$$\exists \; H_{a_1}, \ldots, H_{a_m} \in \{H_a\} \quad \text{such that} \quad A \subset H_{a_1} \cup \cdots \cup H_{a_m}$$

Now let $H = H_{a_1} \cup \cdots \cup H_{a_m}$ and $G = G_{a_1} \cap \cdots \cap G_{a_m}$. H and G are open since they are respectively the union and finite intersection of open sets. Furthermore, $A \subset H$ and $p \in G$ since p belongs to each G_{a_i} individually.

Lastly we claim that $G \cap H = \emptyset$. Note first that $G_{a_i} \cap H_{a_i} = \emptyset$ implies that $G \cap H_{a_i} = \emptyset$. Thus, by the distributive law,

$$G \cap H = G \cap (H_{a_1} \cup \cdots \cup H_{a_m}) = (G \cap H_{a_1}) \cup \cdots \cup (G \cap H_{a_m}) = \emptyset \cup \cdots \cup \emptyset = \emptyset$$

Thus the proof is complete.

8. Let A be a compact subset of a Hausdorff space X. Show that if $p \notin A$, then there is an open set G such that $p \in G \subset A^c$.

Solution:

By Problem 7 there exist open sets G and H such that $p \in G$, $A \subset H$ and $G \cap H = \emptyset$. Hence $G \cap A = \emptyset$, and $p \in G \subset A^c$.

9. **Prove Theorem 11.5:** Let A be a compact subset of a Hausdorff space X. Then A is closed.

Solution:

We prove, equivalently, that A^c is open. Let $p \in A^c$, i.e. $p \notin A$. Then by Problem 8 there exists an open set G_p such that $p \in G_p \subset A^c$. Hence $A^c = \bigcup \{G_p : p \in A^c\}$.

Thus A^c is open as it is the union of open sets, or, A is closed.

10. Prove Theorem 11.6:　Let A and B be disjoint compact subsets of a Hausdorff space X. Then there exist disjoint open sets G and H such that $A \subset G$ and $B \subset H$.

Solution:

　　Let $a \in A$.　Then $a \notin B$, for A and B are disjoint.　By hypothesis, B is compact; hence by Problem 1 there exist open sets G_a and H_a such that

$$a \in G_a, \quad B \subset H_a \quad \text{and} \quad G_a \cap H_a = \emptyset$$

Since $a \in G_a$, $\{G_a : a \in A\}$ is an open cover of A.　Since A is compact, we can select a finite number of the open sets, say $G_{a_1}, \ldots, G_{a_m}$, so that $A \subset G_{a_1} \cup \cdots \cup G_{a_m}$.　Furthermore, $B \subset H_{a_1} \cap \cdots \cap H_{a_m}$ since B is a subset of each individually.

　　Now let $G = G_{a_1} \cup \cdots \cup G_{a_m}$ and $H = H_{a_1} \cap \cdots \cap H_{a_m}$.　Observe, by the above, that $A \subset G$ and $B \subset H$.　In addition, G and H are open as they are the union and finite intersection respectively of open sets.　The theorem is proven if we show that G and H are disjoint.　First observe that, for each i, $G_{a_i} \cap H_{a_i} = \emptyset$ implies $G_{a_i} \cap H = \emptyset$.　Hence, by the distributive law,

$$G \cap H \;=\; (G_{a_1} \cup \cdots \cup G_{a_m}) \cap H \;=\; (G_{a_1} \cap H) \cup \cdots \cup (G_{a_m} \cap H) \;=\; \emptyset \cup \cdots \cup \emptyset \;=\; \emptyset$$

Thus the theorem is proven.

11. Prove Theorem 11.8:　Let f be a one-one continuous function from a compact space X into a Hausdorff space Y.　Then X and $f[X]$ are homeomorphic.

Solution:

　　Now $f : X \to f[X]$ is onto and, by hypothesis, one-one and continuous, so $f^{-1} : f[X] \to X$ exists. We must show that f^{-1} is continuous.　Recall that f^{-1} is continuous if, for every closed subset F of X, $(f^{-1})^{-1}[F] = f[F]$ is a closed subset of $f[X]$.　By Theorem 11.3, the closed subset F of the compact space X is also compact.　Since f is continuous, $f[F]$ is a compact subset of $f[X]$.　But the subspace $f[X]$ of the Hausdorff space Y is also Hausdorff; hence by Theorem 11.5, $f[F]$ is closed. Accordingly, f^{-1} is continuous, so $f : X \to f[X]$ is a homeomorphism, and X and $f[X]$ are homeomorphic.

12. Let $(X, \mathcal{T})$ be compact and let $(X, \mathcal{T}^*)$ be Hausdorff.　Show that if $\mathcal{T}^* \subset \mathcal{T}$, then $\mathcal{T}^* = \mathcal{T}$.

Solution:

　　Consider the function $f : (X, \mathcal{T}) \to (X, \mathcal{T}^*)$ defined by $f(x) = x$, i.e. the identity function on X. Now f is one-one and onto.　Furthermore, f is continuous since $\mathcal{T}^* \subset \mathcal{T}$.　Thus by the preceding problem, f is a homeomorphism and therefore $\mathcal{T}^* = \mathcal{T}$.

SEQUENTIALLY AND COUNTABLY COMPACT SETS

13. Show that a continuous image of a sequentially compact set is sequentially compact.

Solution:

　　Let $f : X \to Y$ be a continuous function and let A be a sequentially compact subset of X.　We want to show that $f[A]$ is a sequentially compact subset of Y.　Let $\langle b_1, b_2, \ldots \rangle$ be a sequence in $f[A]$.　Then

$$\exists\; a_1, a_2, \ldots \in A \qquad \text{such that} \qquad f(a_n) = b_n, \;\; \forall\, n \in \mathbf{N}$$

But A is sequentially compact, so the sequence $\langle a_1, a_2, \ldots \rangle$ contains a subsequence $\langle a_{i_1}, a_{i_2}, \ldots \rangle$ which converges to a point $a_0 \in A$.　Now f is continuous and hence sequentially continuous, so

$$\langle f(a_{i_1}), f(a_{i_2}), \ldots \rangle \;=\; \langle b_{i_1}, b_{i_2}, \ldots \rangle \quad \text{converges to} \quad f(a_0) \in f[A]$$

Thus $f[A]$ is sequentially compact.

14. Let $\mathcal{T}$ be the topology on X which consists of $\emptyset$ and the complements of countable subsets of X. Show that every infinite subset of X is not sequentially compact.

Solution:

Recall (Example 7.3, Page 71) that a sequence in $(X, \mathcal{T})$ converges iff it is of the form

$$\langle a_1, a_2, \ldots, a_{n_0}, p, p, p, \ldots \rangle$$

that is, is constant from some term on. Hence if A is an infinite subset of X, there exists a sequence $\langle b_n \rangle$ in A with distinct terms. Thus $\langle b_n \rangle$ does not contain any convergent subsequence, and A is not sequentially compact.

15. Show that: (i) a continuous image of a countably compact set need not be countably compact; (ii) a closed subset of a countably compact space is countably compact.

Solution:

(i) Let $X = (\mathbf{N}, \mathcal{T})$ where $\mathcal{T}$ is the topology on the positive integers $\mathbf{N}$ generated by the sets $\{1, 2\}, \{3, 4\}, \{5, 6\}, \ldots$. By Example 6.3, X is countably compact. Let $Y = (\mathbf{N}, \mathcal{D})$ where $\mathcal{D}$ is the discrete topology on $\mathbf{N}$. Now Y is not countably compact. On the other hand, the function $f : X \to Y$ which maps $2n$ and $2n - 1$ onto n for $n \in \mathbf{N}$ is continuous and maps the countably compact set X onto the non-countably compact set Y.

(ii) Suppose X is countably compact and suppose F is a closed subset of X. Let A be an infinite subset of F. Since $F \subset X$, A is also an infinite subset of X. By hypothesis, X is countably compact; then A has an accumulation point $p \in X$. Since $A \subset F$, p is also an accumulation point of F. But F is closed and so contains its accumulation points; hence $p \in F$. We have shown that any infinite subset A of F has an accumulation point $p \in F$, that is, that F is countably compact.

16. Prove: Let X be compact. Then X is also countably compact.

Solution:

Let A be a subset of X with no accumulation points in X. Then each point $p \in X$ belongs to an open set G_p which contains at most one point of A. Observe that the class $\{G_p : p \in X\}$ is an open cover of the compact set X and, hence, contains a finite subcover, say $\{G_{p_1}, \ldots, G_{p_m}\}$.

Hence $A \subset X \subset G_{p_1} \cup \cdots \cup G_{p_m}$

But each G_{p_i} contains at most one point of A; hence A, a subset of $G_{p_1} \cup \cdots \cup G_{p_m}$, can contain at most m points, i.e. A is finite. Accordingly, every infinite subset of X contains an accumulation point in X, i.e. X is countably compact.

17. Prove: Let X be sequentially compact. Then X is also countably compact.

Solution:

Let A be any infinite subset of X. Then there exists a sequence $\langle a_1, a_2, \ldots \rangle$ in A with distinct terms. Since X is sequentially compact, the sequence $\langle a_n \rangle$ contains a subsequence $\langle a_{i_1}, a_{i_2}, \ldots \rangle$ (also with distinct terms) which converges to a point $p \in X$. Hence every open neighborhood of p contains an infinite number of the terms of the convergent subsequence $\langle a_{i_n} \rangle$. But the terms are distinct; hence every open neighborhood of p contains an infinite number of points in A. Accordingly, $p \in X$ is an accumulation point of A. In other words, X is countably compact.

Remark: Note that Problems 16 and 17 imply Theorem 11.9.

18. Prove: Let $A \subset X$ be sequentially compact. Then every countable open cover of A is reducible to a finite cover.

Solution:

We may assume A is infinite, for otherwise the proof is trivial. We prove the contrapositive, i.e. assume $\exists$ a countable open cover $\{G_i : i \in \mathbf{N}\}$ with no finite subcover. We define the sequence $\langle a_1, a_2, \ldots \rangle$ as follows.

Let n_1 be the smallest positive integer such that $A \cap G_{n_1} \neq \emptyset$. Choose $a_1 \in A \cap G_{n_1}$. Let n_2 be the least positive integer larger than n_1 such that $A \cap G_{n_2} \neq \emptyset$. Choose

$$a_2 \in (A \cap G_{n_2}) \setminus (A \cap G_{n_1})$$

Such a point always exists, for otherwise G_{n_1} covers A. Continuing in this manner, we obtain the sequence $\langle a_1, a_2, \ldots \rangle$ with the property that, for every $i \in \mathbf{N}$,

$$a_i \in A \cap G_{n_i}, \quad a_i \notin \cup_{j=1}^{n-1} (A \cap G_{n_j}) \quad \text{and} \quad n_i > n_{i-1}$$

We claim that $\langle a_i \rangle$ has no convergent subsequence in A. Let $p \in A$. Then

$$\exists \ G_{i_0} \in \{G_i\} \quad \text{such that} \quad p \in G_{i_0}$$

Now $A \cap G_{i_0} \neq \emptyset$, since $p \in A \cap G_{i_0}$; hence

$$\exists \ j_0 \in \mathbf{N} \quad \text{such that} \quad G_{n_{j_0}} = G_{i_0}$$

But by the choice of the sequence $\langle a_1, a_2, \ldots \rangle$

$$i > j_0 \quad \Rightarrow \quad a_i \notin G_{i_0}$$

Accordingly, since G_{i_0} is an open set containing p, no subsequence of $\langle a_i \rangle$ converges to p. But p was arbitrary, so A is not sequentially compact.

COMPACTNESS IN METRIC SPACES

19. Prove Lemma 11.17: Let A be a sequentially compact subset of a metric space X. Then A is totally bounded.

Solution:

We prove the contrapositive of the above statement, i.e. if A is not totally bounded, then A is not sequentially compact. If A is not totally bounded then there exists an $\epsilon > 0$ such that A possesses no (finite) ϵ-net. Let $a_1 \in A$. Then there exists a point $a_2 \in A$ with $d(a_1, a_2) \geq \epsilon$, for otherwise $\{a_1\}$ would be an ϵ-net for A. Similarly, there exists a point $a_3 \in A$ with $d(a_1, a_3) \geq \epsilon$ and $d(a_2, a_3) \geq \epsilon$, for otherwise $\{a_1, a_2\}$ would be an ϵ-net for A. Continuing in this manner, we arrive at a sequence $\langle a_1, a_2, \ldots \rangle$ with the property that $d(a_i, a_j) \geq \epsilon$ for $i \neq j$. Thus the sequence $\langle a_n \rangle$ cannot contain any subsequence which converges. In other words, A is not sequentially compact.

20. Prove Lemma (Lebesgue) 11.18: Let $\mathcal{A} = \{G_i\}$ be an open cover of a sequentially compact set A. Then $\mathcal{A}$ has a (positive) Lebesgue number.

Solution:

Suppose $\mathcal{A}$ does not have a Lebesgue number. Then for each positive integer $n \in \mathbf{N}$ there exists a subset B_n of A with the property that

$$0 < d(B_n) < 1/n \quad \text{and} \quad B_n \not\subset G_i \quad \text{for every } G_i \text{ in } \mathcal{A}$$

For each $n \in \mathbf{N}$, choose a point $b_n \in B_n$. Since A is sequentially compact, the sequence $\langle b_1, b_2, \ldots \rangle$ contains a subsequence $\langle b_{i_1}, b_{i_2}, \ldots \rangle$ which converges to a point $p \in A$.

Since $p \in A$, p belongs to an open set G_p in the cover $\mathcal{A}$. Hence there exists an open sphere $S(p, \epsilon)$, with center p and radius ϵ, such that $p \in S(p, \epsilon) \subset G_p$. Since $\langle b_{i_n} \rangle$ converges to p, there exists a positive integer i_{n_0} such that

$$d(p, b_{i_{n_0}}) < \tfrac{1}{2}\epsilon, \quad b_{i_{n_0}} \in B_{i_{n_0}} \quad \text{and} \quad d(B_{i_{n_0}}) < \tfrac{1}{2}\epsilon$$

Using the Triangle Inequality we get $B_{i_{n_0}} \subset S(p, \epsilon) \subset G_p$. But this contradicts the fact that $B_{i_{n_0}} \not\subset G_i$ for every G_i in the cover $\mathcal{A}$. Accordingly $\mathcal{A}$ does possess a Lebesgue number.

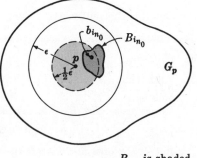

$B_{i_{n_0}}$ is shaded

21. Prove: Let A be a countably compact subset of a metric space X. Then A is also sequentially compact.

Solution:

Let $\langle a_1, a_2, \ldots \rangle$ be a sequence in A. If the set $B = \{a_1, a_2, \ldots\}$ is finite, then one of the points, say a_{i_0}, satisfies $a_{i_0} = a_j$ for infinitely many $j \in \mathbf{N}$. Hence $\langle a_{i_0}, a_{i_0}, \ldots \rangle$ is a subsequence of $\langle a_n \rangle$ which converges to the point a_{i_0} in A.

On the other hand, suppose $B = \{a_1, a_2, \ldots\}$ is infinite. By hypothesis, A is countably compact. Hence the infinite subset B of A contains an accumulation point p in A. But X is a metric space; hence we can choose a subsequence $\langle a_{i_1}, a_{i_2}, \ldots \rangle$ of the sequence $\langle a_n \rangle$ which converges to the point p in A. In other words, A is sequentially compact.

22. Prove Theorem 11.14: Let A be a subset of a metric space X. Then the following are equivalent: (i) A is compact, (ii) A is countably compact, and (iii) A is sequentially compact.

Solution:

Recall (see Theorem 11.8) that (i) implies (ii) in every topological space; hence it is true for a metric space. In the preceding problem we proved that (ii) implies (iii). Accordingly, the theorem is proven if we show that (iii) implies (i).

Let A be sequentially compact, and let $\mathcal{A} = \{G_i\}$ be an open cover of A. We want to show that A is compact, i.e. that $\mathcal{A}$ possesses a finite subcover. By hypothesis, A is sequentially compact; hence, by Lemma 11.18, the cover $\mathcal{A}$ possesses a Lebesgue number $\delta > 0$. In addition, by Lemma 11.17, A is totally bounded. Hence there is a decomposition of A into a finite number of subsets, say $B_1, \ldots, B_m$, with $d(B_i) < \delta$. But δ is a Lebesgue number for $\mathcal{A}$; hence there are open sets $G_{i_1}, \ldots, G_{i_m} \in \mathcal{A}$ such that

$$B_1 \subset G_{i_1}, \quad \ldots, \quad B_m \subset G_{i_m}$$

Accordingly,
$$A \subset B_1 \cup B_2 \cup \cdots \cup B_m \subset G_{i_1} \cup G_{i_2} \cup \cdots \cup G_{i_m}$$

Thus $\mathcal{A}$ possesses a finite subcover $\{G_{i_1}, \ldots, G_{i_m}\}$, i.e. A is compact.

23. Let A be a compact subset of a metric space (X, d). Show that for any $B \subset X$ there is a point $p \in A$ such that $d(p, B) = d(A, B)$.

Solution:

Let $d(A, B) = \epsilon$. Since $d(A, B) = \inf \{d(a, b) : a \in A, \ b \in B\}$, for every positive integer $n \in \mathbf{N}$,

$$\exists \ a_n \in A, \ b_n \in B \quad \text{such that} \quad \epsilon \leq d(a_n, b_n) < \epsilon + 1/n$$

Now A is compact and hence sequentially compact; so the sequence $\langle a_1, a_2, \ldots \rangle$ has a subsequence which converges to a point $p \in A$. We claim that $d(p, B) = d(A, B) = \epsilon$.

Suppose $d(p, B) > \epsilon$, say $d(p, B) = \epsilon + \delta$ where $\delta > 0$. Since a subsequence of $\langle a_n \rangle$ converges to p,

$$\exists \ n_0 \in \mathbf{N} \quad \text{such that} \quad d(p, a_{n_0}) < \tfrac{1}{2}\delta \ \text{ and } \ d(a_{n_0}, b_{n_0}) < \epsilon + 1/n_0 < \epsilon + \tfrac{1}{2}\delta$$

Then
$$d(p, a_{n_0}) + d(a_{n_0}, b_{n_0}) < \tfrac{1}{2}\delta + \epsilon + \tfrac{1}{2}\delta = \epsilon + \delta = d(p, B) \leq d(p, b_{n_0})$$

But this contradicts the Triangle Inequality; hence $d(p, B) = d(A, B)$.

24. Let A be a compact subset of a metric space (X, d) and let B be a closed subset of X such that $A \cap B = \emptyset$. Show that $d(A, B) > 0$.

Solution:

Suppose $d(A, B) = 0$. Then, by the preceding problem,

$$\exists \ p \in A \quad \text{such that} \quad d(p, B) = d(A, B) = 0$$

But B is closed and therefore contains all points whose distance from B is zero. Thus $p \in B$ and so $p \in A \cap B$. But this contradicts the hypothesis; hence $d(A, B) > 0$.

25. Prove: Let f be a continuous function from a compact metric space (X, d) into a metric space (Y, d^*). Then f is *uniformly continuous*, i.e. for every $\epsilon > 0$ there exists a $\delta > 0$ such that

$$d(x, y) < \delta \quad \Rightarrow \quad d^*(f(x), f(y)) < \epsilon$$

(*Remark:* Uniform continuity is a stronger condition than continuity, in that the δ above depends only upon the ϵ and not also on any particular point.)

Solution:

Let $\epsilon > 0$. Since f is continuous, for each point $p \in X$ there exists an open sphere $S(p, \delta_p)$ such that

$$x \in S(p, \delta_p) \quad \Rightarrow \quad f(x) \in S(f(p), \tfrac{1}{2}\epsilon)$$

Observe that the class $\mathcal{A} = \{S(p, \delta_p) : p \in X\}$ is an open cover of X. By hypothesis, X is compact and hence also sequentially compact. Therefore the cover $\mathcal{A}$ possesses a Lebesgue number $\delta > 0$.

Now let $x, y \in X$ with $d(x, y) < \delta$. But $d(x, y) = d\{x, y\} < \delta$ implies $\{x, y\}$ is contained in a member $S(p_0, \delta_{p_0})$ of the cover A. Now

$$x, y \in S(p_0, \delta_{p_0}) \quad \Rightarrow \quad f(x), f(y) \in S(f(p_0), \tfrac{1}{2}\epsilon)$$

But the sphere $S(f(p_0), \tfrac{1}{2}\epsilon)$ has diameter ϵ. Accordingly,

$$d(x, y) < \delta \quad \Rightarrow \quad d^*(f(x), f(y)) < \epsilon$$

In other words, f is uniformly continuous.

Supplementary Problems

COMPACT SPACES

26. Prove: If E is compact and F is closed, then $E \cap F$ is compact.

27. Let $A_1, \ldots, A_m$ be compact subsets of a topological space X. Show that $A_1 \cup \cdots \cup A_m$ is also compact.

28. Prove that compactness is a topological property.

29. Prove Proposition 11.11: The class $\mathcal{T}_\infty$ is a topology on X_∞ and $(X_\infty, \mathcal{T}_\infty)$ is a compactification of $(X, \mathcal{T})$. (Here $(X_\infty, \mathcal{T}_\infty)$ is the Alexandrov one-point compactification of $(X, \mathcal{T})$.)

30. Prove Theorem 11.12: If $(X, \mathcal{T})$ is a locally compact Hausdorff space, then $(X_\infty, \mathcal{T}_\infty)$ is a compact Hausdorff space.

SEQUENTIALLY AND COUNTABLY COMPACT SPACES

31. Show that sequential compactness is a topological property.

32. Prove: A closed subset of a sequentially compact space is sequentially compact.

33. Show that countable compactness is a topological property.

34. Suppose $(X, \mathcal{T})$ is countably compact and $\mathcal{T}^* \subseteq \mathcal{T}$. Show that $(X, \mathcal{T}^*)$ is also countably compact.

35. Prove: Let X be a topological space such that every countable open cover of X is reducible to a finite cover. Then X is countably compact.

36. Prove: Let X be a T_1-space. Then X is countably compact if and only if every countable open cover of X is reducible to a finite cover.

37. Prove: Let X be a second countable T_1-space. Then X is compact if and only if X is countably compact.

TOTALLY BOUNDED SETS

38. Prove Proposition 11.15: A set A is totally bounded if and only if for every $\epsilon > 0$ there exists a decomposition of A into a finite number of sets each with diameter less than ϵ.

39. Prove Proposition 11.16: Totally bounded sets are bounded.

40. Show that every subset of a totally bounded set is totally bounded.

41. Show that if A is totally bounded then $\bar{A}$ is also totally bounded.

42. Prove: Every totally bounded metric space is separable.

COMPACTNESS AND METRIC SPACES

43. Prove: A compact subset of a metric space X is closed and bounded.

44. Prove: Let $f: X \to Y$ be a continuous function from a compact space X into a metric space Y. Then $f[X]$ is a bounded subset of Y.

45. Prove: A subset A of the real line R is compact if and only if A is closed and bounded.

46. Prove: Let A be a compact subset of a metric space X. Then the derived set A' of A is compact.

47. Prove: The Hilbert cube $\mathbf{I} = \{\langle a_n \rangle : 0 \leq a_n \leq 1/n\}$ is a compact subset of $\mathbf{R}^\infty$.

48. Prove: Let A and B be compact subsets of a metric space X. Then there exist $a \in A$ and $b \in B$ such that $d(a, b) = d(A, B)$.

LOCALLY COMPACT SPACES

49. Show that local compactness is a topological property.

50. Show that every discrete space is locally compact.

51. Show that every indiscrete space is locally compact.

52. Show that the plane $\mathbf{R}^2$ with the usual topology is locally compact.

53. Prove: Let A be a closed subset of a locally compact space $(X, \mathcal{T})$. Then A with the relative topology is locally compact.

Chapter 12

Product Spaces

PRODUCT TOPOLOGY

Let $\{X_i : i \in I\}$ be any class of sets and let X denote the Cartesian product of these sets, i.e. $X = \prod_i X_i$. Note that X consists of all points $p = \langle a_i : i \in I \rangle$ where $a_i \in X_i$. Recall that, for each $j_0 \in I$, we defined the *projection* π_{j_0} from the product set X to the coordinate space X_{j_0}, i.e. $\pi_{j_0} : X \to X_{j_0}$, by

$$\pi_{j_0}(\langle a_i : i \in I \rangle) = a_{j_0}$$

These projections are used to define the product topology.

> **Definition:** Let $\{(X_i, \mathcal{T}_i)\}$ be a collection of topological spaces and let X be the product of the sets X_i, i.e. $X = \prod_i X_i$. The coarsest topology $\mathcal{T}$ on X with respect to which all the projections $\pi_i : X \to X_i$ are continuous is called the (Tychonoff) *product topology*. The product set X with the product topology $\mathcal{T}$, i.e. $(X, \mathcal{T})$, is called the *product topological space* or, simply, *product space*.

In other words, the *product topology* $\mathcal{T}$ on the product set $X = \prod_i X_i$ is the topology generated by the projections (see Chapter 7).

Example 1.1: Consider the Cartesian plane $\mathbf{R}^2 = \mathbf{R} \times \mathbf{R}$. Recall that the inverses $\pi_1^{-1}(a, b)$ and $\pi_2^{-1}(a, b)$ are infinite open strips which form a subbase for the usual topology on $\mathbf{R}^2$.

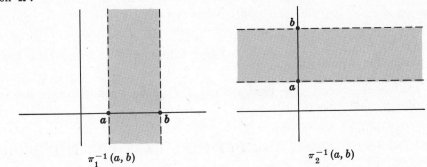

$$\pi_1^{-1}(a, b) \qquad\qquad \pi_2^{-1}(a, b)$$

Thus the usual topology on $\mathbf{R}^2$ is the topology generated by the projections from $\mathbf{R}^2$ into $\mathbf{R}$.

In view of the above definition, we can state the result in Example 1.1 as follows:

Theorem 12.1: The usual topology on $\mathbf{R}^2 = \mathbf{R} \times \mathbf{R}$ is the product topology.

Example 1.2: Let $\{X_i : i \in I\}$ be a collection of Hausdorff spaces and let X be the product space, i.e. $X = \prod_i X_i$. We show that X is also a Hausdorff space. Let $p = \langle a_i : i \in I \rangle$ and $q = \langle b_i : i \in I \rangle$ be distinct points in X. Then p and q must differ in at least one coordinate space, say X_{j_0}, i.e. $a_{j_0} \neq b_{j_0}$. By hypothesis, X_{j_0} is Hausdorff; hence there exist disjoint open subsets G and H of X_{j_0} such that $a_{j_0} \in G$ and $b_{j_0} \in H$. By definition of the product space, the projection $\pi_{j_0} : X \to X_{j_0}$ is continuous.   Accordingly $\pi_{j_0}^{-1}[G]$ and $\pi_{j_0}^{-1}[H]$ are open disjoint subsets of X containing p and q respectively.   Hence X is also a Hausdorff space.

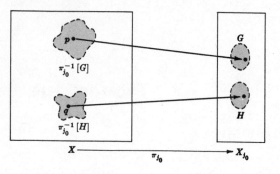

BASE FOR A FINITE PRODUCT TOPOLOGY

The Cartesian product $A \times B$ of two open finite intervals A and B is an open rectangle in $\mathbf{R}^2$ as illustrated below.

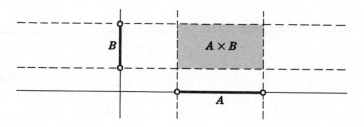

As noted previously, the open rectangles form a base for the usual topology on $\mathbf{R}^2$ which is also the product topology on $\mathbf{R}^2$.   A similar statement is true for every finite product topology.   Namely,

Proposition 12.2:   Let $X_1, \cdots, X_m$ be a finite number of topological spaces, and let X be the product space, i.e. $X = X_1 \times \cdots \times X_m$.   Then the following subsets of the product space X,

$$G_1 \times G_2 \times \cdots \times G_m$$

where G_i is an open subset of X_i, form a base for the product topology on X.

As we shall see in the next section, the above proposition is not true in the case of an infinite product space.

DEFINING SUBBASE AND DEFINING BASE FOR THE PRODUCT TOPOLOGY

Let $\{X_i : i \in I\}$ be a collection of topological spaces and let X denote the product space, i.e. $X = \prod_i X_i$.   If G_{j_0} is an open subset of the coordinate space X_{j_0}, then $\pi_{j_0}^{-1}[G_{j_0}]$ consists of all points $p = \langle a_i : i \in I \rangle$ in X such that $\pi_{j_0}(p) \in G_{j_0}$.   In other words,

$$\pi_{j_0}^{-1}[G_{j_0}] \;=\; \prod \{X_i : i \neq j_0\} \times G_{j_0}$$

In particular, if we have a denumerable collection of topological spaces, say $\{X_1, X_2, \ldots\}$, then the product space

$$X \;=\; \prod_{n=1}^{\infty} X_n \;=\; X_1 \times X_2 \times X_3 \times \cdots$$

consists of all sequences

$$p \;=\; \langle a_1, a_2, a_3, \ldots \rangle \qquad \text{where } a_n \in X_n$$

and, furthermore,

$$\pi_{j_0}^{-1}[G_{j_0}] \;=\; X_1 \times \cdots \times X_{j_0-1} \times G_{j_0} \times X_{j_0+1} \times \cdots$$

By definition, the product topology on X is the "smallest", i.e. coarsest, topology on X with respect to which all the projections are continuous, i.e. the topology generated by the projections. Accordingly, the inverse projections of open sets in the coordinate spaces form a subbase for the product topology (Theorem 7.8). Namely,

Theorem 12.3: The class of subsets of a product space $X = \prod_i X_i$ of the form

$$\pi_{j_0}^{-1}[G_{j_0}] \quad = \quad \prod \{X_i : i \neq j_0\} \times G_{j_0}$$

where G_{j_0} is an open subset of the coordinate space X_{j_0}, is a subbase and is called the *defining subbase* for the product topology.

Furthermore, since finite intersections of the subbase elements form a base for the topology, we also have

Theorem 12.4: The class of subsets of a product space $X = \prod_i X_i$ of the form

$$\pi_{j_1}^{-1}[G_{j_1}] \cap \cdots \cap \pi_{j_m}^{-1}[G_{j_m}] \quad = \quad \prod \{X_i : i \neq j_1, \ldots, j_m\} \times G_{j_1} \times \cdots \times G_{j_m}$$

where G_{j_k} is an open subset of the coordinate space X_{j_k}, is a base and is called the *defining base* for the product topology.

Using the above properties, we can prove the following central facts about product spaces.

Theorem 12.5: A function f from a topological space Y into a product space $X = \prod_i X_i$ is continuous if and only if, for every projection π_i, the composition mapping $\pi_i \circ f : Y \to X_i$ is also continuous.

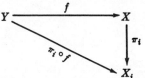

Theorem 12.6: Every projection $\pi_i : X \to X_i$ on a product space $X = \prod_i X_i$ is both open and continuous, i.e. bicontinuous.

Theorem 12.7: A sequence $p_1, p_2, \ldots$ of points in a product space $X = \prod_i X_i$ converges to the point q in X if and only if, for every projection $\pi_i : X \to X_i$, the sequence $\pi_i(p_1), \pi_i(p_2), \ldots$ converges to $\pi_i(q)$ in the coordinate space X_i.

In other words, if $p_1 = \langle a_{1_i} \rangle$, $p_2 = \langle a_{2_i} \rangle$, $\ldots$ and $q = \langle b_i \rangle$ are points in a product space $X = \prod_i X_i$, then

$$p_n \to q \text{ in } X \quad \text{iff} \quad a_{n_i} \to b_i \text{ in every coordinate space } X_i$$

EXAMPLE OF A PRODUCT SPACE

Let R_i denote a copy of **R**, the set of real numbers with the usual topology, indexed by the closed unit interval $I = [0, 1]$. Consider the product space $X = \prod \{R_i : i \in I\}$. We can represent X graphically as in the adjoining figure. Here the horizontal axis denotes the index set $I = [0, 1]$, and each vertical line through a point, say j_0, in I denotes the coordinate space R_{j_0}. Consider an element $p = \langle a_i : i \in I \rangle$ in the product space X. Observe that p assigns to each number $i \in I$ the real number a_i, i.e. p is a real-valued function defined on the index set $I = [0, 1]$. In other words, the product space X is the class of all real-valued functions defined on I, i.e.,

$$X = \{p : p : I \to \mathbf{R}\}$$

Some of the elements of X are also displayed in the figure.

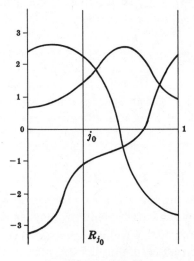

Members of X

We now describe one of the members of the defining subbase $\mathcal{S}$ for the product topology on X. Recall that $\mathcal{S}$ consists of all the subsets of X of the form

$$\pi_{j_0}^{-1}[G_{j_0}] = \prod \{R_i : i \neq j_0\} \times G_{j_0}$$

where G_{j_0} is an open subset of the coordinate space R_{j_0}. Suppose G_{j_0} is the open interval $(1,2)$. Then $\pi_{j_0}^{-1}[G_{j_0}]$ consists of all points $p = \langle a_i : i \in I \rangle$ in X such that $a_{j_0} \in G_{j_0} = (1,2)$, i.e. all functions $p : I \to \mathbf{R}$ such that $1 < p(j_0) < 2$. Graphically, $\pi_{j_0}^{-1}[G_{j_0}]$ consists of all those functions passing through the open interval $G_{j_0} = (1,2)$ on the vertical line representing the coordinate space R_{j_0}, as illustrated in the adjacent diagram.

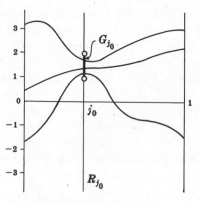

Members of $\pi_{j_0}^{-1}[G_{j_0}]$

Lastly we describe one of the open sets, say B, of the defining base $\mathcal{B}$ for the product topology on X. Recall that B is the intersection of a finite number of the members of the defining subbase $\mathcal{S}$ for the product topology, say,

$$\begin{aligned} B &= \pi_{j_1}^{-1}[G_{j_1}] \cap \pi_{j_2}^{-1}[G_{j_2}] \cap \pi_{j_3}^{-1}[G_{j_3}] \\ &= \prod \{R_i : i \neq j_1, j_2, j_3\} \times G_{j_1} \times G_{j_2} \times G_{j_3} \end{aligned}$$

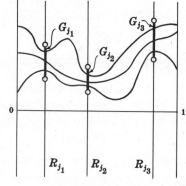

Graphically, then, B consists of all functions passing through the open sets G_{j_1}, G_{j_2} and G_{j_3} which lie on the vertical lines denoting the coordinate spaces R_{j_1}, R_{j_2} and R_{j_3}. Some of the members of B are displayed in the diagram on the right.

Members of B

Consider the following proposition.

Proposition 12.8: Let $\{(X_i, \mathcal{T}_i)\}$ be a collection of topological spaces and let X denote the product of the sets X_i, i.e. $X = \prod_i X_i$. Then the subsets of X of the form

$$\prod \{G_i : i \in I\}$$

where G_i is an open subset of the coordinate space X_i, form a base for a topology on the product set X.

Remark: The topology on the product set $X = \prod_i X_i$ appearing in Proposition 12.8 is not always identical to the product topology on X as defined in this chapter. On the other hand, Proposition 12.2 shows that the two topologies coincide in the case of a finite product space. Historically, the topology in Proposition 12.8 appeared and was investigated first. Tychonoff is credited with defining the (Tychonoff) product topology and proving for it one of the most important and useful theorems in topology, the Tychonoff Product Theorem. It is because of this theorem that the product topology is considered the "right" topology on the product set.

TYCHONOFF'S PRODUCT THEOREM

A property P of a topological space is said to be *product invariant* if a product space $X = \prod_i X_i$ possesses P whenever each coordinate space X_i possesses P. For example, the property P of being a Hausdorff space is product invariant since, in view of Example 1.2, the product of Hausdorff spaces is also Hausdorff. The celebrated Tychonoff Product Theorem states that compactness is also a product invariant property:

Theorem (Tychonoff) 12.9: The product of compact spaces is compact.

The proof of Theorem 12.9 which is given in the solved problem section relies on the set-theoretic lemma which follows; the proof of the lemma requires Zorn's Lemma. This is not too surprising since it has been shown that the Tychonoff Product Theorem is, in fact, equivalent to Zorn's Lemma.

Lemma 12.10: Let $\mathcal{A}$ be a class of subsets of a set X with the finite intersection property. Consider the collection **P** of all superclasses of $\mathcal{A}$ which have the finite intersection property. Then **P**, ordered by class inclusion, contains a maximal element $\mathcal{M}$.

Recall (see Chapter 11) that a class $\mathcal{A} = \{A_i : i \in I\}$ possesses the finite intersection property iff every finite subclass of $\mathcal{A}$, say $\{A_{i_1}, \ldots, A_{i_m}\}$, has a non-empty intersection, i.e. $A_{i_1} \cap \cdots \cap A_{i_m} \neq \emptyset$.

METRIC PRODUCT SPACES

Let $\{(X_i, d_i)\}$ be a collection of metric spaces and let X denote the product of the sets X_i, i.e. $X = \prod_i X_i$. Since the metric spaces (X_i, d_i) are also topological spaces, we can speak of the product topology on X. On the other hand, it is natural to ask whether or not it is possible to define a metric d on the product set X so that the topology on X induced by the metric d is identical to this product topology. The next two propositions give a positive answer to this question in the case of a finite or a denumerable collection of metric spaces. The metrics given are in no way unique.

Proposition 12.11: Let $(X_1, d_1), \ldots, (X_m, d_m)$ be metric spaces and let $p = \langle a_1, \ldots, a_m \rangle$ and $q = \langle b_1, \ldots, b_m \rangle$ be arbitrary points in the product set $X = \prod_{i=1}^{m} X_i$.

Then each of the following functions is a metric on the product set X:

$$d(p, q) = \sqrt{d_1(a_1, b_1)^2 + \cdots + d_m(a_m, b_m)^2}$$
$$d(p, q) = \max\{d_1(a_1, b_1), \ldots, d_m(a_m, b_m)\}$$
$$d(p, q) = d_1(a_1, b_1) + \cdots + d_m(a_m, b_m)$$

Moreover, the topology on X induced by each of the above metrics is the product topology.

Proposition 12.12: Let $\{(X_1, d_1), (X_2, d_2), \ldots\}$ be a denumerable collection of metric spaces and let $p = \langle a_1, a_2, \ldots \rangle$ and $q = \langle b_1, b_2, \ldots \rangle$ be arbitrary points in the product set $X = \prod_{n=1}^{\infty} X_n$. Then the function d defined by

$$d(p, q) = \sum_{n=1}^{\infty} \frac{1}{2^n} \frac{d_n(a_n, b_n)}{1 + d_n(a_n, b_n)}$$

is a metric on the product set X and the topology induced by d is the product topology.

CANTOR SET

We now construct a set T of real numbers, called the *Cantor* or *ternary* set, which has some remarkable properties. Trisect the closed unit interval $I = [0, 1]$ at the points $\frac{1}{3}$ and $\frac{2}{3}$ and then delete the open interval $(\frac{1}{3}, \frac{2}{3})$, called the "middle third". Let T_1 denote the remainder of the points in I, i.e.,

$$T_1 = [0, \tfrac{1}{3}] \cup [\tfrac{2}{3}, 1]$$

We now trisect each of the two segments in T_1 at $\frac{1}{9}$ and $\frac{2}{9}$ and $\frac{7}{9}$ and $\frac{8}{9}$, and then delete the "middle third" from each segment, i.e. $(\frac{1}{9}, \frac{2}{9})$ and $(\frac{7}{9}, \frac{8}{9})$. Let T_2 denote the remainder of the points in T_1, i.e.,

$$T_2 \;=\; [0, \tfrac{1}{9}] \;\cup\; [\tfrac{2}{9}, \tfrac{1}{3}] \;\cup\; [\tfrac{2}{3}, \tfrac{7}{9}] \;\cup\; [\tfrac{8}{9}, 1]$$

If we continue in this manner we obtain a descending sequence of sets

$$T_1 \;\supset\; T_2 \;\supset\; T_3 \;\supset\; \cdots$$

where T_m consists of the points in T_{m-1} excluding the "middle thirds", as shown.

Observe that T_m consists of 2^m disjoint closed intervals and, if we number them consecutively from left to right, we can speak of the odd or even intervals in T_m.

The Cantor set T is the intersection of these sets, i.e. $T = \cap \{T_i : i \in \mathbf{N}\}$.

PROPERTIES OF THE CANTOR SET

We define a function f on the Cantor set T as follows:

$$f(x) \;=\; \langle a_1, a_2, \ldots \rangle$$

where

$$a_m \;=\; \begin{cases} 0 & \text{if } x \text{ belongs to an odd interval in } T_m \\ 2 & \text{if } x \text{ belongs to an even interval in } T_m \end{cases}$$

The above sequence corresponds precisely to the "decimal expansion" of x written to the base 3, i.e. where

$$x \;=\; a_1\left(\frac{1}{3}\right) + a_2\left(\frac{1}{9}\right) + a_3\left(\frac{1}{27}\right) + \cdots + a_m\left(\frac{1}{3^m}\right) + \cdots$$

Consider now a discrete space of two elements, say $A = \{0, 2\}$, and let A_n denote a copy of A indexed by $i \in \mathbf{N}$, the positive integers.

Proposition 12.13:   The Cantor set T is homeomorphic to the product space

$$X \;=\; \prod \{A_i : i \in \mathbf{N}\}$$

In particular, the function $f : T \to X$ defined above is a homeomorphism.

Remark: The Cantor set T possesses the following interesting properties:

(1) T is non-denumerable.   For T is equivalent to the set of sequences $\langle a_1, a_2, \ldots \rangle$, where $a_i = 0$ or 2, which has cardinality $2^{\aleph_0} = \mathbf{c}$.

(2) T has "measure" zero.   For the measure of the complement of T relative to $I = [0, 1]$, i.e. the union of the middle thirds, equals

$$\frac{1}{3} + \frac{2}{9} + \frac{4}{27} + \frac{8}{81} + \cdots \;=\; 1$$

But the measure of $I = [0, 1]$ is also 1.   Hence the measure of T is zero.

Solved Problems

PRODUCT SPACES

1. Consider the topology $\mathcal{T} = \{X, \emptyset, \{a\}, \{b, c\}\}$ on $X = \{a, b, c\}$ and the topology $\mathcal{T}^* = \{Y, \emptyset, \{u\}\}$ on $Y = \{u, v\}$.

   (i) Determine the defining subbase $\mathcal{S}$ of the product topology on $X \times Y$.

   (ii) Determine the defining base $\mathcal{B}$ for the product topology on $X \times Y$.

   Solution:
   Note first that $X \times Y = \{\langle a, u \rangle, \langle a, v \rangle, \langle b, u \rangle, \langle b, v \rangle, \langle c, u \rangle, \langle c, v \rangle\}$

   is the product set on which the product topology is defined.

   (i) The defining subbase $\mathcal{S}$ is the class of inverse sets $\pi_x^{-1}[G]$ and $\pi_y^{-1}[H]$ where G is an open subset of X and H is an open subset of Y. Computing, we have

   $$\pi_x^{-1}[X] = \pi_y^{-1}[Y] = X \times Y$$
   $$\pi_x^{-1}[\emptyset] = \pi_y^{-1}[\emptyset] = \emptyset$$
   $$\pi_x^{-1}[\{a\}] = \{\langle a, u \rangle, \langle a, v \rangle\}$$
   $$\pi_x^{-1}[\{b, c\}] = \{\langle b, u \rangle, \langle b, v \rangle, \langle c, u \rangle, \langle c, v \rangle\}$$
   $$\pi_y^{-1}[\{u\}] = \{\langle a, u \rangle, \langle b, u \rangle, \langle c, u \rangle\}$$

   Hence the defining subbase $\mathcal{S}$ consists of the subsets of $X \times Y$ above.

   (ii) The defining base $\mathcal{B}$ consists of finite intersections of members of the defining subbase $\mathcal{S}$. That is,

   $$\mathcal{B} = \{X \times Y, \ \emptyset, \ \{\langle a, u \rangle\}, \ \{\langle b, u \rangle, \langle c, u \rangle\}, \ \{\langle a, u \rangle, \langle a, v \rangle\},$$
   $$\{\langle b, u \rangle, \langle b, v \rangle, \langle c, u \rangle, \langle c, v \rangle\}, \ \{\langle a, u \rangle, \langle b, u \rangle, \langle c, u \rangle\}\}$$

2. Prove Theorem 12.5: A function $f : Y \to X$ from a topological space Y into a product space $X = \prod_i X_i$ is continuous if and only if, for every projection $\pi_i : X \to X_i$, the composition $\pi_i \circ f : Y \to X_i$ is continuous.

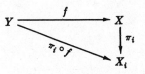

   Solution:
   By definition of the product space, all projections are continuous. So if f is continuous, then $\pi_i \circ f$, the composition of two continuous functions, is also continuous.

   On the other hand, suppose every composition function $\pi_i \circ f : Y \to X_i$ is continuous. Let G be an open subset of X_i. Then, by continuity of $\pi_i \circ f$,

   $$(\pi_i \circ f)^{-1}[G] = f^{-1}[\pi_i^{-1}[G]]$$

   is an open set in Y. But the class of sets of the form

   $$\pi_i^{-1}[G] \quad \text{where } G \text{ is an open subset of } X_i$$

   is the defining subbase for the product topology on X. Since their inverses under f are open subsets of Y, f is a continuous function by Theorem 7.2.

3. Let B be a member of the defining base for a product space $X = \prod_i X_i$. Show that the projection of B into any coordinate space is open.

   Solution:
   Since B belongs to the defining base for X,

   $$B = \prod \{X_i : i \neq j_1, \ldots, j_m\} \times G_{j_1} \times \cdots \times G_{j_m}$$

   where G_{j_k} is an open subset of X_{j_k}. So, for any projection $\pi_\alpha : X \to X_\alpha$,

   $$\pi_\alpha(B) = \begin{cases} X_\alpha & \text{if } \alpha \neq j_1, \ldots, j_m \\ G_\alpha & \text{if } \alpha \in \{j_1, \ldots, j_m\} \end{cases}$$

   In either case, $\pi_\alpha(B)$ is an open set.

4. Prove Theorem 12.6: Every projection $\pi_i : X \to X_i$ on a product space $X = \prod_i X_i$ is both open and continuous, i.e. bicontinuous.

Solution:

By definition of the product space, all projections are continuous. So we need only show that they are open.

Let G be an open subset of the product space $X = \prod_i X_i$. For every point $p \in G$ there is a member B of the defining base of the product topology such that $p \in B \subset G$. Thus, for any projection $\pi_i : X \to X_i$,

$$\pi_i(p) \in \pi_i[B] \subset \pi_i[G]$$

By the preceding problem, $\pi_i[B]$ is an open set. In other words, every point $\pi_i(p)$ in $\pi_i[G]$ belongs to an open set $\pi_i[B]$ which is contained in $\pi_i[G]$. Thus $\pi_i[G]$ is an open set.

5. Prove Theorem 12.7: A sequence $p_1, p_2, \ldots$ of points in a product space $X = \prod_i X_i$ converges to the point $q \in X$ if and only if, for every projection $\pi_i : X \to X_i$, the sequence $\pi_i(p_1), \pi_i(p_2), \ldots$ converges to $\pi_i(q)$ in the coordinate space X_i.

Solution:

Suppose $p_n \to q$. Then, since all projections are continuous, $\pi_i(p_n) \to \pi_i(q)$.

Conversely, suppose $\pi_i(p_n) \to \pi_i(q)$ for every projection π_i. In order to prove that $p_n \to q$, it is sufficient to show that if B is a member of the defining base of the product space $X = \prod_i X_i$ that contains the point $q \in X$, then

$$\exists\ n_0 \in N \quad \text{such that} \quad n > n_0 \ \Rightarrow \ p_n \in B$$

By definition of the defining base for the product space $X = \prod_i X_i$,

$$B \ = \ \pi_{j_1}^{-1}[G_{j_1}] \ \cap \ \cdots \ \cap \ \pi_{j_m}^{-1}[G_{j_m}]$$

where G_{j_k} is an open subset of the coordinate space X_{j_k}. Recall that $q \in B$; hence $\pi_{j_1}(q) \in \pi_{j_1}[B] = G_{j_1}, \ldots, \pi_{j_m}(q) \in \pi_{j_m}[B] = G_{j_m}$. By hypothesis, $\pi_{j_1}(p_n) \to \pi_{j_1}(q)$. Hence, for each $i = 1, \ldots, m$,

$$\exists\ n_i \in \mathbf{N} \quad \text{such that} \quad n > n_i \ \Rightarrow \ \pi_{j_i}(p_n) \in G_{j_i} \ \Rightarrow \ p_n \in \pi_{j_i}^{-1}[G_{j_i}]$$

Let $n_0 = \max(n_1, \ldots, n_m)$. Then

$$n > n_0 \ \Rightarrow \ p_n \in \pi_{j_1}^{-1}[G_{j_1}] \cap \cdots \cap \pi_{j_m}^{-1}[G_{j_m}] \ = \ B$$

Consequently, $p_n \to q$.

TYCHONOFF THEOREM

6. Prove Lemma 12.10: Let $\mathcal{A}$ be a class of subsets of a set X with the finite intersection property. Consider the collection $\mathbf{P}$ of all superclasses of $\mathcal{A}$ each with the finite intersection property. Then $\mathbf{P}$, ordered by class inclusion, contains a maximal element $\mathcal{M}$.

Solution:

Let $\mathbf{T} = \{B_i\}$ be a totally ordered subcollection of $\mathbf{P}$, and let $\mathcal{B} = \cup_i \mathcal{B}_i$. We show that $\mathcal{B}$ belongs to $\mathbf{P}$, i.e. that $\mathcal{B}$ is a superclass of $\mathcal{A}$ with the finite intersection property. It will then follow that $\mathcal{B}$ is an upper bound for $\mathbf{T}$ and so, by Zorn's Lemma, $\mathbf{P}$ contains a maximal element $\mathcal{M}$.

Clearly $\mathcal{B} = \cup_i \mathcal{B}_i$ is a superclass of $\mathcal{A}$ since each $\mathcal{B}_i$ is a superclass of $\mathcal{A}$. To show that $\mathcal{B}$ has the finite intersection property, let $\{A_1, \ldots, A_m\}$ be any finite subclass of $\mathcal{B}$. But $\mathcal{B} = \cup_i \mathcal{B}_i$; hence

$$\exists\ \mathcal{B}_{i_1}, \ldots, \mathcal{B}_{i_m} \in \mathbf{T} \quad \text{such that} \quad A_1 \in \mathcal{B}_{i_1}, \ldots, \ A_m \in \mathcal{B}_{i_m}$$

Recall that $\mathbf{T}$ is totally ordered; hence one of the classes, say $\mathcal{B}_{i_{j_0}}$, contains all the sets A_i and, furthermore, since it has the finite intersection property,

$$A_1 \cap A_2 \cap \cdots \cap A_m \ \neq \ \emptyset$$

We have just shown that each finite subclass $\{A_1, \ldots, A_m\}$ of $\mathcal{B}$ has a non-empty intersection, i.e. $\mathcal{B}$ has the finite intersection property. Consequently, $\mathcal{B}$ belongs to $\mathbf{P}$.

7. **Prove:** The maximal element $\mathcal{M}$ in Lemma 12.10 possesses the following properties:
   (i) Every superset of a member of $\mathcal{M}$ also belongs to $\mathcal{M}$.
   (ii) The intersection of a finite number of members of $\mathcal{M}$ also belongs to $\mathcal{M}$.
   (iii) If $A \cap M \neq \emptyset$, for every $M \in \mathcal{M}$, then A belongs to $\mathcal{M}$.

   Solution:
   We only prove (ii) here. The proofs of (i) and (iii) will be left as supplementary problems.

   (ii) We prove that the intersection of any two sets $A, B \in \mathcal{M}$ also belongs to $\mathcal{M}$. The theorem will then follow by induction. Let $C = A \cap B$. If we show that $\mathcal{M} \cup \{C\}$ has the finite intersection property, then $\mathcal{M} \cup \{C\}$ will belong to **P** and, since $\mathcal{M}$ is maximal in **P**, $\mathcal{M} = \mathcal{M} \cup \{C\}$. Thus C will belong to $\mathcal{M}$, as was to be proven.

   Let $\{A_1, A_2, \ldots, A_m\}$ be a finite subclass of $\mathcal{M} \cup \{C\}$. There are two cases:

   Case I. C does not belong to $\{A_1, \ldots, A_m\} \subset \mathcal{M} \cup \{C\}$. Then $\{A_1, \ldots, A_m\}$ is a finite subclass of $\mathcal{M}$ alone. But $\mathcal{M}$ has the finite intersection property; hence $A_1 \cap \cdots \cap A_m \neq \emptyset$.

   Case II. C does belong to $\{A_1, \ldots, A_m\}$, say $C = A_1$. Then
   $$A_1 \cap \cdots \cap A_m = C \cap A_2 \cap \cdots \cap A_m = A \cap B \cap A_2 \cap \cdots \cap A_m \neq \emptyset$$
   since $A, B, A_2, \ldots, A_m \in \mathcal{M}$.

   In either case, $\{A_1, \ldots, A_m\}$ has a non-empty intersection. So $\mathcal{M} \cup \{C\} \in$ **P** and, for reasons stated above, $C \in \mathcal{M}$.

8. **Prove Theorem (Tychonoff) 12.9:** Let $\{A_i : i \in I\}$ be a collection of compact topological spaces. Then the product space $X = \prod \{A_i : i \in I\}$ is also compact.

   Solution:
   Let $\mathcal{A} = \{F_j\}$ be a class of closed subsets of X with the finite intersection property. The theorem is proven if we show that $\mathcal{A}$ has a non-empty intersection, i.e.,
   $$\exists\, p \in X \quad \text{such that} \quad p \in F_j \text{ for every } F_j \in \mathcal{A}$$

   Let $\mathcal{M} = \{M_k : k \in K\}$ be a maximal superclass of $\mathcal{A}$ with the finite intersection property (see Lemma 12.10). Define $\overline{\mathcal{M}} = \{\overline{M}_k : k \in K\}$. Observe that
   $$F_j \in \mathcal{A} \Rightarrow F_j = \overline{F}_j \quad \text{and} \quad F_j \in \mathcal{M} \Rightarrow F_j \in \overline{\mathcal{M}}$$

   So if we prove that $\overline{\mathcal{M}}$ has a non-empty intersection, then $\mathcal{A}$ will also have a non-empty intersection. In other words, the proof is complete if
   $$\exists\, p \in X \quad \text{such that} \quad p \in \overline{M}_k \text{ for every } k \in K$$
   or, equivalently,
   $$\exists\, p \in X \quad \text{such that} \quad p \in B \Rightarrow B \cap M_k \neq \emptyset \text{ for every } k \in K \tag{1}$$

   where B is a member of the defining base for the product topology on $X = \prod_i A_i$, since p is then an accumulation point of each of the sets M_k and so is contained in $\overline{M}_k$.

   Recall that $\mathcal{M} = \{M_k : k \in K\}$ has the finite intersection property; so, for each projection $\pi_i : X \to A_i$, the class
   $$\{\pi_i[M_k] : k \in K\}$$
   of subsets of the coordinate space A_i also has the finite intersection property. Hence the class of closures
   $$\overline{\{\pi_i[M_k]\}} : k \in K\}$$
   is a class of closed subsets of A_i with the finite intersection property. By hypothesis, A_i is compact; so $\{\overline{\pi_i[M_k]} : k \in K\}$ has a non-empty intersection, i.e.,
   $$\exists\, a_i \in A_i \quad \text{such that} \quad a_i \in \overline{\pi_i[M_k]} \text{ for every } k \in K$$
   or, equivalently,
   $$\exists\, a_i \in A_i \quad \text{such that} \quad a_i \in G_i \Rightarrow G_i \cap \pi_i[M_k] \neq \emptyset \text{ for every } k \in K \tag{2}$$

   where G_i is any open subset of the coordinate space A_i.

Define $p = \langle a_i : i \in I \rangle$. We want to show that p satisfies the condition (1). Let $p \in B$, where B is a member of the defining base for the product topology on $X = \prod_i A_i$, i.e.,

$$B = \pi_{i_1}^{-1}[G_{i_1}] \cap \cdots \cap \pi_{i_m}^{-1}[G_{i_m}]$$

where G_{i_α} is an open subset of A_{i_α}.

Observe that $p \in B$ implies $\pi_{i_1}(p) = a_{i_1}$ belongs to $\pi_{i_1}[B] = G_{i_1}$. So, by (2) above,

$$G_{i_1} \cap \pi_{i_1}[M_k] \neq \varnothing \quad \text{for every } k \in K$$

which implies that

$$\pi_{i_1}^{-1}[G_{i_1}] \cap M_k = \left(\prod \{A_i : i \neq i_1\} \times G_{i_1} \right) \cap M_k \neq \varnothing \quad \text{for every } M_k \in \mathcal{M}$$

By the property (iii) of $\mathcal{M}$, stated in the preceding problem, $\pi_{i_1}^{-1}[G_{i_1}]$ belongs to $\mathcal{M}$. Similarly, $\pi_{i_2}^{-1}[G_{i_2}], \ldots, \pi_{i_m}^{-1}[G_{i_m}]$ also belongs to $\mathcal{M}$. But $\mathcal{M}$ satisfies the finite intersection property; so

$$B \cap M_k = \pi_{i_1}^{-1}[G_{i_1}] \cap \cdots \cap \pi_{i_m}^{-1}[G_{i_m}] \cap M_k \neq \varnothing \quad \text{for every } k \in K$$

Thus (1) is satisfied and the theorem is proven.

CANTOR SET

9. Show that the Cantor set T is a closed subset of **R**.

Solution:

Recall that T_m is the union of 2^m closed intervals. So T_m, the union of a finite number of closed sets, is also closed. But $T = \cap \{T_i : i \in \mathbf{N}\}$; hence T is closed, as it is the intersection of closed sets.

10. Show that T is compact.

Solution:

Since T is a closed and bounded set of real numbers, it is compact.

11. Let $X = \prod \{A_i : i \in \mathbf{N}\}$ where $A_i = \{0, 2\}$ with the discrete topology. Show that X is compact.

Solution:

Observe that A_i is compact since it is finite. So, by the Tychonoff Product Theorem, $X = \prod_i A_i$ is also compact.

12. Let $X = \prod \{A_i : i \in N\}$ where $A_i = \{0, 2\}$ with the discrete topology.

(i) Prove that the function $f : X \to T$ defined by

$$f(\langle a_1, a_2, \ldots \rangle) = a_1(\tfrac{1}{3}) + a_2(\tfrac{1}{9}) + a_3(\tfrac{1}{27}) + \cdots = \sum_{i=1}^{\infty} a_i(\tfrac{1}{3})^i$$

is continuous.

(ii) Prove that X is homeomorphic to T.

Solution:

(i)   Let $p = \langle a_1, a_2, \ldots \rangle \in X$ and let $\epsilon > 0$. We need to show that there is an open subset B of X containing p such that

$$x \in B \quad \text{implies} \quad |f(x) - f(p)| < \epsilon$$

Note that $\sum\limits_{i=1}^{\infty} (\tfrac{2}{3})^i$ converges. Hence

$$\exists \, n_0 \in N \quad \text{such that} \quad \sum_{i=n_0+1}^{\infty} (\tfrac{2}{3})^i < \epsilon$$

Consider the subset

$$B = \{a_1\} \times \{a_2\} \times \cdots \times \{a_{n_0}\} \times A_{n_0+1} \times A_{n_0+2} \times \cdots$$

of X. Observe that $p \in B$ and B is a member of the defining base of the product topology on $X = \prod_i A_i$ and so is open. Furthermore,

$$x \;=\; \langle a_1, \,\ldots,\, a_{n_0}, \; b_{n_0+1}, \; b_{n_0+2}, \,\ldots \rangle \;\in\; B$$

implies
$$|f(x) - f(p)| \;=\; \left| \sum_{i=n_0+1}^{\infty} (b_i - a_i)(\tfrac{1}{3})^i \right| \;\le\; \sum_{i=n_0+1}^{\infty} (\tfrac{2}{3})^i \;<\; \epsilon$$

Thus f is continuous.

(ii)   The function $f : X \to T$ is a one-one continuous function from the compact space X onto the metric space T. By Theorem 11.8, f is a homeomorphism.

Supplementary Problems

PRODUCT SPACES

13.   Show that the property of being a T_1-space is product invariant, i.e. the product of T_1-spaces is a T_1-space.

14.   Show that the property of being a regular space is product invariant.

15.   Show that the property of being a completely regular space is product invariant.

16.   Prove: Let $p = \langle a_i : i \in I \rangle$ be any point in a product space $X = \prod \{X_i : i \in I\}$. Then, for any $j_0 \in I$,
$$X_{j_0} \times \prod \{a_i : i \ne j_0\} \quad \text{is homeomorphic to} \quad X_{j_0}$$

(In the special case of Euclidean 3-space $\mathbf{R}^3$, this theorem states that the line, say
$$Y \;=\; \{a_1\} \times \{a_2\} \times \mathbf{R}_3 \;=\; \{\langle a_1, a_2, x \rangle : x \in \mathbf{R}\}$$
through $p = \langle a_1, a_2, a_3 \rangle$, is homeomorphic to $\mathbf{R}$.) See Fig. (a).

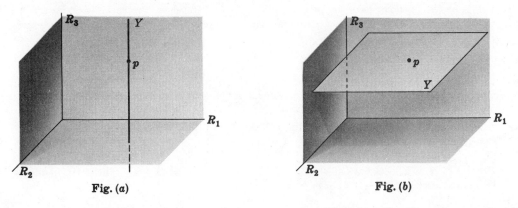

Fig. (a)   Fig. (b)

17.   Prove: Let $p = \langle a_i : i \in I \rangle$ be any point in a product space $X = \prod \{X_i : i \in I\}$. Then, for any $j_0 \in I$,
$$\{a_{j_0}\} \times \prod \{X_i : i \ne j_0\} \quad \text{is homeomorphic to} \quad \prod \{X_i : i \ne j_0\}$$

(In the special case of Euclidean 3-space $\mathbf{R}^3$, this theorem states that the plane, say
$$Y \;=\; \mathbf{R}_1 \times \mathbf{R}_2 \times \{a_3\} \;=\; \{\langle x, y, a_3 \rangle : x, y \in \mathbf{R}\}$$
through $p = \langle a_1, a_2, a_3 \rangle$, is homeomorphic to $\mathbf{R}^2 = \mathbf{R} \times \mathbf{R}$.) See Fig. (b).

18.   Prove the converse of the Tychonoff Product Theorem, i.e. if a product space $X = \prod_i X_i$ is compact, then each coordinate space X_i is also compact.

19.   Let A be a subset of a product space $X = \prod \{X_i : i \in I\}$ and let $\pi_{i_A} : A \to X_i$ denote the restriction of the projection $\pi_i : X \to X_i$ to A. Show that the relative topology on A is the coarsest topology with respect to which the functions π_{i_A} are continuous.

20. (i) Prove that a countable product of first countable spaces is first countable.

    (ii) Show that an arbitrary product of first countable spaces need not be first countable.

21. Show that an uncountable product space $X = \prod_i X_i$ is not metrizable (unless all except a countable number of coordinate spaces are singleton sets).

22. (i) Prove that a countable product of second countable spaces is second countable.

    (ii) Show that an arbitrary product of second countable spaces need not be second countable.

23. Let A_i be an arbitrary subset of a topological space X_i; thus $\prod_i A_i$ is a subset of the product space $X = \prod_i X_i$. Prove that (i) $\prod_i \bar{A}_i = \overline{\prod_i A_i}$, (ii) $\prod_i A_i^\circ \supset (\prod_i A_i)^\circ$. Give an example to show that equality does not hold for (ii) in general.

24. Let A_i be an arbitrary subspace of X_i. Show that the product topology on $\prod_i A_i$ is equal to the relative topology on $\prod_i A_i$ as a subset of the product space $\prod_i X_i$.

ARBITRARY TOPOLOGIES ON PRODUCT SETS

25. Prove Proposition 12.8: Let $\{(X_i, \mathcal{T}_i) : i \in I\}$ be a collection of topological spaces and let X be the product of the sets X_i, i.e. $X = \prod_i X_i$. Then the subsets of X of the form $\prod \{G_i : i \in I\}$ where G_i is an open subset of the coordinate space X_i, form a base for a topology $\mathcal{T}$ on the product set X.

26. Show that the product topology on a product set $X = \prod_i X_i$ is coarser than the topology $\mathcal{T}$ on X defined in the preceding problem (Proposition 12.8).

27. Give an example of a topology $\mathcal{T}$ on a product set $X = \prod_i X_i$ which is coarser than the product topology on X.

28. Let $\mathcal{T}$ be the topology on a product set $X = \prod_i X_i$ defined in Problem 25 (Proposition 12.8). Show that $(X, \mathcal{T})$ is discrete if each coordinate space X_i is discrete.

FINITE PRODUCTS

29. Prove Proposition 12.2: The subsets of a product space $X = X_1 \times \cdots \times X_m$ of the form $G_1 \times \cdots \times G_m$, where G_i is an open subset of X_i, form a base for the product topology on X.

30. Prove: If $\mathcal{B}$ is a base for X and $\mathcal{B}^*$ is a base for Y, then $\{G \times H : G \in \mathcal{B},\ H \in \mathcal{B}^*\}$ is a base for the product space $X \times Y$.

31. Prove: If $\mathcal{B}_a$ is a local base at $a \in X$ and $\mathcal{B}_b$ is a local base at $b \in Y$, then $\{G \times H : G \in \mathcal{B}_a,\ H \in \mathcal{B}_b\}$ is a local base at $p = \langle a, b \rangle \in X \times Y$.

32. Prove that the product of two first countable spaces is first countable.

33. Prove that the product of two second countable spaces is second countable.

34. Prove that the product of two separable spaces is separable.

35. Prove that the product of two compact spaces is compact (without using Zorn's Lemma or its equivalents).

36. Let $\mathcal{B}^*$ be the topology on the plane $\mathbf{R}^2$ generated by the half-open rectangles

$$[a, b) \times [c, d) = \{\langle x, y \rangle : a \leq x < b,\ c \leq y < d\}$$

Furthermore, let $\mathcal{T}$ be the topology on the real line $\mathbf{R}$ generated by the closed-open intervals $[a, b)$. Show that $(\mathbf{R}^2, \mathcal{T}^*)$ is the product of $(\mathbf{R}, \mathcal{T})$ with itself.

37. Show, by a counter-example, that the product of two normal spaces need not be normal.

38. Let $A \subset X$ and $B \subset Y$ and hence $A \times B \subset X \times Y$. Prove that

$$\text{(i)} \quad \bar{A} \times \bar{B} = \overline{A \times B} \qquad \text{(ii)} \quad A^\circ \times B^\circ = (A \times B)^\circ$$

(Recall [see Problem 23] that equality does not hold in general.)

39. Let $f : X \to Y$ and let $F : X \to X \times Y$ be defined by $F(x) = \langle x, f(x) \rangle$. Prove that f is continuous if and only if F is a homeomorphism of X with $F[X]$. (Recall that $F[X]$ is called the *graph* of f.)

40. Let X be a normed vector space over **R**. Show that the function $f : X \times X \to X$ defined by $f(\langle p, q \rangle) = p + q$ is continuous.

41. Let X be a normed vector space over **R**. Show that the function $f : \mathbf{R} \times X \to X$ defined by $f(\langle k, p \rangle) = kp$ is continuous.

METRIC PRODUCT SPACES

42. Prove: Every closed and bounded subset of Euclidean m-space R^m is compact.

43. Prove Proposition 12.11: Let $(X_1, d_1), \ldots, (X_m, d_m)$ be metric spaces and let $p = \langle a_1, \ldots, a_m \rangle$ and $q = \langle b_1, \ldots, b_m \rangle$ be arbitrary points in the product set $X = \prod\limits_{i=1}^{m} X_i$. Then each of the following functions is a metric on X:

(i) $\quad d(p, q) \quad = \quad \sqrt{d_1(a_1, b_1)^2 + \cdots + d_m(a_m, b_m)^2}$

(ii) $\quad d(p, q) \quad = \quad \max \{d_1(a_1, b_1), \ldots, d_m(a_m, b_m)\}$

(iii) $\quad d(p, q) \quad = \quad d_1(a_1, b_1) + \cdots + d_m(a_m, b_m)$

Moreover, the topology on X induced by each of the above metrics is the product topology.

44. Prove Proposition 12.12: Let $\{(X_1, d_1), (X_2, d_2), \ldots\}$ be a denumerable collection of metric spaces and let $p = \langle a_1, a_2, \ldots \rangle$ and $q = \langle b_1, b_2, \ldots \rangle$ be arbitrary points in the product set $X = \prod\limits_{n=1}^{\infty} X_n$. Then the function d defined by

$$d(p, q) \quad = \quad \sum_{n=1}^{\infty} \frac{1}{2^n} \frac{d_n(a_n, b_n)}{1 + d_n(a_n, b_n)}$$

is a metric on X and the topology induced by d is the product topology.

Chapter 13

Connectedness

SEPARATED SETS

Two subsets A and B of a topological space X are said to be *separated* if (i) A and B are disjoint, and (ii) neither contains an accumulation point of the other. In other words, A and B are separated iff

$$A \cap \bar{B} = \emptyset \quad \text{and} \quad \bar{A} \cap B = \emptyset$$

Example 1.1: Consider the following intervals on the real line $\mathbf{R}$:
$$A = (0,1), \quad B = (1,2) \quad \text{and} \quad C = [2,3)$$

Now A and B are separated since $\bar{A} = [0,1]$ and $\bar{B} = [1,2]$, and so $A \cap \bar{B}$ and $\bar{A} \cap B$ are empty. On the other hand, B and C are not separated since $2 \in C$ is a limit point of B; thus:

$$\bar{B} \cap C = [1,2] \cap [2,3) = \{2\} \neq \emptyset$$

Example 1.2: Consider the following subsets of the plane $\mathbf{R}^2$:

$$A = \{\langle 0,y \rangle : \tfrac{1}{2} \le y \le 1\}$$
$$B = \{\langle x,y \rangle : y = \sin(1/x),\ 0 < x \le 1\}$$

Now each point in A is an accumulation point of B; hence A and B are not separated sets.

CONNECTED SETS

Definition: A subset A of a topological space X is *disconnected* if there exist open subsets G and H of X such that $A \cap G$ and $A \cap H$ are disjoint non-empty sets whose union is A. In this case, $G \cup H$ is called a *disconnection* of A. A set is *connected* if it is not disconnected.

Observe that
$$A = (A \cap G) \cup (A \cap H) \quad \text{iff} \quad A \subset G \cup H$$

and
$$\emptyset = (A \cap G) \cap (A \cap H) \quad \text{iff} \quad G \cap H \subset A^c$$

Therefore $G \cup H$ is a disconnection of A if and only if

$$A \cap G \neq \emptyset, \quad A \cap H \neq \emptyset, \quad A \subset G \cup H, \quad \text{and} \quad G \cap H \subset A^c$$

Note that the empty set $\emptyset$ and singleton sets $\{p\}$ are always connected.

Example 2.1: The following subset of the plane $\mathbf{R}^2$ is disconnected:

$$A = \{\langle x, y \rangle : x^2 - y^2 \geq 4\}$$

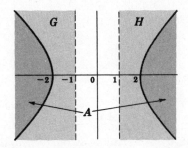

For the two open half-planes

$$G = \{\langle x, y \rangle : x < -1\} \quad \text{and} \quad H = \{\langle x, y \rangle : x > 1\}$$

form a disconnection of A as indicated in the diagram above.

Example 2.2: Consider the following topology on $X = \{a, b, c, d, e\}$:

$$\mathcal{T} = \{X,\ \varnothing,\ \{a, b, c\},\ \{c, d, e\},\ \{c\}\}$$

Now $A = \{a, d, e\}$ is disconnected. For let $G = \{a, b, c\}$ and $H = \{c, d, e\}$; then $A \cap G = \{a\}$ and $A \cap H = \{d, e\}$ are non-empty disjoint sets whose union is A. (Observe that G and H are not disjoint.)

The basic relationship between connectedness and separation follows:

Theorem 13.1: A set is connected if and only if it is not the union of two non-empty separated sets.

The following proposition is very useful.

Proposition 13.2: If A and B are connected sets which are not separated, then $A \cup B$ is connected.

Example 2.3: Let A and B be the subsets of the plane $\mathbf{R}^2$ defined and illustrated in Example 1.2. We show later that A and B are each connected. But A and B are not separated; hence, by the previous proposition, $A \cup B$ is a connected set.

CONNECTED SPACES

Connectedness, like compactness, is an absolute property of a set; namely,

Theorem 13.3: Let A be a subset of a topological space $(X, \mathcal{T})$. Then A is connected with respect to $\mathcal{T}$ if and only if A is connected with respect to the relative topology $\mathcal{T}_A$ on A.

Accordingly, we can frequently limit our investigation of connectedness to those topological spaces which are themselves connected, i.e. to *connected spaces.*

Example 3.1: Let X be a topological space which is disconnected, and let $G \cup H$ be a disconnection of X; then

$$X = (X \cap G) \cup (X \cap H) \quad \text{and} \quad (X \cap G) \cap (X \cap H) = \varnothing$$

But $X \cap G = G$ and $X \cap H = H$; thus X is disconnected if and only if there exist non-empty open sets G and H such that

$$X = G \cup H \quad \text{and} \quad G \cap H = \varnothing$$

In view of the discussion in the above example, we can give a simple characterization of connected spaces.

Theorem 13.4: A topological space X is connected if and only if (i) X is not the union of two non-empty disjoint open sets; or, equivalently, (ii) X and $\varnothing$ are the only subsets of X which are both open and closed.

Example 3.2:   Consider the following topology on $X = \{a, b, c, d, e\}$:

$$\mathcal{T} \;=\; \{X,\ \varnothing,\ \{a\},\ \{c, d\},\ \{a, c, d\},\ \{b, c, d, e\}\}$$

Now X is disconnected; for $\{a\}$ and $\{b, c, d, e\}$ are complements and hence both open and closed. In other words,

$$X \;=\; \{a\} \cup \{b, c, d, e\}$$

is a disconnection of X. Observe that the relative topology on the subset $A = \{b, d, e\}$ is $\{A, \varnothing, \{d\}\}$. Accordingly, A is connected since A and $\varnothing$ are the only subsets of A both open and closed in the relative topology.

Example 3.3:   The real line $\mathbf{R}$ with the usual topology is a connected space since $\mathbf{R}$ and $\varnothing$ are the only subsets of $\mathbf{R}$ which are both open and closed.

Example 3.4:   Let f be a continuous function from a connected space X into a topological space Y. Thus $f : X \to f[X]$ is continuous (where $f[X]$ has the relative topology).

We show that $f[X]$ is connected. Suppose $f[X]$ is disconnected; say G and H form a disconnection of $f[X]$. Then

$$f[X] \;=\; G \cup H \quad \text{and} \quad G \cap H \;=\; \varnothing$$

and so $X \;=\; f^{-1}[G] \cup f^{-1}[H] \quad \text{and} \quad f^{-1}[G] \cap f^{-1}[H] \;=\; \varnothing$

Since f is continuous, $f^{-1}[G]$ and $f^{-1}[H]$ are open subsets of X and hence form a disconnection of X, which is impossible. Thus if X is connected, so is $f[X]$.

We state the result of the preceding example as a theorem.

Theorem 13.5:   Continuous images of connected sets are connected.

Example 3.5:   Let X be a disconnected space; say, $G \cup H$ is a disconnection of X. Then the function $f(x) = \begin{cases} 0 & \text{if } x \in G \\ 1 & \text{if } x \in H \end{cases}$ is a continuous function from X onto the discrete space $Y = \{0, 1\}$.

On the other hand, by Theorem 13.5, a continuous image of a connected space X cannot be the disconnected discrete space $Y = \{0, 1\}$. In other words,

Lemma 13.6:   A topological space X is connected if and only if the only continuous functions from X into $Y = \{0, 1\}$ are the constant functions, $f(x) = 0$ or $f(x) = 1$.

CONNECTEDNESS ON THE REAL LINE

The connected sets of real numbers can be simply described as follows:

Theorem 13.7:   A subset E of the real line $\mathbf{R}$ containing at least two points is connected if and only if E is an interval.

Recall that the intervals on the real line $\mathbf{R}$ are of the following form:

$$(a, b),\ (a, b],\ [a, b),\ [a, b], \qquad \text{finite intervals}$$

$$(-\infty, a),\ (-\infty, a],\ (a, \infty),\ [a, \infty),\ (-\infty, \infty), \qquad \text{infinite intervals}$$

An interval E can be characterized by the following property:

$$a,\ b \in E,\ a < x < b \quad \Rightarrow \quad x \in E$$

Since the continuous image of a connected set is connected, we have the following generalization of the Weierstrass Intermediate Value Theorem (see Page 53, Theorem 4.9):

Theorem 13.8:   Let $f : X \to \mathbf{R}$ be a real continuous function defined on a connected set X. Then f assumes as a value each number between any two of its values.

Example 4.1: An interesting application of the theory of connectedness is the following "fixed-point theorem": Let $I = [0, 1]$ and let $f : I \to I$ be continuous; then $\exists p \in I$ such that $f(p) = p$.

This theorem can be interpreted geometrically. Note first that the graph of $f : I \to I$ lies in the unit square

$$I^2 \quad = \quad \{\langle x, y \rangle : 0 \leqq x \leqq 1,\ 0 \leqq y \leqq 1\}$$

The theorem then states that the graph of f, which connects a point on the left edge of the square to a point on the right edge of the square, must intersect the diagonal line Δ at, say, $\langle p, p \rangle$ as indicated in the diagram.

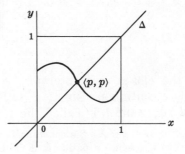

COMPONENTS

A *component* E of a topological space X is a maximal connected subset of X; that is, E is connected and E is not a proper subset of any connected subset of X. Clearly E is non-empty. The central facts about the components of a space are contained in the following theorem.

Theorem 13.9: The components of a topological space X form a partition of X, i.e. they are disjoint and their union is X. Every connected subset of X is contained in some component.

Thus each point $p \in X$ belongs to a unique component of X, called the *component of p*.

Example 5.1: If X is connected, then X has only one component: X itself.

Example 5.2: Consider the following topology on $X = \{a, b, c, d, e\}$:

$$\mathcal{T} \quad = \quad \{X,\ \varnothing,\ \{a\},\ \{c, d\},\ \{a, c, d\},\ \{b, c, d, e\}\}$$

The components of X are $\{a\}$ and $\{b, c, d, e\}$. Any other connected subset of X, such as $\{b, d, e\}$ (see Example 3.2), is a subset of one of the components.

The statement in Example 5.1 is used to prove that connectedness is product invariant; that is,

Theorem 13.10: The product of connected spaces is connected.

Corollary 13.11: Euclidean m-space $\mathbf{R}^m$ is connected.

LOCALLY CONNECTED SPACES

A topological space X is *locally connected at $p \in X$* iff every open set containing p contains a connected open set containing p, i.e. if the open connected sets containing p form a local base at p. X is said to be *locally connected* if it is locally connected at each of its points or, equivalently, if the open connected subsets of X form a base for X.

Example 6.1: Every discrete space X is locally connected. For if $p \in X$, then $\{p\}$ is an open connected set containing p which is contained in every open set containing p. Note that X is not connected if X contains more than one point.

Example 6.2: Let A and B be the subsets of the plane $\mathbf{R}^2$ of Example 1.2. Now $A \cup B$ is a connected set. But $A \cup B$ is not locally connected at $p = \langle 0, 1 \rangle$. For example, the open disc with center p and radius $\frac{1}{4}$ does not contain any connected neighborhood of p.

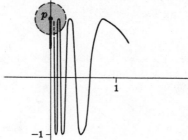

PATHS

Let $I = [0, 1]$, the closed unit interval. A *path* from a point a to a point b in a topological space X is a continuous function $f : I \to X$ with $f(0) = a$ and $f(1) = b$. Here a is called the *initial point* and b is called the *terminal point* of the path.

Example 7.1: For any $p \in X$, the constant function $e_p : I \to X$ defined by $e_p(s) = p$ is continuous and hence a path. It is called the *constant path* at p.

Example 7.2: Let $f : I \to X$ be a path from a to b. Then the function $\hat{f} : I \to X$ defined by $\hat{f}(s) = f(1-s)$ is a path from b to a.

Example 7.3: Let $f : I \to X$ be a path from a to b and let $g : I \to X$ be a path from b to c. Then the juxtaposition of the two paths f and g, denoted by $f * g$, is the function $f * g : I \to X$ defined by

$$(f * g)(s) \;=\; \begin{cases} f(2s) & \text{if } 0 \leq s \leq \frac{1}{2} \\ g(2s-1) & \text{if } \frac{1}{2} \leq s \leq 1 \end{cases}$$

which is a path from a to c obtained by following the path f from a to b and then following g from b to c.

ARCWISE CONNECTED SETS

A subset E of a topological space X is said to be *arcwise connected* if for any two points $a, b \in E$ there is a path $f : I \to X$ from a to b which is contained in E, i.e. $f[I] \subset E$. The maximal arcwise connected subsets of X, called *arcwise connected components*, form a partition of X. The relationship between connectedness and arcwise connectedness follows:

Theorem 13.12: Arcwise connected sets are connected.

The converse of this theorem is not true, as seen in the next example.

Example 8.1: Consider the following subsets of the plane $\mathbf{R}^2$:

$$A \;=\; \{\langle x, y \rangle : 0 \leq x \leq 1, \ y = x/n, \ n \in \mathbf{N}\}$$
$$B \;=\; \{\langle x, 0 \rangle : \tfrac{1}{2} \leq x \leq 1\}$$

Here A consists of the points on the line segments joining the origin $\langle 0, 0 \rangle$ to the points $\langle 1, 1/n \rangle$, $n \in \mathbf{N}$; and B consists of the points on the x-axis between $\frac{1}{2}$ and 1. Now A and B are both arcwise connected, hence also connected. Furthermore, A and B are not separated since each $p \in B$ is a limit point of A; and so $A \cup B$ is connected. But $A \cup B$ is not arcwise connected; in fact, there exists no path from any point in A to any point in B.

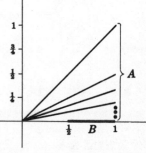

Example 8.2: Let A and B be the subsets of the plane $\mathbf{R}^2$ defined in Example 1.2. Now A and B are continuous images of intervals and are therefore connected. Moreover, A and B are not separated sets and so $A \cup B$ is connected. But $A \cup B$ is not arcwise connected; in fact, there exists no path from a point in A to a point in B.

The topology of the plane $\mathbf{R}^2$ is an essential part of the theory of functions of a complex variable. In this case, a *region* is defined as an open connected subset of the plane. The following theorem plays an important role in this theory.

Theorem 13.13:   An open connected subset of the plane $\mathbf{R}^2$ is arcwise connected.

HOMOTOPIC PATHS

Let $f : I \to X$ and $g : I \to X$ be two paths with the same initial point $p \in X$ and the same terminal point $q \in X$. Then f is said to be *homotopic* to g, written $f \simeq g$, if there exists a continuous function

$$H : I^2 \to X$$

such that

$$H(s, 0) = f(s) \qquad H(0, t) = p$$
$$H(s, 1) = g(s) \qquad H(1, t) = q$$

as indicated in the adjacent diagram. We then say that f can be continuously deformed into g. The function H is called a *homotopy* from f to g.

Example 9.1:   Let X be the set of points between two concentric circles (called an *annulus*). Then the paths f and g in the diagram on the left below are homotopic, whereas the paths f' and g' in the diagram on the right below are not homotopic.

Example 9.2:   Let $f : I \to X$ be any path. Then $f \simeq f$, i.e. f is homotopic to itself. For the function $H : I^2 \to X$ defined by

$$H(s, t) = f(s)$$

is a homotopy from f to f.

Example 9.3:   Let $f \simeq g$ and, say, $H : I^2 \to X$ is a homotopy from f to g. Then the function $\hat{H} : I^2 \to X$ defined by

$$\hat{H}(s, t) = H(s, 1 - t)$$

is a homotopy from g to f, and so $g \simeq f$.

Example 9.4:   Let $f \simeq g$ and $g \simeq h$; say, $F : I^2 \to X$ is a homotopy from f to g and $G : I^2 \to X$ is a homotopy from g to h. The function $H : I^2 \to X$ defined by

$$H(s, t) = \begin{cases} F(s, 2t) & \text{if } 0 \le t \le \tfrac{1}{2} \\ G(s, 2t - 1) & \text{if } \tfrac{1}{2} \le t \le 1 \end{cases}$$

is a homotopy from f to h, and so $f \simeq h$. The homotopy H can be interpreted geometrically as compressing the domains of F and G into one square.

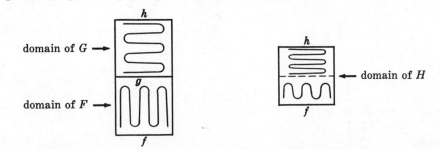

The previous three relations imply the following proposition:

Proposition 13.14: The homotopy relation is an equivalence relation in the collection of all paths from a to b.

SIMPLY CONNECTED SPACES

A path $f : I \to X$ with the same initial and terminal point, say $f(0) = f(1) = p$, is called a *closed path at* $p \in X$. In particular, the constant path $e_p : I \to X$ defined by $e_p(s) = p$ is a closed path at p. A closed path $f : I \to X$ is said to be *contractable to a point* if it is homotopic to the constant path.

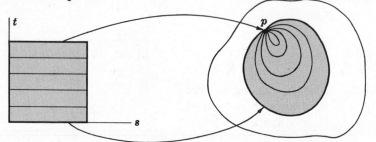

A topological space is *simply connected* iff every closed path in X is contractable to a point.

Example 10.1: An open disc in the plane $\mathbf{R}^2$ is simply connected, whereas an annulus is not simply connected since there are closed curves, as indicated in the diagram, that are not contractable to a point.

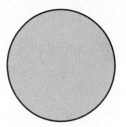

simply connected

not simply connected

Solved Problems

SEPARATED SETS

1. Show that if A and B are non-empty separated sets, then $A \cup B$ is disconnected.

    Solution:

    Since A and B are separated, $A \cap \bar{B} = \emptyset$ and $\bar{A} \cap B = \emptyset$. Let $G = \bar{B}^c$ and $H = \bar{A}^c$. Then G and H are open and
    $$(A \cup B) \cap G = A \quad \text{and} \quad (A \cup B) \cap H = B$$
    are non-empty disjoint sets whose union is $A \cup B$. Thus G and H form a disconnection of $A \cup B$, and so $A \cup B$ is disconnected.

2. Let $G \cup H$ be a disconnection of A. Show that $A \cap G$ and $A \cap H$ are separated sets.

Solution:

Now $A \cap G$ and $A \cap H$ are disjoint; hence we need only show that each set contains no accumulation point of the other. Let p be an accumulation point of $A \cap G$, and suppose $p \in A \cap H$. Then H is an open set containing p and so H contains a point of $A \cap G$ distinct from p, i.e. $(A \cap G) \cap H \neq \emptyset$. But

$$(A \cap G) \cap (A \cap H) = \emptyset = (A \cap G) \cap H$$

Accordingly, $p \notin A \cap H$.

Similarly, if p is an accumulation point of $A \cap H$, then $p \notin A \cap G$. Thus $A \cap G$ and $A \cap H$ are separated sets.

3. Prove Theorem 13.1: A set A is connected if and only if A is not the union of two non-empty separated sets.

Solution:

We show, equivalently, that A is disconnected if and only if A is the union of two non-empty separated sets. Suppose A is disconnected, and let $G \cup H$ be a disconnection of A. Then A is the union of non-empty sets $A \cap G$ and $A \cap H$ which are, by the preceding problem, separated. On the other hand, if A is the union of two non-empty separated sets, then A is disconnected by Problem 1.

CONNECTED SETS

4. Let $G \cup H$ be a disconnection of A and let B be a connected subset of A. Show that either $B \cap H = \emptyset$ or $B \cap G = \emptyset$, and so either $B \subset G$ or $B \subset H$.

Solution:

Now $B \subset A$, and so

$$A \subset G \cup H \quad \Rightarrow \quad B \subset G \cup H \qquad \text{and} \qquad G \cap H \subset A^c \quad \Rightarrow \quad G \cap H \subset B^c$$

Thus if both $B \cap G$ and $B \cap H$ are non-empty, then $G \cup H$ forms a disconnection of B. But B is connected; hence the conclusion follows.

5. Prove Proposition 13.2: If A and B are connected sets which are not separated, then $A \cup B$ is connected.

Solution:

Suppose $A \cup B$ is disconnected and suppose $G \cup H$ is a disconnection of $A \cup B$. Since A is a connected subset of $A \cup B$, either $A \subset G$ or $A \subset H$ by the preceding problem. Similarly, either $B \subset G$ or $B \subset H$.

Now if $A \subset G$ and $B \subset H$ (or $B \subset G$ and $A \subset H$), then, by Problem 2,

$$(A \cup B) \cap G = A \quad \text{and} \quad (A \cup B) \cap H = B$$

are separated sets. But this contradicts the hypothesis; hence either $A \cup B \subset G$ or $A \cup B \subset H$, and so $G \cup H$ is not a disconnection of $A \cup B$. In other words, $A \cup B$ is connected.

6. Prove: Let $\mathcal{A} = \{A_i\}$ be a class of connected subsets of X such that no two members of $\mathcal{A}$ are separated. Then $B = \cup_i A_i$ is connected.

Solution:

Suppose B is not connected and $G \cup H$ is a disconnection of B. Now each $A_i \in \mathcal{A}$ is connected and so (Problem 4) is contained in either G or H and disjoint from the other. Furthermore, any two members $A_{i_1}, A_{i_2} \in \mathcal{A}$ are not separated and so, by Proposition 13.2, $A_{i_1} \cup A_{i_2}$ is connected; then $A_{i_1} \cup A_{i_2}$ is contained in G or H and disjoint from the other. Accordingly, all the members of $\mathcal{A}$, and hence $B = \cup_i A_i$, must be contained in either G or H and disjoint from the other. But this contradicts the fact that $G \cup H$ is a disconnection of B; hence B is connected.

7. **Prove:** Let $\mathcal{A} = \{A_i\}$ be a class of connected subsets of X with a non-empty intersection. Then $B = \cup_i A_i$ is connected.

   Solution:

   Since $\cap_i A_i \neq \emptyset$, any two members of $\mathcal{A}$ are not disjoint and so are not separated; hence, by the preceding problem, $B = \cup_i A_i$ is connected.

8. Let A be a connected subset of X and let $A \subset B \subset \bar{A}$. Show that B is connected and hence, in particular, $\bar{A}$ is connected.

   Solution:

   Suppose B is disconnected and suppose $G \cup H$ is a disconnection of B. Now A is a connected subset of B and so, by Problem 4, either $A \cap H = \emptyset$ or $A \cap G = \emptyset$; say, $A \cap H = \emptyset$. Then H^c is a closed superset of A and therefore $A \subset B \subset \bar{A} \subset H^c$. Consequently, $B \cap H = \emptyset$. But this contradicts the fact that $G \cup H$ is a disconnection of B; hence B is connected.

CONNECTED SPACES

9. Let X be a topological space. Show that the following conditions are equivalent:

   (i) X is disconnected.

   (ii) There exists a non-empty proper subset of X which is both open and closed.

   Solution:

   (i) $\Rightarrow$ (ii): Suppose $X = G \cup H$ where G and H are non-empty and open. Then G is a non-empty proper subset of X and, since $G = H^c$, G is both open and closed.

   (ii) $\Rightarrow$ (i): Suppose A is a non-empty proper subset of X which is both open and closed. Then A^c is also non-empty and open, and $X = A \cup A^c$. Accordingly, X is disconnected.

10. **Prove Theorem 13.3:** Let A be a subset of a topological space $(X, \mathcal{T})$ and let $\mathcal{T}_A$ be the relative topology on A. Then A is $\mathcal{T}$-connected if and only if A is $\mathcal{T}_A$-connected.

   Solution:

   Suppose A is disconnected with $G \cup H$ forming a $\mathcal{T}$-disconnection of A. Now $G, H \in \mathcal{T}$ and so $A \cap G, A \cap H \in \mathcal{T}_A$. Accordingly, $A \cap G$ and $A \cap H$ form a $\mathcal{T}_A$-disconnection of A; hence A is $\mathcal{T}_A$-disconnected.

   On the other hand, suppose A is $\mathcal{T}_A$-disconnected, say G^* and H^* form a $\mathcal{T}_A$-disconnection of A. Then $G^*, H^* \in \mathcal{T}_A$ and so

   $$\exists \; G, H \in \mathcal{T} \quad \text{such that} \quad G^* = A \cap G \;\; \text{and} \;\; H^* = A \cap H$$

   But $\qquad A \cap G^* = A \cap A \cap G = A \cap G \quad \text{and} \quad A \cap H^* = A \cap A \cap H = A \cap H$

   Hence $G \cup H$ is a $\mathcal{T}$-disconnection of A and so A is $\mathcal{T}$-disconnected.

11. Let $p, q \in X$. The subsets $A_1, \ldots, A_m$ of X are said to form a *simple (finite) chain* joining p to q if A_1 (and only A_1) contains p, A_m (and only A_m) contains q, and $A_i \cap A_j = \emptyset$ iff $|i - j| > 1$.

   Prove: Let X be connected and let $\mathcal{A}$ be an open cover of X. Then any pair of points in X can be joined by a simple chain consisting of members of $\mathcal{A}$.

   Solution:

   Let p be any arbitrary point in X and let H consist of those points in X which can be joined to p by some simple chain consisting of members of $\mathcal{A}$. Now $H \neq \emptyset$, since $p \in H$. We claim that H is both open and closed and so $H = X$ since X is connected.

Let $h \in H$. Then $\exists\, G_1, \ldots, G_m \in \mathcal{A}$ which form a simple chain from h to p. But if $x \in G_1 \setminus G_2$, then $G_1, \ldots, G_m$ form a simple chain from x to p; and if $y \in G_1 \cap G_2$, then $G_2, \ldots, G_m$ form a simple chain from y to p, as indicated in the diagram below.

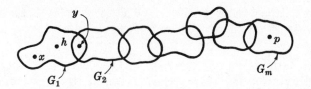

Thus G_1 is a subset of H, i.e. $h \subset G_1 \subset H$. Hence H is a neighborhood of each of its points, and so H is open.

Now let $g \in H^c$. Since $\mathcal{A}$ is a cover of X, $\exists\, G \in \mathcal{A}$ such that $g \in G$, and G is open. If $G \cap H \neq \emptyset$, $\exists\, h \in G \cap H \subset H$ and so $\exists\, G_1, \ldots, G_m \in \mathcal{A}$ forming a simple chain from h to p. But then either $G, G_k, \ldots, G_m$, where we consider the maximum k for which G intersects G_k, or $G_1, \ldots, G_m$ form a simple chain from g to p, and so $g \in H$, a contradiction. Hence $G \cap H = \emptyset$, and so $g \in G \subset H^c$. Thus H^c is an open set, and so $H^{cc} = H$ is closed.

12. **Prove Theorem 13.7:** Let E be a subset of the real line $\mathbf{R}$ containing at least two points. Then E is connected if and only if E is an interval.

Solution:

Suppose E is not an interval; then

$$\exists\, a, b \in E,\ p \notin E \qquad \text{such that} \qquad a < p < b$$

Set $G = (-\infty, p)$ and $H = (p, \infty)$. Then $a \in G$ and $b \in H$, and hence $E \cap G$ and $E \cap H$ are non-empty disjoint sets whose union is E. Thus E is disconnected.

Now suppose E is an interval and, furthermore, assume E is disconnected; say, G and H form a disconnection of E. Set $A = E \cap G$ and $B = E \cap H$; then $E = A \cup B$. Now A and B are non-empty; say, $a \in A$, $b \in B$, $a < b$ and $p = \sup\{A \cap [a, b]\}$. Since $[a, b]$ is a closed set, $p \in [a, b]$ and hence $p \in E$.

Suppose $p \in A = E \cap G$. Then $p < b$ and $p \in G$. Since G is an open set

$$\exists\, \delta > 0 \qquad \text{such that} \qquad p + \delta \in G \quad \text{and} \quad p + \delta < b$$

Hence $p + \delta \in E$ and so $p + \delta \in A$. But this contradicts the definition of p, i.e. $p = \sup\{A \cap [a, b]\}$. Therefore $p \notin A$.

On the other hand, suppose $p \in B = E \cap H$. Then, in particular, $p \in H$. Since H is an open set,

$$\exists\, \delta^* > 0 \qquad \text{such that} \qquad [p - \delta^*, p] \subset H \quad \text{and} \quad a < p - \delta^*$$

Hence $[p - \delta^*, p] \subset E$ and so $[p - \delta^*, p] \subset B$. Accordingly, $[p - \delta^*, p] \cap A = \emptyset$. But then $p - \delta^*$ is an upper bound for $A \cap [a, b]$, which is impossible since $p = \sup\{A \cap [a, b]\}$. Hence $p \notin B$. But this contradicts the fact that $p \in E$, and so E is connected.

13. **Prove (see Example 4.1):** Let $I = [0, 1]$ and let $f : I \to I$ be continuous. Then $\exists\, p \in I$ such that $f(p) = p$.

Solution:

If $f(0) = 0$ or $f(1) = 1$, the theorem follows; hence we can assume that $f(0) > 0$ and $f(1) < 1$. Since f is continuous, the graph of the function

$$F : I \to \mathbf{R}^2 \qquad \text{defined by} \qquad F(x) = \langle x, f(x) \rangle$$

is also continuous.

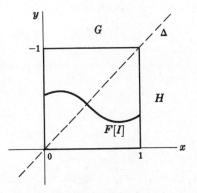

Set $G = \{\langle x, y \rangle : x < y\}$, $H = \{\langle x, y \rangle : y < x\}$; then $\langle 0, f(0) \rangle \in G$, $\langle 1, f(1) \rangle \in H$. Hence if $F[I]$ does not contain a point of the diagonal

$$\Delta = \{\langle x, y \rangle : x = y\} = \mathbf{R}^2 \setminus (G \cup H)$$

then $G \cup H$ is a disconnection of $F[I]$. But this contradicts the fact that $F[I]$, the continuous image of a connected set, is connected; hence $F[I]$ contains a point $\langle p, p \rangle \in \Delta$, and so $f(p) = p$.

COMPONENTS

14. Show that every component E is closed.

> **Solution:**
>   Now E is connected and so, by Problem 6, $\bar{E}$ is connected, $E \subset \bar{E}$. But E, a component, is a maximal connected set; hence $E = \bar{E}$, and so E is closed.

15. Prove: Let $p \in X$ and let $\mathcal{A}_p = \{A_i\}$ be the class of connected subsets of X containing p. Furthermore, let $C_p = \cup_i A_i$. Then: (i) C_p is connected. (ii) If B is a connected subset of X containing p, then $B \subset C_p$. (iii) C_p is a maximal connected subset of X, i.e. a component.

> **Solution:**
> (i) Since each $A_i \in \mathcal{A}_p$ contains p, $p \in \cap_i A_i$ and so, by Problem 7, $C_p = \cup_i A_i$ is connected.
>
> (ii) If B is a connected subset of X containing p, then $B \in \mathcal{A}_p$ and so $B \subset C_p = \cup \{A_i : A_i \in \mathcal{A}_p\}$.
>
> (iii) Let $C_p \subset D$, where D is connected. Then $p \in D$ and hence, by (ii), $D \subset C_p$; that is, $C_p = D$. Therefore C_p is a component.

16. Prove Theorem 13.9: The components of X form a partition of X. Every connected subset of X is contained in some component.

> **Solution:**
>   Consider the class $\mathcal{C} = \{C_p : p \in X\}$ where C_p is defined as in the preceding problem. We claim that $\mathcal{C}$ consists of the components of X. By the preceding problem, each $C_p \in \mathcal{C}$ is a component. On the other hand, if D is a component, then D contains some point $p_0 \in X$ and so $D \subset C_{p_0}$. But D is a component; hence $D = C_{p_0}$.
>
>   We now show that $\mathcal{C}$ is a partition of X. Clearly, $X = \cup \{C_p : p \in X\}$; hence we need only show that distinct components are disjoint or, equivalently, if $C_p \cap C_q \neq \varnothing$, then $C_p = C_q$. Let $a \in C_p \cap C_q$. Then $C_p \subset C_a$ and $C_q \subset C_a$, since C_p and C_q are connected sets containing a. But C_p and C_q are components; hence $C_p = C_a = C_q$.
>
>   Lastly, if E is a non-empty connected subset of X, then E contains a point $p_0 \in X$ and so $E \subset C_{p_0}$ by the preceding problem. If $E = \varnothing$, then E is contained in every component.

17. Show that if X and Y are connected spaces, then $X \times Y$ is connected. Hence a finite product of connected spaces is connected.

> **Solution:**
>   Let $p = \langle x_1, y_1 \rangle$ and $q = \langle x_2, y_2 \rangle$ be any pair of points in $X \times Y$. Now $\{x_1\} \times Y$ is homeomorphic to Y and is therefore connected. Similarly, $X \times \{y_2\}$ is connected.
>
>   But $\{x_1\} \times Y \cap X \times \{y_2\} = \{\langle x_1, y_2 \rangle\}$; hence $\{x_1\} \times Y \cup X \times \{y_2\}$ is connected. Accordingly, p and q belong to the same component. But p and q were arbitrary; hence $X \times Y$ has one component and is therefore connected.

18. Prove Theorem 13.10: The product of connected spaces is connected, i.e. connectedness is a product invariant property.

> **Solution:**
>   Let $\{X_i : i \in I\}$ be a collection of connected spaces and let $X = \prod_i X_i$ be the product space. Furthermore, let $p = \langle a_i : i \in I \rangle \in X$ and let $E \subset X$ be the component of p. We claim that every point $x = \langle x_i : i \in I \rangle \in X$ belongs to the closure of E and hence belongs to E since E is closed. Now let
> $$ G \;=\; \prod \{X_i : i \neq i_1, \ldots, i_m\} \times G_{i_1} \times \cdots \times G_{i_m} $$
> be any basic open set containing $x \in X$. Now
> $$ H \;=\; \prod \{\{a_i\} : i \neq i_1, \ldots, i_m\} \times X_{i_1} \times \cdots \times X_{i_m} $$
> is homeomorphic to $X_{i_1} \times \cdots \times X_{i_m}$ and hence connected. Furthermore, $p \in H$ and so H is a subset of E, the component of p. But $G \cap H$ is non-empty; hence G contains a point of E. Accordingly, $x \in \bar{E} = E$. Thus X has one component and is therefore connected.

ARCWISE CONNECTED SETS

19. Let $f : I \to X$ be any path in X. Show that $f[I]$, the range of f, is connected.

Solution:

$I = [0, 1]$ is connected and f is continuous; hence, by Theorem 13.5, $f[I]$ is connected.

20. Prove: Continuous images of arcwise connected sets are arcwise connected.

Solution:

Let $E \subset X$ be arcwise connected and let $f : X \to Y$ be continuous. We claim that $f[E]$ is arcwise connected. For let $p, q \in f[E]$. Then $\exists\, p^*, q^* \in E$ such that $f(p^*) = p$ and $f(q^*) = q$. But E is arcwise connected and so

$$\exists \text{ a path } g : I \to X \quad \text{such that} \quad g(0) = p^*, \ g(1) = q^* \ \text{and} \ g[I] \subset E$$

Now the composition of continuous functions is continuous and so $f \circ g : I \to Y$ is continuous. Furthermore,

$$f \circ g(0) = f(p^*) = p, \quad f \circ g(1) = f(q^*) = q \quad \text{and} \quad f \circ g[I] = f[g[I]] \subset f[E]$$

Thus $f[E]$ is arcwise connected.

21. Prove Theorem 13.12: Every arcwise connected set A is connected.

Solution:

If A is empty, then A is connected. Suppose A is not empty; say, $p \in A$. Now A is arcwise connected and so, for each $a \in A$, there is a path $f_a : I \to A$ from p to a. Furthermore,

$$a \in f_a[I] \subset A \quad \text{and so} \quad A = \bigcup \{ f_a[I] : a \in A \}$$

But $p \in f_a[I]$, for every $a \in A$; hence $\bigcap \{ f_a[I] : a \in A \}$ is non-empty. Moreover, each $f_a[I]$ is connected and so, by Problem 7, A is connected.

22. Prove: Let $\mathcal{A}$ be a class of arcwise connected subsets of X with a non-empty intersection. Then $B = \bigcup \{ A : A \in \mathcal{A} \}$ is arcwise connected.

Solution:

Let $a, b \in B$. Then

$$\exists\, A_a, A_b \in \mathcal{A} \quad \text{such that} \quad a \in A_a, \ b \in A_b$$

Now $\mathcal{A}$ has a non-empty intersection; say, $p \in \bigcap \{ A : A \in \mathcal{A} \}$. Then $p \in A_a$ and, since A_a is arcwise connected, there is a path $f : I \to A_a \subset B$ from a to p. Similarly, there is a path $g : I \to A_b \subset B$ from p to b. The juxtaposition of the two paths (see Example 7.3) is a path from a to b contained in B. Hence B is arcwise connected.

23. Show that an open disc D in the plane $\mathbf{R}^2$ is arcwise connected.

Solution:

Let $p = \langle a_1, b_1 \rangle$, $q = \langle a_2, b_2 \rangle \in D$. The function $f : I \to \mathbf{R}^2$ defined by

$$f(t) = \langle a_1 + t(a_2 - a_1), \ b_1 + t(b_2 - b_1) \rangle$$

is a path from p to q which is contained in D. (Geometrically, $f[I]$ is the line segment connecting p and q.) Hence D is arcwise connected.

24. Prove Theorem 13.13: Let E be a non-empty open connected subset of the plane $\mathbf{R}^2$. Then E is arcwise connected.

Solution:

Method 1.

Let $p \in E$ and let G consist of those points in E which can be joined to p by a path in E. We claim that G is open. For let $q \in G \subset E$. Now E is open and so $\exists$ an open disc D with center q such that $q \in D \subset E$. But D is arcwise connected; hence each point $x \in D$ can be joined to q which can be joined to p. Hence each point $x \in D$ can be joined to p, and so $q \in D \subset G$. Accordingly, G is open.

Now set $H = E \setminus G$, i.e. H consists of those points in E which cannot be joined to E by a path in E. We claim that H is open. For let $q^* \in H \subset E$. Since E is open, $\exists$ an open disc D^* with center q^* such that $q^* \in D^* \subset E$. Since D^* is arcwise connected, each $x \in D^*$ cannot be joined to p with a path in E, and so $q^* \in D^* \subset H$. Hence H is open.

But E is connected and therefore E cannot be the union of two non-empty disjoint open sets. Then $H = \emptyset$, and so $E = G$ is arcwise connected.

Method 2.

Since E is open, E is the union of open discs. But E is connected; hence, by Problem 11, $\exists$ open discs $S_1, \ldots, S_m \subset E$ which form a simple chain joining any $p \in E$ to any $q \in E$. Let a_i be the center of S_i and let $b_i \in S_i \cap S_{i+1}$. Then the polygonal arc joining p to a_1 to b_1 to a_2, etc., is contained in the union of the discs and hence is contained in E. Thus E is arcwise connected.

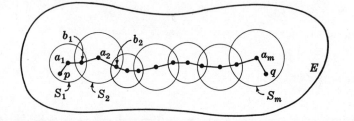

TOTALLY DISCONNECTED SPACES

25. A topological space X is said to be *totally disconnected* if for each pair of points $p, q \in X$ there exists a disconnection $G \cup H$ of X with $p \in G$ and $q \in H$. Show that the real line $\mathbf{R}$ with the topology $\mathcal{T}$ generated by the open closed intervals $(a, b]$ is totally disconnected.

Solution:

Let $p, q \in R$; say, $p < q$. Then $G = (-\infty, p]$ and $H = (p, \infty)$ are open disjoint sets whose union is $\mathbf{R}$, i.e. $G \cup H$ is a disconnection of $\mathbf{R}$. But $p \in G$ and $q \in H$; hence $(\mathbf{R}, \mathcal{T})$ is totally disconnected.

26. Show that the set $\mathbf{Q}$ of rational numbers with the relative usual topology is totally disconnected.

Solution:

Let $p, q \in \mathbf{Q}$; say, $p < q$. Now there exists an irrational number a such that $p < a < q$.

Set $G = \{x \in Q : x < a\}$ and $H = \{x \in Q : x > a\}$. Then $G \cup H$ is a disconnection of $\mathbf{Q}$, and $p \in G$ and $q \in H$. Thus $\mathbf{Q}$ is totally disconnected.

27. Prove: The components of a totally disconnected space X are the singleton subsets of X.

Solution:

Let E be a component of X and suppose $p, q \in E$ with $p \neq q$. Since X is totally disconnected, there exists a disconnection $G \cup H$ of X such that $p \in G$ and $q \in H$. Consequently, $E \cap G$ and $E \cap H$ are non-empty and so $G \cup H$ is a disconnection of E. But this contradicts the fact that E is a component and so is connected. Hence E consists of exactly one point.

LOCALLY CONNECTED SPACES

28. Prove: Let E be a component in a locally connected space X. Then E is open.

Solution:

Let $p \in E$. Since X is locally connected, p belongs to at least one open connected set G_p. But E is the component of p; hence

$$p \in G_p \subset E \quad \text{and so} \quad E = \bigcup\{G_p : p \in E\}$$

Therefore E is open, as it is the union of open sets.

29. Prove: Let X and Y be locally connected. Then $X \times Y$ is locally connected.

Solution:

Now X is locally connected iff X possesses a base $\mathcal{B}$ consisting of connected sets. Similarly, Y possesses a base $\mathcal{B}^*$ consisting of connected sets. But $X \times Y$ is a finite product; hence

$$\{G \times H : G \in \mathcal{B},\ H \in \mathcal{B}^*\}$$

is a base for the product space $X \times Y$. Now each $G \times H$ is connected since G and H are connected. In other words, $X \times Y$ possesses a base consisting of connected sets and so $X \times Y$ is locally connected.

30. Prove: Let $\{X_i\}$ be a collection of connected locally connected spaces. Then the product space $X = \prod_i X_i$ is locally connected.

Solution:

Let G be an open subset of X containing $p = \langle a_i : i \in I \rangle \in X$. Then there exists a member of the defining base

$$B \;=\; G_{i_1} \times \cdots \times G_{i_m} \times \prod \{X_i : i \neq i_1, \ldots, i_m\}$$

such that $p \in B \subset G$, and so $a_{i_k} \in G_{i_k}$. Now each coordinate space is locally connected, and so there exists connected open subsets $H_{i_k} \subset X_{i_k}$ such that

$$a_{i_1} \in H_{i_1} \subset G_{i_1}, \;\; \ldots, \;\; a_{i_m} \in H_{i_m} \subset G_{i_m}$$

Set
$$H \;=\; H_{i_1} \times \cdots \times H_{i_m} \times \prod \{X_i : i \neq i_1, \ldots, i_m\}$$

Since each X_i is connected and each H_{i_k} is connected, H is also connected. Furthermore, H is open and $p \in H \subset B \subset G$. Accordingly, X is locally connected.

Supplementary Problems

CONNECTED SPACES

31. Show that if $(X, \mathcal{T})$ is connected and $\mathcal{T}^* \precsim \mathcal{T}$, then $(X, \mathcal{T}^*)$ is connected.

32. Show that if $(X, \mathcal{T})$ is disconnected and $\mathcal{T} \precsim \mathcal{T}^*$, then $(X, \mathcal{T}^*)$ is disconnected.

33. Show that every indiscrete space is connected.

34. Show, by a counterexample, that connectedness is not a hereditary property.

35. Prove: If $A_1, A_2, \ldots$ is a sequence of connected sets such that A_1 and A_2 are not separated, A_2 and A_3 are not separated, etc., then $A_1 \cup A_2 \cup \cdots$ is connected.

36. Prove: Let E be a connected subset of a T_1-space containing more than one element. Then E is infinite.

37. Prove: A topological space X is connected if and only if every non-empty proper subset of X has a non-empty boundary.

COMPONENTS

38. Determine the components of a discrete space.

39. Determine the components of a cofinite space.

40. Show that any pair of components are separated.

41. Prove: If X has a finite number of components, then each component is both open and closed.

42. Prove: If E is a non-empty connected subset of X which is both open and closed, then E is a component.

43. Prove: Let E be a component of Y and let $f : X \to Y$ be continuous. Then $f^{-1}[E]$ is a union of components of X.

44. Prove: Let X be a compact space. If the components of X are open, then there are only a finite number of them.

ARCWISE CONNECTED SETS

45. Show that an indiscrete space is arcwise connected.

46. Prove: The arcwise connected components of X form a partition of X.

47. Prove: Every component of X is partitioned by arcwise connected components.

MISCELLANEOUS PROBLEMS

48. Show that an indiscrete space is simply connected.

49. Show that a totally disconnected space is Hausdorff.

50. Prove: Let G be an open subset of a locally connected space X. Then G is locally connected.

51. Let $A = \{a, b\}$ be discrete and let $I = [0, 1]$. Show that the product space $X = \prod \{A_i : A_i = A, i \in I\}$ is not locally connected. Hence locally connectedness is not product invariant.

52. Show that "simply connected" is a topological property.

53. Prove: Let X be locally connected. Then X is connected if and only if there exists a simple chain of connected sets joining any pair of points in X.

Chapter 14

Complete Metric Spaces

CAUCHY SEQUENCES

Let X be a metric space. A sequence $\langle a_1, a_2, \ldots \rangle$ in X is a *Cauchy sequence* iff for every $\epsilon > 0$,

$$\exists\, n_0 \in \mathbf{N} \quad \text{such that} \quad n, m > n_0 \;\Rightarrow\; d(a_n, a_m) < \epsilon$$

Hence, in the case that X is a normed space, $\langle a_n \rangle$ is a Cauchy sequence iff for every $\epsilon > 0$,

$$\exists\, n_0 \in \mathbf{N} \quad \text{such that} \quad n, m > n_0 \;\Rightarrow\; \|a_n - a_m\| < \epsilon$$

Example 1.1: Let $\langle a_n \rangle$ be a convergent sequence; say $a_n \to p$. Then $\langle a_n \rangle$ is necessarily a Cauchy sequence since, for every $\epsilon > 0$,

$$\exists\, n_0 \in \mathbf{N} \quad \text{such that} \quad n > n_0 \;\Rightarrow\; d(a_n, p) < \tfrac{1}{2}\epsilon$$

Hence, by the Triangle Inequality,

$$n, m > n_0 \;\Rightarrow\; d(a_n, a_m) \le d(a_n, p) + d(a_m, p) < \tfrac{1}{2}\epsilon + \tfrac{1}{2}\epsilon = \epsilon$$

In other words, $\langle a_n \rangle$ is a Cauchy sequence.

We state the result of Example 1.1 as a proposition.

Proposition 14.1: Every convergent sequence in a metric space is a Cauchy sequence.

The converse of Proposition 14.1 is not true, as seen in the next example.

Example 1.2: Let $X = (0, 1)$ with the usual metric. Then $\langle \tfrac{1}{2}, \tfrac{1}{3}, \tfrac{1}{4}, \ldots \rangle$ is a sequence in X which is Cauchy but which does not converge in X.

Example 1.3: Let d be the trivial metric on any set X and let $\langle a_n \rangle$ be a Cauchy sequence in (X, d). Recall that d is defined by

$$d(a, b) = \begin{cases} 0 & \text{if } a = b \\ 1 & \text{if } a \ne b \end{cases}$$

Let $\epsilon = \tfrac{1}{2}$. Then, since $\langle a_n \rangle$ is Cauchy, $\exists\, n_0 \in \mathbf{N}$ such that

$$n, m > n_0 \;\Rightarrow\; d(a_n, a_m) < \tfrac{1}{2} \;\Rightarrow\; a_n = a_m$$

In other words, $\langle a_n \rangle$ is of the form $\langle a_1, a_2, \ldots, a_{n_0}, p, p, p, \ldots \rangle$, i.e. constant from some term on.

Example 1.4: Let $\langle p_1, p_2, \ldots \rangle$ be a Cauchy sequence in Euclidean m-space $\mathbf{R}^m$; say,

$$p_1 = \langle a_1^{(1)}, \ldots, a_1^{(m)} \rangle, \quad p_2 = \langle a_2^{(1)}, \ldots, a_2^{(m)} \rangle, \quad \ldots$$

The projections of $\langle p_n \rangle$ into each of the m coordinate spaces, i.e.,

$$\langle a_1^{(1)}, a_2^{(1)}, a_3^{(1)}, \ldots \rangle, \quad \ldots, \quad \langle a_1^{(m)}, a_2^{(m)}, a_3^{(m)}, \ldots \rangle \tag{1}$$

are Cauchy sequences in $\mathbf{R}$, for, let $\epsilon > 0$. Since $\langle p_n \rangle$ is Cauchy, $\exists\, n_0 \in \mathbf{N}$ such that

$$r, s > n_0 \;\Rightarrow\; d(p_r, p_s)^2 = |a_r^{(1)} - a_s^{(1)}|^2 + \cdots + |a_r^{(m)} - a_s^{(m)}|^2 < \epsilon^2$$

Hence, in particular,

$$r, s > n_0 \;\Rightarrow\; |a_r^{(1)} - a_s^{(1)}|^2 < \epsilon^2, \quad \ldots, \quad |a_r^{(m)} - a_s^{(m)}|^2 < \epsilon^2$$

In other words, each of the m sequences in (1) is a Cauchy sequence.

COMPLETE METRIC SPACES

Definition: A metric space (X, d) is *complete* if every Cauchy sequence $\langle a_n \rangle$ in X converges to a point $p \in X$.

Example 2.1: By the fundamental Cauchy Convergence Theorem (see Page 52), the real line **R** with the usual metric is complete.

Example 2.2: Let d be the trivial metric on any set X. Now (see Example 1.3) a sequence $\langle a_n \rangle$ in X is Cauchy iff it is of the form $\langle a_1, a_2, \ldots, a_{n_0}, p, p, p, \ldots \rangle$, which clearly converges to $p \in X$. Thus every trivial metric space is complete.

Example 2.3: The open unit interval $X = (0, 1)$ with the usual metric is not complete since (see Example 1.2) the sequence $\langle \frac{1}{2}, \frac{1}{3}, \frac{1}{4}, \ldots \rangle$ in X is Cauchy but does not converge to a point in X.

Remark: Examples 2.1 and 2.3 show that completeness is not a topological property; for **R** is homeomorphic to $(0, 1)$ even though **R** is complete and $(0, 1)$ is not.

Example 2.4: Euclidean m-space $\mathbf{R}^m$ is complete. For, let $\langle p_1, p_2, \ldots \rangle$ be a Cauchy sequence in $\mathbf{R}^m$ where

$$p_1 = \langle a_1^{(1)}, \ldots, a_1^{(m)} \rangle, \quad p_2 = \langle a_2^{(1)}, \ldots, a_2^{(m)} \rangle, \quad \ldots$$

Then (see Example 1.4) the projections of $\langle p_n \rangle$ into the m coordinate spaces are Cauchy; and since **R** is complete, they converge:

$$\langle a_1^{(1)}, a_2^{(1)}, \ldots \rangle \to b_1, \quad \ldots, \quad \langle a_1^{(m)}, a_2^{(m)}, \ldots \rangle \to b_m$$

Thus $\langle p_n \rangle$ converges to the point $q = \langle b_1, \ldots, b_m \rangle \in \mathbf{R}^m$, since each of the m projections converges to the projection of q (see Page 169, Theorem 12.7).

PRINCIPLE OF NESTED CLOSED SETS

Recall that the diameter of a subset A of a metric space X, denoted by $d(A)$, is defined by $d(A) = \sup \{d(a, a') : a, a' \in A\}$ and that a sequence of sets, $A_1, A_2, \ldots$, is said to be nested if $A_1 \supset A_2 \supset \cdots$.

The next theorem gives a characterization of complete metric spaces analogous to the Nested Interval Theorem for the real numbers.

Theorem 14.2: A metric space X is complete if and only if every nested sequence of non-empty closed sets whose diameters tend to zero has a non-empty intersection.

In other words, if $A_1 \supset A_2 \supset \cdots$ are non-empty closed subsets of a complete metric space X such that $\lim_{n \to \infty} d(A_n) = 0$, then $\cap_{n=0}^{\infty} A_n \neq \emptyset$; and vice versa.

The next examples show that the conditions $\lim_{n \to \infty} d(A_n) = 0$ and that the A_i are closed, are both necessary in Theorem 14.2.

Example 3.1: Let X be the real line **R** and let $A_n = [n, \infty)$. Now X is complete, the A_n are closed, and $A_1 \supset A_2 \supset \cdots$. But $\cap_{n=1}^{\infty} A_n$ is empty. Observe that $\lim_{n \to \infty} d(A_n) \neq 0$.

Example 3.2: Let X be the real line **R** and let $A_n = (0, 1/n]$. Now X is complete, $A_1 \supset A_2 \supset \cdots$, and $\lim_{n \to \infty} d(A_n) = 0$. But $\cap_{n=1}^{\infty} A_n$ is empty. Observe that the A_n are not closed.

COMPLETENESS AND CONTRACTING MAPPINGS

Let X be a metric space. A function $f : X \to X$ is called a *contracting mapping* if there exists a real number α, $0 \leq \alpha < 1$, such that, for every $p, q \in X$,

$$d(f(p), f(q)) \leq \alpha \, d(p, q) < d(p, q)$$

Thus, in a contracting mapping, the distance between the images of any two points is less than the distance between the points.

Example 4.1:   Let f be the function on Euclidean 2-space $\mathbf{R}^2$, i.e. $f: \mathbf{R}^2 \to \mathbf{R}^2$, defined by $f(p) = \frac{1}{2}p$. Then f is contracting, for

$$d(f(p), f(q)) \;=\; \|f(p) - f(q)\| \;=\; \|\tfrac{1}{2}p - \tfrac{1}{2}q\|$$

$$=\; \tfrac{1}{2}\|p - q\| \;=\; \tfrac{1}{2}\, d(p, q)$$

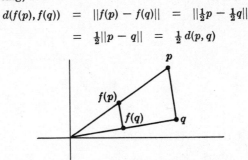

If X is a complete metric space, then we have the following "fixed point" theorem which has many applications in analysis.

Theorem 14.3:   If f is a contracting mapping on a complete metric space X, then there exists a unique point $p \in X$ such that $f(p) = p$.

COMPLETIONS

A metric space X^* is called a *completion* of a metric space X if X^* is complete and X is isometric to a dense subset of X^*.

Example 5.1:   The set $\mathbf{R}$ of real numbers is a completion of the set $\mathbf{Q}$ of rational numbers, since $\mathbf{R}$ is complete and $\mathbf{Q}$ is a dense subset of $\mathbf{R}$.

We now outline one particular construction of a completion of an arbitrary metric space X. Let $C[X]$ denote the collection of all Cauchy sequences in X and let $\sim$ be the relation in $C[X]$ defined by

$$\langle a_n \rangle \sim \langle b_n \rangle \qquad \text{iff} \qquad \lim_{n \to \infty} d(a_n, b_n) = 0$$

Thus, under "$\sim$" we identify those Cauchy sequences which "should" have the same "limit".

Lemma 14.4:   The relation $\sim$ is an equivalence relation in $C[X]$.

Now let X^* denote the quotient set $C[X]/\sim$, i.e. X^* consists of equivalence classes $[\langle a_n \rangle]$ of Cauchy sequences $\langle a_n \rangle \in C[X]$. Let e be the function defined by

$$e([\langle a_n \rangle],\, [\langle b_n \rangle]) \;=\; \lim_{n \to \infty} d(a_n, b_n)$$

where $[\langle a_n \rangle], [\langle b_n \rangle] \in X^*$.

Lemma 14.5:   The function e is well-defined, i.e. $\langle a_n \rangle \sim \langle a_n^* \rangle$ and $\langle b_n \rangle \sim \langle b_n^* \rangle$ implies $\lim\limits_{n \to \infty} d(a_n, b_n) = \lim\limits_{n \to \infty} d(a_n^*, b_n^*)$.

In other words, e does not depend upon the particular Cauchy sequence chosen to represent any equivalence class. Furthermore,

Lemma 14.6:   The function e is a metric on X^*.

Now for each $p \in X$, the sequence $\langle p, p, p, \ldots \rangle \in C[X]$, i.e. is Cauchy. Set

$$\hat{p} \;=\; [\langle p, p, \ldots \rangle] \quad \text{and} \quad \hat{X} = \{\hat{p} : p \in X\}$$

Then $\hat{X}$ is a subset of X^*.

Lemma 14.7: X is isometric to $\hat{X}$, and $\hat{X}$ is dense in X^*.

Lemma 14.8:   Every Cauchy sequence in X^* converges, and so X^* is a completion of X.

Lastly, we show that

Lemma 14.9:   If Y^* is any completion of X, then Y^* is isometric to X^*.

The previous lemmas imply the following fundamental result.

Theorem 14.10:   Every metric space X has a completion and all completions of X are isometric.

In other words, up to isometry, there exists a unique completion of any metric space.

BAIRE'S CATEGORY THEOREM

Recall that a subset A of a topological space X is *nowhere dense* in X iff the interior of the closure of A is empty:

$$\operatorname{int}(\bar{A}) = \varnothing$$

> **Example 6.1:**   The set **Z** of integers is a nowhere dense subset of the real line **R**. For **Z** is closed, i.e. $\mathbf{Z} = \bar{\mathbf{Z}}$, and its interior is empty; hence
> $$\operatorname{int}(\bar{\mathbf{Z}}) = \operatorname{int}(\mathbf{Z}) = \varnothing$$
> Similarly every finite subset of **R** is nowhere dense in **R**.
>
> On the other hand, the set **Q** of rational numbers is not nowhere dense in **R** since the closure of **Q** is **R** and so
> $$\operatorname{int}(\bar{\mathbf{Q}}) = \operatorname{int}(\mathbf{R}) = \mathbf{R} \neq \varnothing$$

A topological space X is said to be of *first category* (or *meager* or *thin*) if X is the countable union of nowhere dense subsets of X. Otherwise X is said to be of *second category* (or *non-meager* or *thick*).

> **Example 6.2:**   The set **Q** of rational numbers is of first category since the singleton subsets $\{p\}$ of **Q** are nowhere dense in **Q**, and **Q** is the countable union of singleton sets.

In view of Baire's Category Theorem, which follows, the real line **R** is of second category.

Theorem (Baire) 14.11:   Every complete metric space X is of second category.

COMPLETENESS AND COMPACTNESS

Let A be a subset of a metric space X. Now A is compact iff A is sequentially compact iff every sequence $\langle a_n \rangle$ in A has a convergent subsequence $\langle a_{i_n} \rangle$. But, by Example 1.1, $\langle a_{i_n} \rangle$ is a Cauchy sequence. Hence it is reasonable to expect that the notion of completeness is related to the notion of compactness and its related concept: total boundedness.

We state two such relationships:

Theorem 14.12:   A metric space X is compact if and only if it is complete and totally bounded.

Theorem 14.13:   Let X be a complete metric space. Then $A \subset X$ is compact if and only if A is closed and totally bounded.

CONSTRUCTION OF THE REAL NUMBERS

The real numbers can be constructed from the rational numbers by the method described in this chapter. Specifically, let **Q** be the set of rational numbers and let **R** be the collection of equivalence classes of Cauchy sequences in **Q**:

$$\mathbf{R} \;=\; \{[\langle a_n \rangle] : \langle a_n \rangle \text{ is a Cauchy sequence in } \mathbf{Q}\}$$

Now $\mathbf{R}$ with the appropriate metric is a complete metric space.

Remark: Let X be a normed vector space. The construction in this chapter gives us a complete metric space X^*. We can then define the following operations of vector addition, scalar multiplication and norm in X^* so that X^* is, in fact, a complete normed vector space, called a *Banach space*:

    (i) $[\langle a_n \rangle] + [\langle b_n \rangle] \equiv [\langle a_n + b_n \rangle]$ (ii) $k\,[\langle a_n \rangle] \equiv [\langle ka_n \rangle]$ (iii) $\|[\langle a_n \rangle]\| = \lim\limits_{n \to \infty} \|a_n\|$

Solved Problems

CAUCHY SEQUENCES

1. Show that every Cauchy sequence $\langle a_n \rangle$ in a metric space X is totally bounded (hence also bounded).

   Solution:
       Let $\epsilon > 0$. We want to show that there is a decomposition of $\{a_n\}$ into a finite number of sets, each with diameter less than ϵ. Since $\langle a_n \rangle$ is Cauchy, $\exists\, n_0 \in \mathbf{N}$ such that

   $$n, m > n_0 \;\Rightarrow\; d(a_n, a_m) < \epsilon$$

   Accordingly, $B = \{a_{n_0+1}, a_{n_0+2}, \ldots\}$ has diameter at most ϵ. Thus $\{a_1\}, \ldots, \{a_{n_0}\}, B$ is a finite decomposition of $\{a_n\}$ into sets with diameter less than ϵ, and so $\langle a_n \rangle$ is totally bounded.

2. Let $\langle a_1, a_2, \ldots \rangle$ be a sequence in a metric space X, and let

   $$A_1 = \{a_1, a_2, \ldots\}, \quad A_2 = \{a_2, a_3, \ldots\}, \quad A_3 = \{a_3, a_4, \ldots\}, \quad \ldots$$

   Show that $\langle a_n \rangle$ is a Cauchy sequence if and only if the diameters of the A_n tend to zero, i.e. $\lim\limits_{n \to \infty} d(A_n) = 0$.

   Solution:
       Suppose $\langle a_n \rangle$ is a Cauchy sequence. Let $\epsilon > 0$. Then

   $$\exists\, n_0 \in \mathbf{N} \quad \text{such that} \quad n, m > n_0 \Rightarrow d(a_n, a_m) < \epsilon$$

   Accordingly, $n > n_0 \;\Rightarrow\; d(A_n) < \epsilon$ and so $\lim\limits_{n \to \infty} (A_n) = 0$

       On the other hand, suppose $\lim\limits_{n \to \infty} d(A_n) = 0$. Let $\epsilon > 0$. Then

   $$\exists\, n_0 \in \mathbf{N} \quad \text{such that} \quad d(A_{n_0+1}) < \epsilon$$

   Hence $n, m > n_0 \;\Rightarrow\; a_n, a_m \in A_{n_0+1} \;\Rightarrow\; d(a_n, a_m) < \epsilon$

   and so $\langle a_n \rangle$ is a Cauchy sequence.

3. Let $\langle a_1, a_2, \ldots \rangle$ be a Cauchy sequence in X and let $\langle a_{i_1}, a_{i_2}, \ldots \rangle$ be a subsequence of $\langle a_n \rangle$. Show that $\lim\limits_{n \to \infty} d(a_n, a_{i_n}) = 0$.

   Solution:
       Let $\epsilon > 0$. Since $\langle a_n \rangle$ is a Cauchy sequence,

   $$\exists\, n_0 \in \mathbf{N} \quad \text{such that} \quad n, m > n_0 - 1 \;\Rightarrow\; d(a_n, a_m) < \epsilon$$

   Now $i_{n_0} \geq n_0 > n_0 - 1$ and therefore $d(a_{n_0}, a_{i_{n_0}}) < \epsilon$. In other words, $\lim\limits_{n \to \infty} d(a_n, a_{i_n}) = 0$.

4. Let $\langle a_1, a_2, \ldots \rangle$ be a Cauchy sequence in X and let $\langle a_{i_1}, a_{i_2}, \ldots \rangle$ be a subsequence of $\langle a_n \rangle$ converging to $p \in X$. Show that $\langle a_n \rangle$ also converges to p.

Solution:

By the Triangle Inequality, $d(a_n, p) \le d(a_n, a_{i_n}) + d(a_{i_n}, p)$ and therefore

$$\lim_{n \to \infty} d(a_n, p) \le \lim_{n \to \infty} d(a_n, a_{i_n}) + \lim_{n \to \infty} d(a_{i_n}, p)$$

Since $a_{i_n} \to p$, $\lim_{n \to \infty} d(a_{i_n}, p) = 0$ and, by the preceding problem, $\lim_{n \to \infty} d(a_n, a_{i_n}) = 0$. Then

$$\lim_{n \to \infty} d(a_n, p) = 0 \quad \text{and so} \quad a_n \to p$$

5. Let $\langle b_1, b_2, \ldots \rangle$ be a Cauchy sequence in a metric space X, and let $\langle a_1, a_2, \ldots \rangle$ be a sequence in X such that $d(a_n, b_n) < 1/n$ for every $n \in \mathbf{N}$.

(i) Show that $\langle a_n \rangle$ is also a Cauchy sequence in X.

(ii) Show that $\langle a_n \rangle$ converges to, say, $p \in X$ if and only if $\langle b_n \rangle$ converges to p.

Solution:

(i)   By the Triangle Inequality,

$$d(a_m, a_n) \le d(a_m, b_m) + d(b_m, b_n) + d(b_n, a_n)$$

Let $\epsilon > 0$. Then $\exists\, n_1 \in \mathbf{N}$ such that $1/n_1 < \epsilon/3$. Hence

$$n, m > n_1 \;\Rightarrow\; d(a_m, a_n) < \epsilon/3 + d(b_m, b_n) + \epsilon/3$$

By hypothesis, $\langle b_1, b_2, \ldots \rangle$ is a Cauchy sequence; hence

$$\exists\, n_2 \in \mathbf{N} \quad \text{such that} \quad n, m > n_2 \;\Rightarrow\; d(b_m, b_n) < \epsilon/3$$

Set $n_0 = \max\{n_1, n_2\}$. Then

$$n, m > n_0 \;\Rightarrow\; d(a_m, a_n) < \epsilon/3 + \epsilon/3 + \epsilon/3 = \epsilon$$

Thus $\langle a_n \rangle$ is a Cauchy sequence.

(ii)   By the Triangle Inequality, $d(b_n, p) \le d(b_n, a_n) + d(a_n, p)$; hence

$$\lim_{n \to \infty} d(b_n, p) \le \lim_{n \to \infty} d(b_n, a_n) + \lim_{n \to \infty} (a_n, p)$$

But $\lim_{n \to \infty} d(b_n, a_n) \le \lim_{n \to \infty} (1/n) = 0$. Hence, if $a_n \to p$, $\lim_{n \to \infty} d(b_n, p) \le \lim_{n \to \infty} (a_n, p) = 0$ and so $\langle b_n \rangle$ also converges to p.

Similarly, if $b_n \to p$ then $a_n \to p$.

COMPLETE SPACES

6. Prove Theorem 14.2: The following are equivalent: (i) X is a complete metric space. (ii) Every nested sequence of non-empty closed sets whose diameters tend to zero has a non-empty intersection.

Solution:

(i) $\Rightarrow$ (ii):

Let $A_1 \supset A_2 \supset \cdots$ be non-empty closed subsets of X such that $\lim_{n \to \infty} d(A_n) = 0$. We want to prove that $\cap_n A_n \ne \emptyset$. Since each A_i is non-empty, we can choose a sequence

$$\langle a_1, a_2, \ldots \rangle \quad \text{such that} \quad a_1 \in A_1,\; a_2 \in A_2,\; \ldots$$

We claim that $\langle a_n \rangle$ is a Cauchy sequence. Let $\epsilon > 0$. Since $\lim_{n \to \infty} d(A_n) = 0$,

$$\exists\, n_0 \in \mathbf{N} \quad \text{such that} \quad d(A_{n_0}) < \epsilon$$

But the A_i are nested; hence

$$n, m > n_0 \;\Rightarrow\; A_n, A_m \subset A_{n_0} \;\Rightarrow\; a_n, a_m \in A_{n_0} \;\Rightarrow\; d(a_n, a_m) < \epsilon$$

Thus $\langle a_n \rangle$ is Cauchy.

Now X is complete and so $\langle a_n \rangle$ converges to, say, $p \in X$. We claim that $p \in \cap_n A_n$. Suppose not, i.e. suppose

$$\exists \; k \in \mathbf{N} \quad \text{such that} \quad p \notin A_k$$

Since A_k is a closed set, the distance between p and A_k is non-zero; say, $d(p, A_k) = \delta > 0$. Then A_k and the open sphere $S = S(p, \tfrac{1}{2}\delta)$ are disjoint. Hence

$$n > k \quad \Rightarrow \quad a_n \in A_k \quad \Rightarrow \quad a_n \notin S(p, \tfrac{1}{2}\delta)$$

This is impossible since $a_n \to p$. In other words, $p \in \cap_n A_n$ and so $\cap_n A_n$ is non-empty.

(ii) $\Rightarrow$ (i):

Let $\langle a_1, a_2, \ldots \rangle$ be a Cauchy sequence in X. We want to show that $\langle a_n \rangle$ converges. Set

$$A_1 = \{a_1, a_2, \ldots\}, \quad A_2 = \{a_2, a_3, \ldots\}, \quad \ldots$$

i.e. $A_k = \{a_n : n \geq k\}$. Then $A_1 \supset A_2 \supset \cdots$ and, by Problem 2, $\lim_{n \to \infty} d(A_n) = 0$. Furthermore, since $d(\bar{A}) = d(A)$, where $\bar{A}$ is the closure of A, $\bar{A}_1 \supset \bar{A}_2 \supset \cdots$ is a sequence of non-empty closed sets whose diameters tend to zero. Therefore, by hypothesis, $\cap_n \bar{A}_n \neq \emptyset$; say, $p \in \cap_n \bar{A}_n$. We claim that the Cauchy sequence $\langle a_n \rangle$ converges to p.

Let $\epsilon > 0$. Since $\lim_{n \to \infty} d(\bar{A}) = 0$,

$$\exists \; n_0 \in \mathbf{N} \quad \text{such that} \quad d(\bar{A}_{n_0}) < \epsilon$$

and so

$$n > n_0 \quad \Rightarrow \quad a_n, p \in \bar{A}_{n_0} \quad \Rightarrow \quad d(a_n, p) < \epsilon$$

In other words, $\langle a_n \rangle$ converges to p.

---

7. Let X be a metric space and let $f: X \to X$ be a contracting mapping on X, i.e. there exists $\alpha \in \mathbf{R}$, $0 \leq \alpha < 1$, such that, for every $p, q \in X$, $d(f(p), f(q)) \leq \alpha \, d(p, q)$. Show that f is continuous.

Solution:

We show that f is continuous at each point $x_0 \in X$. Let $\epsilon > 0$. Then

$$d(x, x_0) < \epsilon \quad \Rightarrow \quad d(f(x), f(x_0)) \leq \alpha \, d(x, x_0) \leq \alpha\epsilon < \epsilon$$

and so f is continuous.

---

8. Prove Theorem 14.3: Let f be a contracting mapping on a complete metric space X, say

$$d(f(a), f(b)) \leq \alpha \, d(a, b), \qquad 0 \leq \alpha < 1$$

Then there exists one and only one point $p \in X$ such that $f(p) = p$.

Solution:

Let a_0 be any point in X. Set

$$a_1 = f(a_0), \; a_2 = f(a_1) = f^2(a_0), \; \ldots, \; a_n = f(a_{n-1}) = f^n(a_0), \; \ldots$$

We claim that $\langle a_1, a_2, \ldots \rangle$ is a Cauchy sequence. First notice that

$$d(f^{s+t}(a_0), f^t(a_0)) \leq \alpha \, d(f^{s+t-1}(a_0), f^{t-1}(a_0)) \leq \cdots \leq \alpha^t \, d(f^s(a_0), a_0)$$

$$\leq \alpha^t [d(a_0, f(a_0)) + d(f(a_0), f^2(a_0)) + \cdots + d(f^{s-1}(a_0), f^s(a_0))]$$

But $d(f^{i+1}(a_0), f^i(a_0)) \leq \alpha^i \, d(f(a_0), a_0)$ and so

$$d(f^{s+t}(a_0), f^t(a_0)) \leq \alpha^t \, d(f(a_0), a_0)(1 + \alpha + \alpha^2 + \cdots + \alpha^{s-1})$$

$$\leq \alpha^t \, d(f(a_0), a_0) \, [1/(1 - \alpha)]$$

since $(1 + \alpha + \alpha^2 + \cdots + \alpha^{s-1}) \leq 1/(1 - \alpha)$.

Now let $\epsilon > 0$ and set

$$\delta = \begin{cases} \epsilon(1 - \alpha) & \text{if } d(f(a_0), a_0) = 0 \\ \epsilon(1 - \alpha)/d(f(a_0), a_0) & \text{if } d(f(a_0), a_0) \neq 0 \end{cases}$$

Since $\alpha < 1$, $\exists\ n_0 \in \mathbf{N}$ such that $\alpha^{n_0} < \delta$

Hence if $r \geq s > n_0$,

$$d(a_s, a_r) \leq \alpha^s [1/(1 - \alpha)]\, d(f(a_0), a_0) < \delta [1/(1 - \alpha)]\, d(f(a_0), a_0) \leq \epsilon$$

and so $\langle a_n \rangle$ is a Cauchy sequence.

Now X is complete and so $\langle a_n \rangle$ converges to, say, $p \in X$. We claim that $f(p) = p$; for f is continuous and hence sequentially continuous, and so

$$f(p) = f\left(\lim_{n \to \infty} a_n \right) = \lim_{n \to \infty} f(a_n) = \lim_{n \to \infty} a_{n+1} = p$$

Lastly, we show that p is unique. Suppose $f(p) = p$ and $f(q) = q$; then

$$d(p, q) = d(f(p), f(q)) \leq \alpha\, d(p, q)$$

But $\alpha < 1$; hence $d(p, q) = 0$, i.e. $p = q$.

COMPLETIONS

9. Show that $\langle a_n \rangle \sim \langle b_n \rangle$ if and only if they are both subsequences of some Cauchy sequence $\langle c_n \rangle$.

Solution:

Suppose $\langle a_n \rangle \sim \langle b_n \rangle$, i.e. $\lim\limits_{n \to \infty} d(a_n, b_n) = 0$. Define $\langle c_n \rangle$ by

$$c_n = \begin{cases} a_{\frac{1}{2}n} & \text{if } n \text{ is even} \\ b_{\frac{1}{2}(n+1)} & \text{if } n \text{ is odd} \end{cases}$$

Thus $\langle c_n \rangle = \langle b_1, a_1, b_2, a_2, \ldots \rangle$. We claim $\langle c_n \rangle$ is a Cauchy sequence. For, let $\epsilon > 0$; now

$$\exists\ n_1 \in \mathbf{N} \quad \text{such that} \quad m, n > n_1 \ \Rightarrow\ d(a_m, a_n) < \tfrac{1}{2}\epsilon$$

$$\exists\ n_2 \in \mathbf{N} \quad \text{such that} \quad m, n > n_2 \ \Rightarrow\ d(b_m, b_n) < \tfrac{1}{2}\epsilon$$

$$\exists\ n_3 \in \mathbf{N} \quad \text{such that} \qquad n > n_3 \ \Rightarrow\ d(a_n, b_n) < \tfrac{1}{2}\epsilon$$

Set $n_0 = \max(n_1, n_2, n_3)$. We claim that

$$m, n > 2n_0 \ \Rightarrow\ d(c_m, c_n) < \epsilon$$

Note that $m > 2n_0 \ \Rightarrow\ \tfrac{1}{2}m > n_1, n_3;\ \tfrac{1}{2}(m+1) > n_2, n_3$

Thus m, n even $\ \Rightarrow\ c_m = a_{\frac{1}{2}m},\ c_n = a_{\frac{1}{2}n} \ \Rightarrow\ d(c_m, c_n) < \tfrac{1}{2}\epsilon < \epsilon$

m, n odd $\ \Rightarrow\ c_m = b_{\frac{1}{2}(m+1)},\ c_n = b_{\frac{1}{2}(n+1)} \ \Rightarrow\ d(c_m, c_n) < \tfrac{1}{2}\epsilon < \epsilon$

m even, n odd $\ \Rightarrow\ c_m = a_{\frac{1}{2}m},\ c_n = b_{\frac{1}{2}(n+1)} \ \Rightarrow$

$$d(c_m, c_n) \leq d(a_{\frac{1}{2}m}, b_{\frac{1}{2}m}) + d(b_{\frac{1}{2}m}, b_{\frac{1}{2}(n+1)}) < \tfrac{1}{2}\epsilon + \tfrac{1}{2}\epsilon = \epsilon$$

and so $\langle c_n \rangle$ is a Cauchy sequence.

Conversely, if there exists a Cauchy sequence $\langle c_n \rangle$ for which $\langle a_n \rangle = \langle c_{j_n} \rangle$ and $\langle b_n \rangle = \langle c_{k_n} \rangle$, then

$$\lim_{n \to \infty} d(a_n, b_n) = \lim_{n \to \infty} d(c_{j_n}, c_{k_n}) = 0$$

since $\langle c_n \rangle$ is Cauchy and $n \to \infty$ implies $j_n, k_n \to \infty$.

10. Prove Lemma 14.5: The function e is well-defined, i.e. $\langle a_n \rangle \sim \langle a_n^* \rangle$ and $\langle b_n \rangle \sim \langle b_n^* \rangle$ implies $\lim\limits_{n \to \infty} d(a_n, b_n) = \lim\limits_{n \to \infty} d(a_n^*, b_n^*)$.

Solution:

Set $r = \lim d(a_n, b_n)$ and $r^* = \lim d(a_n^*, b_n^*)$, and let $\epsilon > 0$. Note that

$$d(a_n, b_n) \ \leq \ d(a_n, a_n^*) + d(a_n^*, b_n^*) + d(b_n^*, b_n)$$

Now

$$\exists \ n_1 \in \mathbf{N} \quad \text{such that} \quad n > n_1 \ \Rightarrow \ d(a_n, a_n^*) < \epsilon/3$$
$$\exists \ n_2 \in \mathbf{N} \quad \text{such that} \quad n > n_2 \ \Rightarrow \ d(b_n, b_n^*) < \epsilon/3$$
$$\exists \ n_3 \in \mathbf{N} \quad \text{such that} \quad n > n_3 \ \Rightarrow \ |d(a_n^*, b_n^*) - r^*| < \epsilon/3$$

Accordingly, if $n > \max(n_1, n_2, n_3)$, then

$$d(a_n, b_n) < r^* + \epsilon \quad \text{and so} \quad \lim_{n \to \infty} d(a_n, b_n) = r \leq r^* + \epsilon$$

But this inequality holds for every $\epsilon > 0$; hence $r \leq r^*$. In the same manner we may show that $r^* \leq r$; thus $r = r^*$.

11. Let $\langle a_n \rangle$ be a Cauchy sequence in X. Show that $\alpha = [\langle a_n \rangle] \in X^*$ is the limit of the sequence $\langle \hat{a}_1, \hat{a}_2, \ldots \rangle$ in $\hat{X}$. (Here $\hat{X} = \{\hat{p} = [\langle p, p, p, \ldots \rangle] : p \in X\}$.)

Solution:

Since $\langle a_n \rangle$ is a Cauchy sequence in X,

$$\lim_{m \to \infty} e(\hat{a}_m, \alpha) \ = \ \lim_{m \to \infty} \left(\lim_{n \to \infty} d(a_m, a_n) \right) \ = \ \lim_{\substack{m \to \infty \\ n \to \infty}} d(a_m, a_n) \ = \ 0$$

Accordingly, $\langle \hat{a}_n \rangle \to \alpha$.

12. Prove Lemma 14.7: X is isometric to $\hat{X}$, and $\hat{X}$ is dense in X^*.

Solution:

For every $p, q \in X$,

$$e(\hat{p}, \hat{q}) \ = \ \lim_{n \to \infty} d(p, q) \ = \ d(p, q)$$

and so X is isometric to $\hat{X}$. We show that $\hat{X}$ is dense in X^* by showing that every point in X^* is the limit of a sequence in $\hat{X}$. Let $\alpha = [\langle a_1, a_2, \ldots \rangle]$ be an arbitrary point in X^*. Then $\langle a_n \rangle$ is a Cauchy sequence in X and so, by the preceding problem, α is the limit of the sequence $\langle \hat{a}_1, \hat{a}_2, \ldots \rangle$ in $\hat{X}$. Thus $\hat{X}$ is dense in X^*.

13. Prove Lemma 14.8: Every Cauchy sequence in (X^*, e) converges, and so (X^*, e) is a completion of X.

Solution:

Let $\langle \alpha_1, \alpha_2, \ldots \rangle$ be a Cauchy sequence in X^*. Since $\hat{X}$ is dense in X^*, for every $n \in \mathbf{N}$,

$$\exists \ \hat{a}_n \in \hat{X} \quad \text{such that} \quad e(\hat{a}_n, \alpha_n) < 1/n$$

Then (Problem 5) $\langle \hat{a}_1, \hat{a}_2, \ldots \rangle$ is also a Cauchy sequence and, by Problem 12, $\langle \hat{a}_1, \hat{a}_2, \ldots \rangle$ converges to $\beta = [\langle a_1, a_2, \ldots \rangle] \in X^*$. Hence (Problem 5) $\langle \alpha_n \rangle$ also converges to β and therefore (X^*, e) is complete.

14. Prove Lemma 14.9: If Y^* is a completion of X, then Y^* is isometric to X^*.

Solution:

We can assume X is a subspace of Y^*. Hence, for every $p \in Y^*$, there exists a sequence $\langle a_1, a_2, \ldots \rangle$ in X converging to p; and in particular, $\langle a_n \rangle$ is a Cauchy sequence. Let $f : Y^* \to X^*$ be defined by

$$f(p) \ = \ [\langle a_1, a_2, \ldots \rangle]$$

Now if $\langle a_1^*, a_2^*, \ldots \rangle \in X$ also converges to p, then

$$\lim_{n \to \infty} d(a_n, a_n^*) \ = \ 0 \quad \text{and so} \quad [\langle a_n \rangle] \ = \ [\langle a_n^* \rangle]$$

In other words, f is well-defined.

Furthermore, f is onto. For if $[\langle b_1, b_2, \ldots \rangle] \in X^*$, then $\langle b_1, b_2, \ldots \rangle$ is a Cauchy sequence in $X \subset Y^*$ and, since Y^* is complete, $\langle b_n \rangle$ converges to, say, $q \in Y^*$. Accordingly, $f(q) = [\langle b_n \rangle]$.

Now let $p, q \in Y^*$ with, say, sequences $\langle a_n \rangle$ and $\langle b_n \rangle$ in X converging, respectively, to p and q. Then

$$e(f(p), f(q)) = e([\langle a_n \rangle], [\langle b_n \rangle]) = \lim_{n \to \infty} d(a_n, b_n) = d\left(\lim_{n \to \infty} a_n, \lim_{n \to \infty} b_n \right) = d(p, q)$$

Consequently, f is an isometry between Y^* and X^*.

BAIRE'S CATEGORY THEOREM

15. Let N be a nowhere dense subset of X. Show that $\overline{N}^c$ is dense in X.

Solution:

Suppose $\overline{N}^c$ is not dense in X, i.e. $\exists \; p \in X$ and an open set G such that

$$p \in G \quad \text{and} \quad G \cap \overline{N}^c = \emptyset$$

Then $p \in G \subset \overline{N}$ and so $p \in \text{int}\,(\overline{N})$. But this is impossible since N is nowhere dense in X, i.e. $\text{int}\,(\overline{N}) = \emptyset$. Therefore $\overline{N}^c$ is dense in X.

16. Let G be an open subset of the metric space X and let N be nowhere dense in X. Show that there exist $p \in X$ and $\delta > 0$ such that $S(p, \delta) \subset G$ and $S(p, \delta) \cap N = \emptyset$.

Solution:

Set $H = G \cap \overline{N}^c$. Then $H \subset G$ and $H \cap N = \emptyset$. Furthermore, H is non-empty since G is open and $\overline{N}^c$ is dense in X; say, $p \in H$. But H is open since G and $\overline{N}^c$ are open; hence $\exists \; \delta > 0$ such that $S(p, \delta) \subset H$. Consequently, $S(p, \delta) \subset G$ and $S(p, \delta) \cap N = \emptyset$.

17. Prove Theorem 14.11: Every complete metric space X is of second category.

Solution:

Let $M \subset X$ and let M be of first category. We want to show that $M \neq X$, i.e. $\exists \; p \in X$ such that $p \notin M$. Since M is of first category, $M = N_1 \cup N_2 \cup \cdots$ where each N_i is nowhere dense in X.

Since N_1 is nowhere dense in X, $\exists \; a_1 \in X$ and $\delta_1 > 0$ such that $S(a_1, \delta_1) \cap N_1 = \emptyset$. Set $\epsilon_1 = \delta_1/2$. Then

$$\overline{S(a_1, \epsilon_1)} \cap N_1 = \emptyset$$

Now $S(a_1, \epsilon_1)$ is open and N_2 is nowhere dense in X, and so, by Problem 16,

$$\exists \; a_2 \in X \text{ and } \delta_2 > 0 \quad \text{such that} \quad S(a_2, \delta_2) \subset S(a_1, \epsilon_1) \subset \overline{S(a_1, \epsilon_1)} \quad \text{and} \quad S(a_2, \delta_2) \cap N_2 = \emptyset$$

Set $\epsilon_2 = \delta_2/2 \leq \epsilon_1/2 = \delta_1/4$. Then

$$\overline{S(a_2, \epsilon_2)} \subset \overline{S(a_1, \epsilon_1)} \quad \text{and} \quad \overline{S(a_2, \epsilon_2)} \cap N_2 = \emptyset$$

Continuing in this manner, we obtain a nested sequence of closed sets

$$\overline{S(a_1, \epsilon_1)} \supset \overline{S(a_2, \epsilon_2)} \supset \overline{S(a_3, \epsilon_3)} \supset \cdots$$

such that, for every $n \in N$, $\quad \overline{S(a_n, \epsilon_n)} \cap N_n = \emptyset \quad$ and $\quad \epsilon_n \leq \delta_1/2^n$

Thus $\lim_{n \to \infty} \epsilon_n \leq \lim_{n \to \infty} \delta_1/2^n = 0$ and so, by Theorem 14.2,

$$\exists \; p \in X \quad \text{such that} \quad p \in \cap_{n=1}^{\infty} \overline{S(a_n, \epsilon_n)}$$

Furthermore, for every $n \in N$, $p \notin N_n$ and so $p \notin M$.

COMPLETENESS AND COMPACTNESS

18. Show that every compact metric space X is complete.

Solution:

Let $\langle a_1, a_2, \ldots \rangle$ be a Cauchy sequence in X. Now X is compact and so sequentially compact; hence $\langle a_n \rangle$ contains a subsequence $\langle a_{i_1}, a_{i_2}, \ldots \rangle$ which converges to, say, $p \in X$. But (Problem 4) $\langle a_n \rangle$ also converges to p. Hence X is complete.

19. Let E be a totally bounded subset of a metric space X. Show that every sequence $\langle a_n \rangle$ in E contains a Cauchy subsequence.

Solution:

Since E is totally bounded, we can decompose E into a finite number of subsets of diameter less than $\epsilon_1 = 1$. One of these sets, call it A_1, must contain an infinite number of the terms of the sequence; hence

$$\exists\ i_1 \in \mathbf{N} \qquad \text{such that} \qquad a_{i_1} \in A_1$$

Now A_1 is totally bounded and can be decomposed into a finite number of subsets of diameter less than $\epsilon_2 = \frac{1}{2}$. Similarly, one of these sets, call it A_2, must contain an infinite number of the terms of the sequence; hence

$$\exists\ i_2 \in \mathbf{N} \qquad \text{such that} \qquad i_2 > i_1 \text{ and } a_{i_2} \in A_2$$

Furthermore, $A_2 \subset A_1$.

We continue in this manner and obtain a nested sequence of sets

$$E \supset A_1 \supset A_2 \supset \cdots \qquad \text{with} \qquad d(A_n) < 1/n$$

and a subsequence $\langle a_{i_1}, a_{i_2}, \ldots \rangle$ of $\langle a_n \rangle$ with $a_{i_n} \in A_n$. We claim that $\langle a_{i_n} \rangle$ is a Cauchy sequence. For, let $\epsilon > 0$; then

$$\exists\ n_0 \in \mathbf{N} \qquad \text{such that} \qquad 1/n_0 < \epsilon \text{ and so } d(A_{n_0}) < \epsilon$$

Therefore $\qquad\qquad\qquad i_n, i_m > i_{n_0} \quad \Rightarrow \quad a_{i_n}, a_{i_m} \in A_{n_0} \quad \Rightarrow \quad d(a_{i_m}, d_{i_n}) < \epsilon$

20. Prove Theorem 14.12: A metric space X is compact if and only if X is complete and totally bounded.

Solution:

Suppose X is compact. Then, by Problem 15, X is complete and, by Lemma 11.17, Page 158, X is totally bounded.

On the other hand, suppose X is complete and totally bounded. Let $\langle a_1, a_2, \ldots \rangle$ be a sequence in X. Then, by the preceding problem, $\langle a_n \rangle$ contains a Cauchy subsequence $\langle a_{i_n} \rangle$ which converges since X is complete. Thus X is sequentially compact and therefore compact.

21. Prove Theorem 14.13: Let A be a subset of a complete metric space X. Then the following are equivalent: (i) A is compact. (ii) A is closed and totally bounded.

Solution:

If A is compact, then by Theorem 11.5 and Lemma 11.17 it is closed and totally bounded.

Conversely, suppose A is closed and totally bounded. Now a closed subset of a complete space is complete, and so A is complete and totally bounded. Hence, by the preceding problem, A is compact.

Supplementary Problems

COMPLETE METRIC SPACES

22. Let (X, d) be a metric space and let e be the metric on X defined by $e(a, b) = \min \{1, d(a, b)\}$. Show that $\langle a_n \rangle$ is a Cauchy sequence in (X, d) if and only if $\langle a_n \rangle$ is a Cauchy sequence in (X, e).

23. Show that every finite metric space is complete.

24. Prove: Every closed subspace of a complete metric space is complete.

25. Prove that Hilbert Space (l_2-space) is complete.

26. Prove: Let $\mathcal{B}(X, \mathbf{R})$ be the collection of bounded real-valued functions defined on X with norm

$$\|f\| = \sup \{|f(x)| : x \in X\}$$

Then $\mathcal{B}(X, \mathbf{R})$ is complete.

27. Prove: A metric space X is complete if and only if every infinite totally bounded subset of X has an accumulation point.

28. Show that a countable union of first category sets is of first category.

29. Show that a metric space X is totally bounded if and only if every sequence in X contains a Cauchy subsequence.

30. Show that if X is isometric to Y and X is complete, then Y is complete.

MISCELLANEOUS PROBLEM

31. Prove: Every normed vector space X can be densely embedded in a Banach space, i.e. a complete normed vector space. (*Hint*: See Remark on Page 199).

Chapter 15

Function Spaces

FUNCTION SPACES

Let X and Y be arbitrary sets, and let $\mathcal{F}(X, Y)$ denote the collection of all functions from X into Y. Any subcollection of $\mathcal{F}(X, Y)$ with some topology $\mathcal{T}$ is called a *function space*.

We can identify $\mathcal{F}(X, Y)$ with a product set as follows: Let Y_x denote a copy of Y indexed by $x \in X$, and let **F** denote the product of the sets Y_x, i.e.,

$$\textbf{F} = \prod \{Y_x : x \in X\}$$

Recall that **F** consists of all points $p = \langle a_x : x \in X \rangle$ which assign to each $x \in X$ the element $a_x \in Y_x = Y$, i.e. **F** consists of all functions from X into Y, and so $\textbf{F} = \mathcal{F}(X, Y)$.

Now for each element $x \in X$, the mapping e_x from the function set $\mathcal{F}(X, Y)$ into Y defined by

$$e_x(f) = f(x)$$

is called the *evaluation mapping* at x. (Here f is any function in $\mathcal{F}(X, Y)$, i.e. $f : X \to Y$.) Under our identification of $\mathcal{F}(X, Y)$ with **F**, the evaluation mapping e_x is precisely the projection mapping π_x from **F** into the coordinate space $Y_x = Y$.

Example 1.1: Let $\mathcal{F}(I, \mathbf{R})$ be the collection of all real-valued functions defined on $I = [0, 1]$, and let $f, g, h \in \mathcal{F}(I, \mathbf{R})$ be the functions

$$f(x) = x^2, \quad g(x) = 2x + 1, \quad h(x) = \sin \pi x$$

Consider the evaluation function $e_j : \mathcal{F}(I, \mathbf{R}) \to \mathbf{R}$ at, say, $j = \frac{1}{2}$. Then

$$e_j(f) = f(j) = f(\tfrac{1}{2}) = \tfrac{1}{4}$$
$$e_j(g) = g(j) = g(\tfrac{1}{2}) = 2$$
$$e_j(h) = h(j) = h(\tfrac{1}{2}) = 1$$

Graphically, $e_j(f)$, $e_j(g)$ and $e_j(h)$ are the points where the graphs of f, g and h intersect the vertical line R_j through $x = j$.

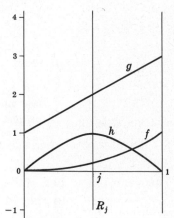

POINT OPEN TOPOLOGY

Let X be an arbitrary set and let Y be a topological space. We first investigate the product topology $\mathcal{T}$ on $\mathcal{F}(X, Y)$ where we identify $\mathcal{F}(X, Y)$ with the product set $\textbf{F} = \prod \{Y_x : x \in X\}$ as above. Recall that the defining subbase $\mathcal{S}$ of the product topology on **F** consists of all subsets of **F** of the form

$$\pi_{x_0}^{-1}[G] = \{f : \pi_{x_0}(f) \in G\}$$

where $x_0 \in X$ and G is an open subset of the coordinate space $Y_{x_0} = Y$. But $\pi_{x_0}(f) = e_{x_0}(f) = f(x_0)$, where e_{x_0} is the evaluation mapping at $x_0 \in X$. Hence the defining subbase $\mathcal{S}$ of the product topology $\mathcal{T}$ on $\mathcal{F}(X, Y)$ consists of all subsets of $\mathcal{F}(X, Y)$ of the form

$\{f : f(x_0) \in G\}$, i.e. all functions which map an arbitrary point $x_0 \in X$ into an arbitrary open set G of Y. We call this product topology on $\mathcal{F}(X, Y)$, appropriately, the *point open topology*.

Alternatively, we can define the point open topology on $\mathcal{F}(X, Y)$ to be the coarsest topology on $\mathcal{F}(X, Y)$ with respect to which the evaluation functions $e_x : \mathcal{F}(X, Y) \to Y$ are continuous. This definition corresponds directly to the definition of the product topology.

Example 2.1: Let $\mathcal{T}$ be the point open topology on $\mathcal{F}(I, \mathbf{R})$ where $I = [0, 1]$. As above, members of the defining subbase of $\mathcal{T}$ are of the form

$$\{f : f(j_0) \in G\}$$

where $j_0 \in I$ and G is an open subset of $\mathbf{R}$. Graphically, the above subbase element consists of all functions passing through the open set G on the vertical real line $\mathbf{R}$ through the point j_0 on the horizontal axis. Recall that this is identical to the subbase element of the product space

$$X = \prod \{R_i : i \in I\}$$

illustrated in Chapter 12, Page 170.

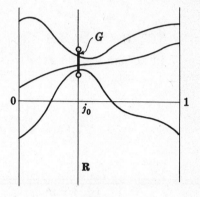

Example 2.2: If A is a subset of a product space $\prod \{X_i : i \in I\}$, then A is a subset of the product of its projections, i.e.

$$A \subset \prod \{\pi_i[A] : i \in I\}$$

(as indicated in the diagram).

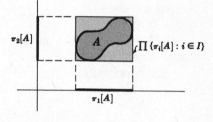

Thus $A \subset \prod \{\overline{\pi_i[A]} : i \in I\}$ where $\overline{\pi_i[A]}$ is the closure of $\pi_i[A]$. Accordingly, if $\mathcal{A} = \mathcal{A}(X, Y)$ is a subcollection of $\mathcal{F}(X, Y)$, then

$$\mathcal{A} \subset \prod \{\overline{\pi_x[\mathcal{A}]} : x \in X\} = \prod \{\overline{e_x[\mathcal{A}]} : x \in X\}$$

and $\overline{e_x[\mathcal{A}]} = \overline{\{f(x) : f \in \mathcal{A}\}}$. By the Tychonoff Product Theorem, if $\overline{\{f(x) : x \in X\}}$ is compact for every $x \in X$, then $\prod \{\overline{\pi_x[\mathcal{A}]} : x \in X\}$ is a compact subset of the product space $\prod \{Y_x : x \in X\}$.

Recall that a closed subset of a compact set is compact. Hence the result of Example 2.2 implies

Theorem 15.1: Let $\mathcal{A}$ be a subcollection of $\mathcal{F}(X, Y)$. Then $\mathcal{A}$ is compact with respect to the point open topology on $\mathcal{F}(X, Y)$ if (i) $\mathcal{A}$ is a closed subset of $\mathcal{F}(X, Y)$ and (ii) for every $x \in X$, $\overline{\{f(x) : f \in \mathcal{A}\}}$ is compact in Y.

In the case that Y is Hausdorff we have the following stronger result:

Theorem 15.2: Let Y be a Hausdorff space and let $\mathcal{A} \subset \mathcal{F}(X, Y)$. Then $\mathcal{A}$ is compact with respect to the point open topology if and only if $\mathcal{A}$ is closed and, for every $x \in X$, $\overline{\{f(x) : f \in \mathcal{A}\}}$ is compact.

POINTWISE CONVERGENCE

Let $\langle f_1, f_2, \ldots \rangle$ be a sequence of functions from an arbitrary set X into a topological space Y. The sequence $\langle f_n \rangle$ is said to converge *pointwise* to a function $g : X \to Y$ if, for every $x_0 \in X$,

$$\langle f_1(x_0), f_2(x_0), \ldots \rangle \text{ converges to } g(x_0), \quad \text{i.e. } \lim_{n \to \infty} f_n(x_0) = g(x_0)$$

In particular, if Y is a metric space then $\langle f_n \rangle$ converges pointwise to g iff for every $\epsilon > 0$ and every $x_0 \in X$,

$$\exists \, n_0 = n_0(x_0, \epsilon) \in \mathbf{N} \quad \text{such that} \quad n > n_0 \;\Rightarrow\; d(f_n(x_0), g(x_0)) < \epsilon$$

Note that the n_0 depends upon the ϵ and also upon the point x_0.

Example 3.1: Let $\langle f_1, f_2, \ldots \rangle$ be the sequence of functions from $I = [0,1]$ into $\mathbf{R}$ defined by

$$f_1(x) = x, \;\; f_2(x) = x^2, \;\; f_3(x) = x^3, \;\; \ldots$$

Then $\langle f_n \rangle$ converges pointwise to the function $g : I \to \mathbf{R}$ defined by

$$g(x) \;=\; \begin{cases} 0 & \text{if } 0 \leq x < 1 \\ 1 & \text{if } x = 1 \end{cases}$$

Observe that the limit function g is not continuous even though each of the functions f_i is continuous.

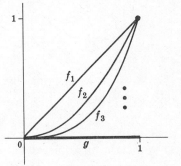

The notion of pointwise convergence is related to the point open topology as follows:

Theorem 15.3: A sequence of functions $\langle f_1, f_2, \ldots \rangle$ in $\mathcal{F}(X, Y)$ converges to $g \in \mathcal{F}(X, Y)$ with respect to the point open topology on $\mathcal{F}(X, Y)$ if and only if $\langle f_n \rangle$ converges pointwise to g.

In view of the above theorem, the point open topology on $\mathcal{F}(X, Y)$ is also called the *topology of pointwise convergence*.

Remark: Recall that metrizability is not invariant under uncountable products; therefore, the topology of pointwise convergence of real-valued functions defined on $[0,1]$ is not a metric topology. The theory of topological spaces, as a generalization of metric spaces, was first motivated by the study of pointwise convergence of functions.

UNIFORM CONVERGENCE

Let $\langle f_1, f_2, \ldots \rangle$ be a sequence of functions from an arbitrary set X into a metric space (Y, d). Then $\langle f_n \rangle$ is said to converge uniformly to a function $g : X \to Y$ if, for every $\epsilon > 0$,

$$\exists \, n_0 = n_0(\epsilon) \in \mathbf{N} \quad \text{such that} \quad n > n_0 \;\Rightarrow\; d(f_n(x), g(x)) < \epsilon, \;\; \forall \, x \in X$$

In particular, $\langle f_n \rangle$ converges pointwise to g; that is, uniform convergence implies pointwise convergence. Observe that the n_0 depends only on the ϵ, whereas, in pointwise convergence, the n_0 depends on both the ϵ and the point x.

In the case where X is a topological space, we have the following classical result:

Proposition 15.4: Let $\langle f_1, f_2, \ldots \rangle$ be a sequence of continuous functions from a topological space X into a metric space Y. If $\langle f_n \rangle$ converges uniformly to $g : X \to Y$, then g is continuous.

Example 4.1: Let $f_1, f_2, \ldots$ be the following continuous functions from $I = [0,1]$ into $\mathbf{R}$:

$$f_1(x) = x, \;\; f_2(x) = x^2, \;\; f_3(x) = x^3, \;\; \ldots$$

Now, by Example 3.1, $\langle f_n \rangle$ converges pointwise to $g : I \to \mathbf{R}$ defined by

$$g(x) \;=\; \begin{cases} 0 & \text{if } 0 \leq x < 1 \\ 1 & \text{if } x = 1 \end{cases}$$

Since g is not continuous, $\langle f_n \rangle$ does not converge uniformly to g.

Example 4.2:   Let $\langle f_1, f_2, \ldots \rangle$ be the following sequence of functions in $\mathcal{F}(\mathbf{R}, \mathbf{R})$:

$$f_n(x) \;=\; \begin{cases} 1 - \dfrac{1}{n}|x| & \text{if } |x| < n \\ 0 & \text{if } |x| \geqq n \end{cases}$$

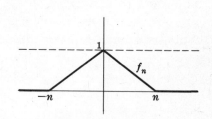

Now $\langle f_n \rangle$ converges pointwise to the constant function $g(x) = 1$. But $\langle f_n \rangle$ does not converge uniformly to g. For, let $\epsilon = \frac{1}{2}$. Note that, for every $n \in \mathbf{N}$, there exist points $x_0 \in \mathbf{R}$ with $f_n(x_0) = 0$ and so $|f_n(x_0) - g(x_0)| = 1 > \epsilon$.

Let $\mathcal{B}(X, Y)$ denote the collection of all bounded functions from an arbitrary set X into a metric space (Y, d), and let e be the metric on $\mathcal{B}(X, Y)$ defined by

$$e(f, g) \;=\; \sup \{d(f(x), g(x)) : x \in X\}$$

This metric has the following property:

Theorem 15.5:   Let $\langle f_1, f_2, \ldots \rangle$ be a sequence of functions in $\mathcal{B}(X, Y)$. Then $\langle f_n \rangle$ converges to $g \in \mathcal{B}(X, Y)$ with respect to the metric e if and only if $\langle f_n \rangle$ converges uniformly to g.

In view of the above theorem, the topology on $\mathcal{B}(X, Y)$ induced by the above metric is called the *topology of uniform convergence*.

Remark: The concept of uniform convergence defined in the case of a metric space Y cannot be defined for a general topological space. However, the notion of uniform convergence can be generalized to a collection of spaces, called *uniform spaces*, which lie between topological spaces and metric spaces.

THE FUNCTION SPACE $C[0, 1]$

The vector space $C[0, 1]$ of all continuous functions from $I = [0, 1]$ into $\mathbf{R}$ with norm defined by

$$\|f\| \;=\; \sup \{|f(x)| : x \in I\}$$

is one of the most important function spaces in analysis. Note that the above norm induces the topology of uniform convergence.

Since $I = [0, 1]$ is compact, each $f \in C[0, 1]$ is *uniformly continuous*; that is,

Proposition 15.6:   Let $f : [0, 1] \to \mathbf{R}$ be continuous. Then for every $\epsilon > 0$,

$$\exists \, \delta = \delta(\epsilon) > 0 \quad \text{such that} \quad |x_0 - x_1| < \delta \;\Rightarrow\; |f(x_0) - f(x_1)| < \epsilon$$

Uniform continuity (like uniform convergence) is stronger than continuity in that the δ depends only on ϵ and not on any particular point.

One consequence of Proposition 15.4 follows:

Theorem 15.7: $C[0, 1]$ is a complete normed vector space.

We shall use the Baire Category Theorem for complete metric spaces to prove the following interesting result:

Proposition 15.8:   There exists a continuous function $f : [0, 1] \to \mathbf{R}$ which is nowhere differentiable.

Remark: All the results proven here for $C[0, 1]$ are also true for the space $C[a, b]$ of all continuous functions on the closed interval $[a, b]$.

UNIFORM BOUNDEDNESS

In establishing necessary and sufficient conditions for subsets of function spaces to be compact, we are led to the concepts of *uniform boundedness* and *equicontinuity* which are interesting in their own right.

A collection of real valued functions $\mathcal{A} = \{f_i : X \to \mathbf{R}\}$ defined on an arbitrary set X is said to be *uniformly bounded* if

$$\exists\, M \in \mathbf{R} \quad \text{such that} \quad |f(x)| \leq M,\ \forall f \in \mathcal{A},\ \forall x \in X$$

That is, each function $f \in \mathcal{A}$ is bounded and there is one bound which holds for all of the functions.

In particular if $\mathcal{A} \subset C[0,1]$, then uniform boundedness is equivalent to

$$\exists\, M \in \mathbf{R} \quad \text{such that} \quad \|f\| \leq M,\ \forall f \in \mathcal{A}$$

or, $\mathcal{A}$ is a bounded subset of $C[0,1]$.

> **Example 5.1:** Let $\mathcal{A}$ be the following subset of $\mathcal{F}(\mathbf{R}, \mathbf{R})$:
> $$\mathcal{A} = \{f_1(x) = \sin x,\ f_2(x) = \sin 2x,\ \ldots\}$$
> Then $\mathcal{A}$ is uniformly bounded. For, let $M = 1$; then, for every $f \in \mathcal{A}$ and every $x \in \mathbf{R}$, $|f(x)| \leq M$. See Fig. (a) below.

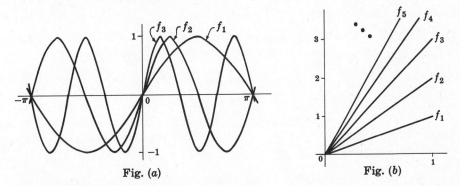

Fig. (a)   Fig. (b)

> **Example 5.2:** Let $\mathcal{A} \subset C[0,1]$ be defined as follows (see Fig. (b) above):
> $$\mathcal{A} = \{f_1(x) = x,\ f_2(x) = 2x,\ f_3(x) = 3x,\ \ldots\}$$
> Although each function in $C[0,1]$, and in particular in $\mathcal{A}$, is bounded, $\mathcal{A}$ is not uniformly bounded. For if M is any real number, however large, $\exists\, n_0 \in \mathbf{N}$ with $n_0 > M$ and hence $f_{n_0}(1) = n_0 > M$.

EQUICONTINUITY. ASCOLI'S THEOREM

A collection of real-valued functions $\mathcal{A} = \{f_i : X \to \mathbf{R}\}$ defined on an arbitrary metric space X is said to be *equicontinuous* if for every $\epsilon > 0$,

$$\exists\, \delta = \delta(\epsilon) > 0 \quad \text{such that} \quad d(x_0, x_1) < \delta \ \Rightarrow\ |f(x_0) - f(x_1)| < \epsilon,\ \forall f \in \mathcal{A}$$

Note that δ depends only on ϵ and not on any particular point or function. It is clear that each $f \in \mathcal{A}$ is uniformly continuous.

> **Theorem (Ascoli) 15.9:** Let $\mathcal{A}$ be a closed subset of the function space $C[0,1]$. Then $\mathcal{A}$ is compact if and only if $\mathcal{A}$ is uniformly bounded and equicontinuous.

COMPACT OPEN TOPOLOGY

Let X and Y be arbitrary sets and let $A \subset X$ and $B \subset Y$. We shall write $F(A, B)$ for the class of functions from X into Y which carry A into B:

$$F(A, B) = \{f \in \mathcal{F}(X, Y) : f[A] \subset B\}$$

Example 6.1: Let $\mathcal{S}$ be the defining subbase for the point open topology on $\mathcal{F}(X,Y)$. Recall that the members of $\mathcal{S}$ are of the form

$$\{f \in \mathcal{F}(X,Y) : f(x) \in G\}, \quad \text{where } x \in X, \ G \text{ an open subset of } Y$$

Following the above notation, we denote this set by $F(x,G)$ and we can then define $\mathcal{S}$ by

$$\mathcal{S} = \{F(x,G) : x \in X, \ G \subset Y \text{ open}\}$$

Now let X and Y be topological spaces and let $\mathcal{A}$ be the class of compact subsets of X and $\mathcal{G}$ be the class of open subsets of Y. The topology $\mathcal{T}$ on $\mathcal{F}(X,Y)$ generated by

$$\mathcal{S} = \{F(A,G) : A \in \mathcal{A}, \ G \in \mathcal{G}\}$$

is called the *compact open topology* on $\mathcal{F}(X,Y)$, and $\mathcal{S}$ is a defining subbase for $\mathcal{T}$.

Since singleton subsets of X are compact, $\mathcal{S}$ contains the members of the defining subbase for the point open topology on $\mathcal{F}(X,Y)$. Thus:

Theorem 15.10: The point open topology on $\mathcal{F}(X,Y)$ is coarser than the compact open topology on $\mathcal{F}(X,Y)$.

Recall that the point open topology is the coarsest topology with respect to which the evaluation mappings are continuous. Hence,

Corollary 15.11: The evaluation functions $e_x : \mathcal{F}(X,Y) \to Y$ are continuous relative to the compact open topology on $\mathcal{F}(X,Y)$.

TOPOLOGY OF COMPACT CONVERGENCE

Let $\langle f_1, f_2, \ldots \rangle$ be a sequence of functions from a topological space X into a metric space (Y,d). The sequence $\langle f_n \rangle$ is said to *converge uniformly on compacta* to $g : X \to Y$ if for every compact subset $E \subset X$ and every $\epsilon > 0$,

$$\exists \ n_0 = n_0(E, \epsilon) \in \mathbf{N} \quad \text{such that} \quad n > n_0 \ \Rightarrow \ d(f_n(x), g(x)) < \epsilon, \ \forall \, x \in E$$

In other words, $\langle f_n \rangle$ converges uniformly on compacta to g iff, for every compact subset $E \subset X$, the restriction of $\langle f_n \rangle$ to E converges uniformly to the restriction of g to E, i.e.,

$$\langle f_1|E, f_2|E, \ldots \rangle \quad \text{converges uniformly to} \quad g|E$$

Now uniform convergence implies uniform convergence on compacta and, since singleton sets are compact, uniform convergence on compacta implies pointwise convergence.

Example 7.1: Let $\langle f_1, f_2, \ldots \rangle$ be the sequence in $\mathcal{F}(\mathbf{R}, \mathbf{R})$ defined by

$$f_n(x) = \begin{cases} 1 - \dfrac{1}{n}|x| & \text{if } |x| < n \\ 0 & \text{if } |x| \geqq n \end{cases}$$

Now $\langle f_n \rangle$ converges pointwise to the constant function $g(x) = 1$ but $\langle f_n \rangle$ does not converge uniformly to g (see Example 4.2). However, since every compact subset E of R is bounded, $\langle f_n \rangle$ does converge uniformly on compacta to g.

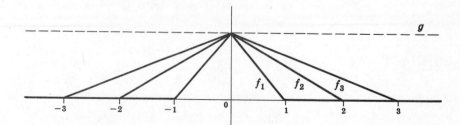

Theorem 15.12: Let $C(X, Y)$ be the collection of continuous functions from a topological space X into a metric space (Y, d). Then a sequence of functions $\langle f_n \rangle$ in $C(X, Y)$ converges to $g \in C(X, Y)$ with respect to the compact open topology if and only if $\langle f_n \rangle$ converges uniformly on compacta to g.

In view of the preceding theorem, the compact open topology is also called the *topology of compact convergence*.

FUNCTIONALS ON NORMED SPACES

Let X be a normed vector space (over $\mathbf{R}$). A real-valued function f with domain X, i.e. $f : X \to \mathbf{R}$, is called a *functional*.

Definition: A functional f on X is *linear* if

(i) $f(x + y) = f(x) + f(y)$, $\forall\, x, y \in X$, and (ii) $f(kx) = k[f(x)]$, $\forall\, x \in X$, $k \in \mathbf{R}$

A linear functional f on X is *bounded* if

$$\exists\, M > 0 \quad \text{such that} \quad |f(x)| \leq M\,\|x\|, \quad \forall\, x \in X$$

Here M is called a *bound* for f.

Example 8.1: Let X be the space of all continuous real-valued functions on $[a, b]$ with norm $\|f\| = \sup\{|f(x)| : x \in [a, b]\}$, i.e. $X = C[a, b]$. Let $\mathbf{I} : X \to \mathbf{R}$ be defined by

$$\mathbf{I}(f) = \int_a^b f(t)\, dt$$

Then $\mathbf{I}$ is a linear functional; for

$$\mathbf{I}(f + g) = \int_a^b (f(t) + g(t))\, dt = \int_a^b f(t)\, dt + \int_a^b g(t)\, dt = \mathbf{I}(f) + \mathbf{I}(g)$$

$$\mathbf{I}(kf) = \int_a^b (kf)(t)\, dt = \int_a^b k[f(t)]\, dt = k \int_a^b f(t)\, dt = k\,\mathbf{I}(f)$$

Furthermore, $M = b - a$ is a bound for $\mathbf{I}$ since

$$\mathbf{I}(f) = \int_a^b f(t)\, dt \leq M \sup\{|f(t)|\} = M\,\|f\|$$

Proposition 15.13: Let f and g be bounded linear functionals on X and let $k \in \mathbf{R}$. Then $f + g$ and $k \cdot f$ are also bounded linear functionals on X.

Thus (by Proposition 8.14, Page 119) the collection X^* of all bounded linear functionals on X is a linear vector space.

Proposition 15.14: The following function on X^* is a norm:

$$\|f\| = \sup\{|f(x)|/\|x\| : x \neq 0\}$$

Observe that if M is a bound for f, i.e. $|f(x)| \leq M\,\|x\|$, $\forall\, x \in X$, then in particular, for $x \neq 0$, $|f(x)|/\|x\| \leq M$ and so $\|f\| \leq M$. In fact, $\|f\|$ could have been defined equivalently by

$$\|f\| = \inf\{M : M \text{ is a bound for } f\}$$

Remark: The normed space of all bounded linear functionals on X is called the *dual space* of X.

Solved Problems

POINTWISE CONVERGENCE, POINT OPEN TOPOLOGY

1. Let $\langle f_1, f_2, \ldots \rangle$ be the sequence of functions in $\mathcal{F}(I, \mathbf{R})$, where $I = [0, 1]$, defined by

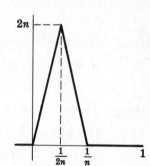

$$f_n(x) = \begin{cases} 4n^2x & \text{if } 0 \leq x \leq 1/2n \\ -4n^2x + 4n & \text{if } 1/2n < x < 1/n \\ 0 & \text{if } 1/n \leq x \leq 1 \end{cases}$$

Show that $\langle f_n \rangle$ converges pointwise to the constant function $g(x) = 0$.

Solution:

Now $f_n(0) = 0$ for every $n \in \mathbf{N}$, and so $\lim_{n \to \infty} f_n(0) = g(0) = 0$. On the other hand, if $x_0 > 0$, then $\exists\, n_0 \in \mathbf{N}$ such that $1/n_0 < x_0$; hence

$$n > n_0 \;\Rightarrow\; f_n(x_0) = 0 \;\Rightarrow\; \lim_{n \to \infty} f_n(x_0) = g(x_0) = 0$$

Thus $\langle f_n \rangle$ converges pointwise to the zero function.

Observe that $\displaystyle\int_0^1 f_n(x)\, dx = 1$, for every $n \in \mathbf{N}$, and $\displaystyle\int_0^1 g(x)\, dx = 0$

Thus, in this case, the limit of the integrals does not equal the integral of the limit, i.e.,

$$\lim_{n \to \infty} \int_0^1 f_n(x)\, dx \;\neq\; \int_0^1 \lim_{n \to \infty} f_n(x)\, dx$$

2. Let $\mathcal{C}(I, \mathbf{R})$ denote the class of continuous real valued functions on $I = [0, 1]$ with norm

$$\|f\| = \int_0^1 |f(x)|\, dx$$

Give an example of a sequence $\langle f_1, f_2, \ldots \rangle$ in $\mathcal{C}(I, \mathbf{R})$ such that $f_n \to g$ in the above norm but $\langle f_n \rangle$ does not converge to g pointwise.

Solution:

Let $\langle f_n \rangle$ be defined by $f_n(x) = x^n$. Then

$$\lim_{n \to \infty} \|f_n\| = \lim_{n \to \infty} \int_0^1 x^n\, dx = \lim_{n \to \infty} 1/(n+1) = 0$$

Hence $\langle f_n \rangle$ converges to the zero function $g(x) = 0$ in the above norm. On the other hand, $\langle f_n \rangle$ converges pointwise (see Example 3.1) to the function f defined by $f(x) = 0$ if $0 \leq x < 1$ and $f(x) = 1$ if $x = 1$. Note $f \neq g$.

3. Show that if Y is T_1, T_2, regular, or connected, then $\mathcal{F}(X, Y)$ with the point open topology also has that property.

Solution:

Since the point open topology on $\mathcal{F}(X, Y)$ is the product topology, $\mathcal{F}(X, Y)$ inherits any product invariant property of Y. By previous results, the above properties are product invariant.

4. Prove Theorem 15.2: Let Y be Hausdorff and let $\mathcal{A}$ be a subset of $\mathcal{F}(X, Y)$ with the point open topology. Then the following are equivalent: (i) $\mathcal{A}$ is compact. (ii) $\mathcal{A}$ is closed and $\overline{\{f(x) : f \in \mathcal{A}\}}$ is compact in Y, for every $x \in X$.

Solution:

By Theorem 15.1, (ii) $\Rightarrow$ (i) and so we need only show that (i) $\Rightarrow$ (ii). Since Y is Hausdorff and T_2 is product invariant, $\mathcal{F}(X, Y)$ is also Hausdorff. Now by Theorem 11.5 a compact subset of a Hausdorff space is closed; hence $\mathcal{A}$ is closed. Furthermore, each evaluation map $e_x : \mathcal{F}(X, Y) \to Y$ is continuous with respect to the point open topology; hence, for each $x \in X$,

$$e_x[\mathcal{A}] \;=\; \{f(x) : f \in \mathcal{A}\}$$

is compact in Y and, since Y is Hausdorff, closed. In other words, $\overline{\{f(x) : f \in \mathcal{A}\}} = \{f(x) : f \in \mathcal{A}\}$ is compact.

5. Prove Theorem 15.3: Let $\mathcal{T}$ be the point open topology on $\mathcal{F}(X, Y)$ and let $\langle f_1, f_2, \ldots \rangle$ be a sequence in $\mathcal{F}(X, Y)$. Then the following are equivalent: (i) $\langle f_n \rangle$ converges to $g \in \mathcal{F}(X, Y)$ with respect to $\mathcal{T}$. (ii) $\langle f_n \rangle$ converges pointwise to g.

Solution:

Method 1.

We identify $\mathcal{F}(X, Y)$ with the product set $\mathbf{F} = \prod \{Y_x : x \in X\}$ and $\mathcal{T}$ with the product topology. Then by Theorem 12.7 the sequence $\langle f_n \rangle$ in $\mathbf{F}$ converges to $g \in \mathbf{F}$ if and only if, for every projection π_x,

$$\langle \pi_x(f_n) \rangle = \langle e_x(f_n) \rangle = \langle f_n(x) \rangle \quad \text{converges to} \quad \pi_x(g) = e_x(g) = g(x)$$

In other words, $\qquad f_n \to g$ with respect to $\mathcal{T}$ $\quad$ iff $\quad \lim f_n(x) = g(x), \; \forall \, x \in X$

i.e. iff $\langle f_n \rangle$ converges pointwise to g.

Method 2.

(i) $\Rightarrow$ (ii): Let x_0 be an arbitrary point in X and let G be an open subset of Y containing $g(x_0)$, i.e. $g(x_0) \in G$. Then

$$g \in F(x_0, G) \;=\; \{f \in \mathcal{F}(X, Y) : f(x_0) \in G\}$$

and so $F(x_0, G)$ is a $\mathcal{T}$-open subset of $\mathcal{F}(X, Y)$ containing g. By (i), $\langle f_n \rangle$ converges to g with respect to $\mathcal{T}$; hence

$$\exists \; n_0 \in \mathbf{N} \quad \text{such that} \quad n > n_0 \;\Rightarrow\; f_n \in F(x_0, G)$$

Accordingly, $\qquad\qquad n > n_0 \;\Rightarrow\; f_n(x_0) \in G \;\Rightarrow\; \lim_{n \to \infty} f_n(x_0) = g(x_0)$

But x_0 was arbitrary; hence $\langle f_n \rangle$ converges pointwise to g.

(ii) $\Rightarrow$ (i): Let $F(x_0, G) = \{f : f(x_0) \in G\}$ be any member of the defining subbase for $\mathcal{T}$ which contains g. Then $g(x_0) \in G$. By (ii), $\langle f_n \rangle$ converges pointwise to g; hence

$$\exists \; n_0 \in \mathbf{N} \quad \text{such that} \quad n > n_0 \;\Rightarrow\; f_n(x_0) \in G$$

and so $\qquad\qquad n > n_0 \;\Rightarrow\; f_n \in F(x_0, G) \;\Rightarrow\; \langle f_n \rangle \; \mathcal{T}\text{-converges to } g$

UNIFORM CONVERGENCE

6. Prove Proposition 15.4: Let $\langle f_1, f_2, \ldots \rangle$ be a sequence of continuous functions from a topological space X into a metric space Y, and let $\langle f_n \rangle$ converge uniformly to $g : X \to Y$. Then g is continuous.

Solution:

Let $x_0 \in X$ and let $\epsilon > 0$. Then g is continuous at x_0 if $\exists$ an open set $G \subset X$ containing x_0 such that

$$x \in G \;\Rightarrow\; d(g(x), g(x_0)) < \epsilon$$

Now $\langle f_n \rangle$ converges uniformly to g, and so

$$\exists\ m \in \mathbf{N} \quad \text{such that} \quad d(f_m(x), g(x)) < \tfrac{1}{3}\epsilon, \quad \blacktriangledown x \in X$$

Hence, by the Triangle Inequality,

$$d(g(x), g(x_0)) \;\leq\; d(g(x), f_m(x)) + d(f_m(x), f_m(x_0)) + d(f_m(x_0), g(x_0)) \;<\; d(f_m(x), f_m(x_0)) + \tfrac{2}{3}\epsilon$$

Since f_m is continuous, $\exists$ an open set $G \subset X$ containing x_0 such that

$$x \in G \;\Rightarrow\; d(f_m(x), f_m(x_0)) < \tfrac{1}{3}\epsilon \quad \text{and so} \quad x \in G \;\Rightarrow\; d(g(x), g(x_0)) < \epsilon$$

Thus g is continuous.

7. Let $\langle f_1, f_2, \ldots \rangle$ be a sequence of real, continuous functions defined on $[a, b]$ and converging uniformly to $g : [a, b] \to \mathbf{R}$. Show that

$$\lim_{n \to \infty} \int_a^b f_n(x)\, dx \;=\; \int_a^b g(x)\, dx$$

Observe (Problem 1) that this statement is not true in the case of pointwise convergence.

Solution:

Let $\epsilon > 0$. We need to show that

$$\exists\ n_0 \in \mathbf{N} \quad \text{such that} \quad n > n_0 \;\Rightarrow\; \left| \int_a^b f_n(x)\, dx - \int_a^b g(x)\, dx \right| < \epsilon$$

Now $\langle f_n \rangle$ converges uniformly to g, and so $\exists\ n_0 \in \mathbf{N}$ such that

$$n > n_0 \;\Rightarrow\; |f_n(x) - g(x)| < \epsilon/(b-a), \quad \blacktriangledown x \in [a, b]$$

Hence, if $n > n_0$,

$$\left| \int_a^b f_n(x)\, dx - \int_a^b g(x)\, dx \right| \;=\; \left| \int_a^b (f_n(x) - g(x))\, dx \right|$$

$$\leq\; \int_a^b |f_n(x) - g(x)|\, dx$$

$$<\; \int_a^b \epsilon/(b-a)\, dx \;=\; \epsilon$$

8. Prove Theorem 15.5: Let $\langle f_1, f_2, \ldots \rangle$ be a sequence in $\mathcal{B}(X, Y)$ with metric

$$e(f, g) \;=\; \sup\ \{d(f(x), g(x)) : x \in X\}$$

Then the following are equivalent: (i) $\langle f_n \rangle$ converges to $g \in \mathcal{F}(X, Y)$ with respect to e. (ii) $\langle f_n \rangle$ converges uniformly to g.

Solution:

(i) $\Rightarrow$ (ii): Let $\epsilon > 0$. Since $\langle f_n \rangle$ converges to g with respect to e,

$$\exists\ n_0 \in \mathbf{N} \quad \text{such that} \quad n > n_0 \;\Rightarrow\; e(f_n, g) < \epsilon$$

Therefore,

$$n > n_0 \;\Rightarrow\; d(f_n(x), g(x)) \leq \sup\ \{d(f_n(x), g(x)) : x \in X\} = e(f_n, g) < \epsilon, \quad \blacktriangledown x \in X$$

that is, $\langle f_n \rangle$ converges uniformly to g.

(ii) $\Rightarrow$ (i): Let $\epsilon > 0$. Since $\langle f_n \rangle$ converges uniformly to g,

$$\exists\ n_0 \in \mathbf{N} \quad \text{such that} \quad n > n_0 \;\Rightarrow\; d(f_n(x), g(x)) < \epsilon/2, \quad \blacktriangledown x \in X$$

Therefore,

$$n > n_0 \;\Rightarrow\; \sup\ \{d(f_n(x), g(x)) : x \in X\} \leq \epsilon/2 < \epsilon$$

that is, $n > n_0$ implies $e(f_n, g) < \epsilon$, and so $\langle f_n \rangle$ converges to g with respect to e.

THE FUNCTION SPACE $C[0,1]$

9. Prove Proposition 15.6: Let $f : I \to \mathbf{R}$ be continuous on $I = [0,1]$. Then for every $\epsilon > 0$,

$$\exists\ \delta = \delta(\epsilon) > 0 \quad \text{such that} \quad |x - y| < \delta \ \Rightarrow\ |f(x) - f(y)| < \epsilon$$

i.e. f is uniformly continuous.

Solution:

Let $\epsilon > 0$. Since f is continuous, for every $p \in I$,

$$\exists\ \delta_p > 0 \quad \text{such that} \quad |x - p| < \delta_p \ \Rightarrow\ |f(x) - f(p)| < \tfrac{1}{2}\epsilon \tag{1}$$

For each $p \in I$, set $S_p = I \cap (p - \tfrac{1}{2}\delta_p,\ p + \tfrac{1}{2}\delta_p)$. Then $\{S_p : p \in I\}$ is an open cover of I and, since I is compact, a finite number of the S_p also cover I; say, $I = S_{p_1} \cup \cdots \cup S_{p_m}$. Set

$$\delta = \tfrac{1}{2} \min (\delta_{p_1}, \ldots, \delta_{p_m})$$

Suppose $|x - y| < \delta$. Then $x \in S_{p_k}$ for some k, and so $|x - p_k| < \tfrac{1}{2}\delta_{p_k} < \delta_{p_k}$ and

$$|y - p_k| \ \leq\ |y - x| + |x - p_k| \ <\ \delta + \tfrac{1}{2}\delta_{p_k} \ \leq\ \tfrac{1}{2}\delta_{p_k} + \tfrac{1}{2}\delta_{p_k} \ =\ \delta_{p_k}$$

Hence by (1), $\qquad\qquad |f(x) - f(p_k)| < \tfrac{1}{2}\epsilon \quad \text{and} \quad |f(y) - f(p_k)| < \tfrac{1}{2}\epsilon$

Thus by the Triangle Inequality,

$$|f(x) - f(y)| \ \leq\ |f(x) - f(p_k)| + |f(p_k) - f(y)| \ <\ \tfrac{1}{2}\epsilon + \tfrac{1}{2}\epsilon \ =\ \epsilon$$

10. Let $\langle f_1, f_2, \ldots \rangle$ be a Cauchy sequence in $C[0,1]$. Show that, for each $x_0 \in I = [0,1]$, $\langle f_1(x_0), f_2(x_0), \ldots \rangle$ is a Cauchy sequence in $\mathbf{R}$.

Solution:

Let $x_0 \in I$ and let $\epsilon > 0$. Since $\langle f_n \rangle$ is Cauchy, $\exists\ n_0 \in \mathbf{N}$ such that

$$m, n > n_0 \ \Rightarrow\ \|f_n - f_m\| \ =\ \sup \{|f_n(x) - f_m(x)| : x \in I\} \ <\ \epsilon$$
$$\Rightarrow\ |f_n(x_0) - f_m(x_0)| \ <\ \epsilon$$

Hence $\langle f_n(x_0) \rangle$ is a Cauchy sequence.

11. Prove Theorem 15.7: $C[0,1]$ is a complete normed vector space.

Solution:

Let $\langle f_1, f_2, \ldots \rangle$ be a Cauchy sequence in $C[0,1]$. Then, for every $x_0 \in I$, $\langle f_n(x_0) \rangle$ is a Cauchy sequence in $\mathbf{R}$ and, since $\mathbf{R}$ is complete, converges. Define $g : I \to \mathbf{R}$ by $g(x) = \lim_{n \to \infty} f_n(x)$. Then (see Problem 32) $\langle f_n \rangle$ converges uniformly to g. But, by Proposition 15.4, g is continuous, i.e. $g \in C[0,1]$; hence $C[0,1]$ is complete.

12. Let $f \in C[0,1]$ and let $\epsilon > 0$. Show that $\exists\ n_0 \in \mathbf{N}$ and points

$$p_0 = (0,\ \epsilon k_0/5),\ \ldots,$$
$$p_i = (i/n_0,\ \epsilon k_i/5),\ \ldots,$$
$$p_{n_0} = (1,\ \epsilon k_{n_0}/5)$$

where $k_0, \ldots, k_{n_0}$ are integers such that, if g is the polygonal arc connecting the p_i, then $\|f - g\| < \epsilon$ (see adjacent diagram). In other words, the piecewise linear (or polygonal) functions are dense in $C[0,1]$.

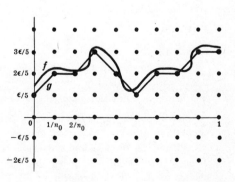

Solution:

Now f is uniformly continuous on $[0,1]$ and so

$$\exists\ n_0 \in \mathbf{N} \quad \text{such that} \quad |a - b| \leq 1/n_0 \ \Rightarrow\ |f(a) - f(b)| < \epsilon/5 \tag{1}$$

Consider the following subset of $I \times \mathbf{R}$:

$$A = \{\langle x, y \rangle : x = i/n_0, \; y = k\epsilon/5 \text{ where } i = 0, \ldots, n_0; \; k \in \mathbf{Z}\}$$

Choose $p_i = \langle x_i, y_i \rangle \in A$ such that $\qquad y_i \leq f(x_i) < y_i + \epsilon/5$

Then $\qquad |f(x_i) - g(x_i)| = |f(x_i) - y_i| < \epsilon/5 \qquad$ and by (1), $\quad |f(x_i) - f(x_{i+1})| < \epsilon/5$

as indicated in the diagram above.

Observe that

$$|g(x_i) - g(x_{i+1})| \leq |g(x_i) - f(x_i)| + |f(x_i) - f(x_{i+1})| + |f(x_{i+1}) - g(x_{i+1})| < \epsilon/5 + \epsilon/5 + \epsilon/5 = 3\epsilon/5$$

Since g is linear between x_i and x_{i+1},

$$x_i \leq z \leq x_{i+1} \;\Rightarrow\; |g(x_i) - g(z)| \leq |g(x_i) - g(x_{i+1})| < 3\epsilon/5$$

Now for any point $z \in I$, $\exists\, x_k$ satisfying $x_k \leq z \leq x_{k+1}$. Hence

$$|f(z) - g(z)| \leq |f(z) - f(x_k)| + |f(x_k) - g(x_k)| + |g(x_k) - g(z)| < \epsilon/5 + \epsilon/5 + 3\epsilon/5 = \epsilon$$

But z was an arbitrary point in I; hence $\|f - g\| < \epsilon$.

13. Let m be an arbitrary positive integer and let $A_m \subset C[0,1]$ consist of those functions f with the property that

$$\exists\, x_0 \in \left[0, 1 - \frac{1}{m}\right] \quad \text{such that} \quad \left|\frac{f(x_0 + h) - f(x_0)}{h}\right| \leq m, \; \forall\, h \in \left(0, \frac{1}{m}\right)$$

Show that A_m is a closed subset of $C[0,1]$. (Notice that every function f in $C[0,1]$ which is differentiable at a point belongs to some A_m for m sufficiently large.)

Solution:

Let $g \in \bar{A}_m$. We want to show that $g \in A_m$, i.e. $\bar{A}_m = A_m$. Since $g \in \bar{A}_m$, there exists a sequence $\langle f_1, f_2, \ldots \rangle$ in A_m converging to g. Now for each f_i there exists a point x_i such that

$$x_i \in \left[0, 1 - \frac{1}{m}\right] \quad \text{and} \quad \left|\frac{f_i(x_i + h) - f_i(x_i)}{h}\right| \leq m, \; \forall\, h \in \left(0, \frac{1}{m}\right) \tag{1}$$

But $\langle x_n \rangle$ is a sequence in a compact set $\left[0, 1 - \frac{1}{m}\right]$ and so has a subsequence $\langle x_{i_n} \rangle$ which converges to, say, $x_0 \in \left[0, 1 - \frac{1}{m}\right]$.

Now $f_n \to g$ implies $f_{i_n} \to g$, and so (Problem 30), passing to the limit in (1), gives

$$\left|\frac{g(x_0 + h) - g(x_0)}{h}\right| \leq m, \; \forall\, h \in \left(0, \frac{1}{m}\right)$$

Hence $g \in A_m$, and A_m is closed.

14. Let $A_m \subset C[0,1]$ be defined as in Problem 13. Show that A_m is nowhere dense in $C[0,1]$.

Solution:

A_m is nowhere dense in $C[0,1]$ iff $\operatorname{int}(\bar{A}_m) = \operatorname{int}(A_m) = \emptyset$. Let $S = S(f, \delta)$ be any open sphere in $C[0,1]$. We claim that S contains a point not belonging to A_m, and so $\operatorname{int}(\bar{A}_m) = \emptyset$.

By Problem 12, there exists a polygonal arc $p \in C[0,1]$ such that $\|f - p\| < \frac{1}{2}\delta$. Let g be a saw-tooth function with magnitude less than $\frac{1}{2}\delta$ and slope sufficiently large (Problem 33). Then the function $h = p + g$ belongs to $C[0,1]$ but does not belong to A_m. Furthermore,

$$\|f - h\| \leq \|f - p\| + \|g\| < \tfrac{1}{2}\delta + \tfrac{1}{2}\delta = \delta$$

so $h \in S$ and the proof is complete.

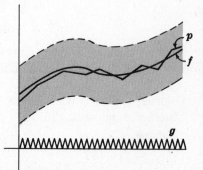

15. Let $A_m \subset C[0,1]$ be defined as in Problem 13. Show that $C[0,1] \neq \cup_{m=1}^{\infty} A_m$.

Solution:

Since A_m is nowhere dense in $C[0,1]$, $B = \cup_{m=1}^{\infty} A_m$ is of the first category. But, by Baire's Category Theorem, $C[0,1]$, a complete space, is of the second category. Hence $C[0,1] \neq B$.

16. Prove Proposition 15.8: There exists a continuous function $f : [0,1] \to \mathbf{R}$ which is nowhere differentiable.

Solution:

Let $f \in C[0,1]$ have a derivative at, say, x_0 and suppose $|f'(x_0)| = t$. Then

$$\exists \, \epsilon > 0 \quad \text{such that} \quad \left| \frac{f(x_0 + h) - f(x_0)}{h} \right| \le t + 1, \quad \forall \, h \in (-\epsilon, \epsilon)$$

Now choose $m_0 \in N$ so that $t + 1 \le m_0$ and $1/m_0 < \epsilon$. Then $f \in A_{m_0}$. Thus $\cup_{m=1}^{\infty} A_m$ contains all functions which are differentiable at some point of I.

But by the preceding problem, $C[0,1] \neq \cup_{m=1}^{\infty} A_m$ and so there exists a function in $C[0,1]$ which is nowhere differentiable.

17. Prove Theorem (Ascoli) 15.9: Let $\mathcal{A}$ be a closed subset of $C[0,1]$. Then the following are equivalent: (i) $\mathcal{A}$ is compact. (ii) $\mathcal{A}$ is uniformly bounded and equicontinuous.

Solution:

(i) $\Rightarrow$ (ii): Since $\mathcal{A}$ is compact it is a bounded subset of $C[0,1]$ and is thus uniformly bounded as a set of functions. Now we need only show that $\mathcal{A}$ is equicontinuous.

Let $\epsilon > 0$. Since $\mathcal{A}$ is compact, it has a finite $\epsilon/3$-net, say, $\mathcal{B} = \{f_1, \ldots, f_t\}$. Hence, for any $f \in \mathcal{A}$,

$$\exists \, f_{i_0} \in \mathcal{B} \quad \text{such that} \quad \|f - f_{i_0}\| \, = \, \sup \, \{|f(x) - f_{i_0}(x)| : x \in I\} \, \le \, \epsilon/3$$

Therefore, for any $x, y \in I = [0,1]$,

$$
\begin{aligned}
|f(x) - f(y)| \, &= \, |f(x) - f_{i_0}(x) + f_{i_0}(x) - f_{i_0}(y) + f_{i_0}(y) - f(y)| \\
&\le \, |f(x) - f_{i_0}(x)| \, + \, |f_{i_0}(x) - f_{i_0}(y)| \, + \, |f_{i_0}(y) - f(y)| \\
&\le \, \epsilon/3 \, + \, |f_{i_0}(x) - f_{i_0}(y)| \, + \, \epsilon/3 \, = \, |f_{i_0}(x) - f_{i_0}(y)| \, + \, 2\epsilon/3
\end{aligned}
$$

Now each $f_i \in \mathcal{B}$ is uniformly continuous and so

$$\exists \, \delta_i > 0 \quad \text{such that} \quad |x - y| < \delta_i \, \Rightarrow \, |f_i(x) - f_i(y)| < \epsilon/3$$

Set $\delta = \min \{\delta_1, \ldots, \delta_t\}$. Then, for any $f \in \mathcal{A}$,

$$|x - y| < \delta \, \Rightarrow \, |f(x) - f(y)| \, \le \, |f_{i_0}(x) + f_{i_0}(y)| + 2\epsilon/3 \, < \, \epsilon/3 + 2\epsilon/3 \, = \, \epsilon$$

Thus $\mathcal{A}$ is equicontinuous.

(ii) $\Rightarrow$ (i): Since $\mathcal{A}$ is a closed subset of the complete space $C[0,1]$, we need only show that $\mathcal{A}$ is totally bounded. Let $\epsilon > 0$. Since $\mathcal{A}$ is equicontinuous,

$$\exists \, n_0 \in \mathbf{N} \quad \text{such that} \quad |a - b| < 1/n_0 \, \Rightarrow \, |f(a) - f(b)| < \epsilon/5, \quad \forall \, f \in \mathcal{A}$$

Now for each $f \in \mathcal{A}$, we can construct, by Problem 12, a polygonal arc p_f such that $\|f - p_f\| < \epsilon$ and p_f connects points belonging to

$$A \, = \, \{\langle x, y \rangle : x = 0, 1/n_0, 2/n_0, \ldots, 1; \, y = n\epsilon/5, \, n \in \mathbf{Z}\}$$

We claim that $\mathcal{B} = \{p_f : f \in \mathcal{A}\}$ is finite and hence a finite ϵ-net for $\mathcal{A}$.

Now $\mathcal{A}$ is uniformly bounded, and so $\mathcal{B}$ is uniformly bounded. Therefore only a finite number of the points in A will appear in the polygonal arcs in $\mathcal{B}$. Hence there can only be a finite number of arcs in $\mathcal{B}$. Thus $\mathcal{B}$ is a finite ϵ-net for $\mathcal{A}$, and so $\mathcal{A}$ is totally bounded.

COMPACT CONVERGENCE

18. Let $\langle f_1, f_2, \ldots \rangle$ in $\mathcal{F}(\mathbf{R}, \mathbf{R})$ be defined by

$$f_n(x) \;=\; \begin{cases} 1 - \dfrac{1}{n}|x| & \text{if } |x| < n \\ 0 & \text{if } |x| \geqq n \end{cases}$$

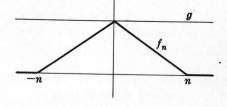

Show that $\langle f_n \rangle$ converges uniformly on compacta
to the constant function $g(x) = 1$.

Solution:

Let E be a compact subset of $\mathbf{R}$ and let $0 < \epsilon < 1$. Since E is compact, it is bounded; say, $E \subset (-M, M)$ for $M > 0$. Now

$$\exists\; n_0 \in \mathbf{N} \qquad \text{such that} \qquad n_0 > M/\epsilon, \text{ or, } M/n_0 < \epsilon$$

Therefore, $\qquad n > n_0 \;\Rightarrow\; |f_n(x) - g(x)| = \dfrac{1}{n}|x| < M/n_0 < \epsilon, \;\; \blacktriangledown\, x \in E$

Hence $\langle f_n \rangle$ converges uniformly to g on E.

19. Show: If Y is Hausdorff, then the compact open topology on $\mathcal{F}(X, Y)$ is also Hausdorff.

Solution:

Method 1. Let $f, g \in \mathcal{F}(X, Y)$ with $f \neq g$. Then $\exists\, p \in X$ such that $f(p) \neq g(p)$. Now Y is Hausdorff, hence $\exists$ open subsets G and H of Y such that $f(p) \in G$, $g(p) \in H$ and $G \cap H = \emptyset$. Hence

$$f \in F(p, G), \quad g \in F(p, H) \quad \text{and} \quad F(p, G) \cap F(p, H) = \emptyset$$

But the singleton set $\{p\}$ is compact, and so $F(p, G)$ and $F(p, H)$ belong to the compact open topology on $\mathcal{F}(X, Y)$. Accordingly, $\mathcal{F}(X, Y)$ is Hausdorff.

Method 2. The compact open topology is finer than the point open topology, which is Hausdorff since T_2 is a product invariant property. Hence the compact open topology is also Hausdorff.

20. Prove Theorem 15.12: Let $\langle f_1, f_2, \ldots \rangle$ be a sequence in $\mathcal{C}(X, Y)$, the collection of all continuous functions from a topological space X into a metric space (Y, d). Then the following are equivalent:

(i) $\langle f_n \rangle$ converges uniformly on compacta to $g \in \mathcal{C}(X, Y)$.

(ii) $\langle f_n \rangle$ converges to g with respect to the compact open topology $\mathcal{T}$ on $\mathcal{C}(X, Y)$.

Solution:

(i) $\Rightarrow$ (ii):

Let $F(E, G)$ be an open subbase element of $\mathcal{T}$ containing g; hence $g[E] \subset G$ where E is compact and G is open. Since g is continuous, $g[E]$ is compact. Furthermore, $g[E] \cap G^c = \emptyset$ and so (see Page 164) the distance between the compact set $g[E]$ and the closed set G^c is greater than zero; say, $d(g[E], G^c) = \epsilon > 0$. Since $\langle f_n \rangle$ converges uniformly on compacta to g,

$$\exists\; n_0 \in \mathbf{N} \qquad \text{such that} \qquad n > n_0 \;\Rightarrow\; d(f_n(x), g(x)) < \epsilon, \;\; \blacktriangledown\, x \in E$$

Therefore, $\qquad d(f_n(x), g[E]) \;\leq\; d(f_n(x), g(x)) \;<\; \epsilon, \quad \blacktriangledown\, x \in E$

and so, for every $x \in E$, $f_n(x) \notin G^c$. In other words,

$$n > n_0 \;\Rightarrow\; f_n[E] \subset G \;\Rightarrow\; f_n \in F(E, G)$$

Accordingly, $\langle f_n \rangle$ converges to g with respect to the compact open topology $\mathcal{T}$.

(ii) $\Rightarrow$ (i):

Let E be a compact subset of X and let $\epsilon > 0$. We want to show that $\langle f_n \rangle$ converges uniformly on E to g, i.e.,

$$\exists\; n_0 \in \mathbf{N} \qquad \text{such that} \qquad n > n_0 \;\Rightarrow\; d(f_n(x), g(x)) < \epsilon, \;\; \blacktriangledown\, x \in E$$

Since E is compact and g is continuous, $g[E]$ is compact. Let $\mathcal{B} = \{p_1, \ldots, p_t\}$ be a finite $\epsilon/3$-net for $g[E]$. Consider the open spheres

$$S_1 = S(p_1, \epsilon/3), \ldots, S_t = S(p_t, \epsilon/3) \quad \text{and} \quad G_1 = S(p_1, 2\epsilon/3), \ldots, G_t = S(p_t, 2\epsilon/3)$$

Hence $\bar{S}_1 \subset G_1, \ldots, \bar{S}_t \subset G_t$. Furthermore, since $\mathcal{B}$ is an $\epsilon/3$-net for $g[E]$,

$$g[E] \subset \bar{S}_1 \cup \cdots \cup \bar{S}_t \quad \text{and so} \quad E \subset g^{-1}[\bar{S}_1] \cup \cdots \cup g^{-1}[\bar{S}_t]$$

Now set $\quad E_i = E \cap g^{-1}[\bar{S}_i] \quad$ and so $\quad E = E_1 \cup \cdots \cup E_t \quad$ and $\quad g[E_i] \subset \bar{S}_i \subset G_i$

We claim that the E_i are compact. For g is continuous and so $g^{-1}[\bar{S}_i]$, the inverse of a closed set, is closed; hence $E_i = E \cap g^{-1}[\bar{S}_i]$, the intersection of a compact and a closed set, is compact.

Now $g[E_i] \subset G_i$ and so the $F(E_i, G_i)$ are $\mathcal{T}$-open subsets of $\mathcal{F}(X, Y)$ containing g; hence $\cap_{i=1}^t F(E_i, G_i)$ is also a $\mathcal{T}$-open set containing g. But $\langle f_n \rangle$ converges to g with respect to $\mathcal{T}$; hence

$$\exists\, n_0 \in \mathbf{N} \quad \text{such that} \quad n > n_0 \;\Rightarrow\; f_n \in \cap_{i=1}^t F(E_i, G_i) \;\Rightarrow\; f_n[E_1] \subset G_1, \ldots, f_n[E_t] \subset G_t$$

Now let $x \in E$. Then $x \in E_{i_0}$ and so, for $n > n_0$,

$$f_n(x) \in f_n[E_{i_0}] \subset G_{i_0} \;\Rightarrow\; d(f_n(x), p_{i_0}) < 2\epsilon/3$$

and $$g(x) \in g[E_{i_0}] \subset \bar{S}_{i_0} \;\Rightarrow\; d(g(x), p_{i_0}) \leq \epsilon/3$$

Therefore, by the Triangle Inequality,

$$n > n_0 \;\Rightarrow\; d(f_n(x), g(x)) \leq d(f_n(x), p_{i_0}) + d(p_{i_0}, g(x)) < 2\epsilon/3 + \epsilon/3 = \epsilon, \quad \forall\, x \in E$$

FUNCTIONALS ON NORMED SPACES

21. Show that if f is a linear functional on X, then $f(0) = 0$.

Solution:
Since f is linear and $0 = 0 + 0$,

$$f(0) = f(0 + 0) = f(0) + f(0)$$

Adding $-f(0)$ to both sides gives $f(0) = 0$.

22. Show that a bounded linear functional f on X is uniformly continuous.

Solution:
Let M be a bound for f and let $\epsilon > 0$. Set $\delta = \epsilon/M$. Then

$$||x - y|| < \delta \;\Rightarrow\; |f(x) - f(y)| = |f(x - y)| \leq M\,||x - y|| < \epsilon$$

23. Prove Proposition 15.13: Let f and g be bounded linear functionals on X and let $c \in \mathbf{R}$. Then $f + g$ and $c \cdot f$ are also bounded linear functionals on X.

Solution:
Let M and M^* be bounds for f and g respectively. Then

$$(f + g)(x + y) = f(x + y) + g(x + y) = f(x) + f(y) + g(x) + g(y) = (f + g)(x) + (f + g)(y)$$

$$(f + g)(kx) = f(kx) + g(kx) = k\,f(x) + k\,g(x) = k[f(x) + g(x)] = k\,(f + g)(x)$$

$$|(f + g)(x)| = |f(x) + g(x)| \leq |f(x)| + |g(x)| \leq M\,||x|| + M^*\,||x|| = (M + M^*)\,||x||$$

Thus $f + g$ is a bounded linear functional.

Furthermore,

$$(c \cdot f)(x + y) \; = \; c \, f(x + y) \; = \; c \, [f(x) + f(y)] \; = \; c \, f(x) + c \, f(y) \; = \; (c \cdot f)(x) + (c \cdot f)(y)$$

$$(c \cdot f)(kx) \; = \; c \, f(kx) \; = \; ck \, f(x) \; = \; kc \, f(x) \; = \; k \, (c \cdot f)(x)$$

$$|(c \cdot f)(x)| \; = \; |c \, f(x)| \; = \; |c| \; |f(x)| \; \leqq \; |c| \, (M \, ||x||) \; = \; (|c| \, M) \, ||x||$$

and so $c \cdot f$ is a bounded linear functional.

24. Prove Proposition 15.14: The following function on X^* is a norm:

$$||f|| \; = \; \sup \, \{|f(x)|/||x|| : x \neq 0\}$$

Solution:

 If $f = 0$, then $f(x) = 0$, $\forall \, x \in X$, and so $||f|| = \sup \{0\} = 0$. If $f \neq 0$, then $\exists \, x_0 \neq 0$ such that $f(x_0) \neq 0$, and so

$$||f|| \; = \; \sup \, \{|f(x)|/||x||\} \; \geqq \; |f(x_0)|/||x_0|| \; > \; 0$$

Thus the axiom $[N_1]$ (see Page 118) is satisfied.

 Now $\qquad ||k \cdot f|| \quad = \quad \sup \{|(k \cdot f)(x)|/||x||\} \quad = \quad \sup \{|k[f(x)]|/||x||\}$

$$= \quad \sup \{|k| \, |f(x)|/||x||\} \quad = \quad |k| \, \sup \{|f(x)|/||x||\} \quad = \quad |k| \, ||f||$$

Hence axiom $[N_2]$ is satisfied.

 Furthermore,

$$||f + g|| \quad = \quad \sup \{|f(x) + g(x)|/||x||\} \quad \leqq \quad \sup \{(|f(x)| + |g(x)|)/||x||\}$$

$$\leqq \quad \sup \{|f(x)|/||x||\} + \sup \{|g(x)|/||x||\} \quad = \quad ||f|| + ||g||$$

and so axiom $[N_3]$ is satisfied.

Supplementary Problems

CONVERGENCE OF SEQUENCES OF FUNCTIONS

25. Let $\langle f_1, f_2, \ldots \rangle$ be the sequence of real-valued functions with domain $I = [0, 1]$ defined by $f_n(x) = x^n/n$.

 (i) Show that $\langle f_n \rangle$ converges pointwise to the constant function $g(x) = 0$, i.e. for every $x \in I$, $\displaystyle \lim_{n \to \infty} f_n(x) = 0$.

 (ii) Show that
$$\lim_{n \to \infty} \frac{d}{dx} f_n(x) \;\; \neq \;\; \frac{d}{dx} \lim_{n \to \infty} f_n(x)$$

26. Let $\langle f_1, f_2, \ldots \rangle$ be a sequence of real-valued differentiable functions with domain $[a, b]$ which converge uniformly to g. Prove:
$$\frac{d}{dx} \lim_{n \to \infty} f_n(x) \;\; = \;\; \lim_{n \to \infty} \frac{d}{dx} f_n(x)$$

(Observe, by the preceding problem, that this result does not hold in the case of pointwise convergence.)

27. Let $f_n : \mathbf{R} \to \mathbf{R}$ be defined by

$$f_n(x) \;\; = \;\; \begin{cases} \dfrac{1}{n} \sqrt{n^2 - x^2} & \text{if } |x| < n \\[2mm] 0 & \text{if } |x| \geqq n \end{cases}$$

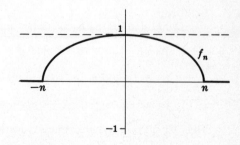

 (i) Show that $\langle f_n \rangle$ does not converge uniformly to the constant function $g(x) = 1$.

 (ii) Prove that $\langle f_n \rangle$ converges uniformly on compacta to the constant function $g(x) = 1$.

28. Let $\langle f_1, f_2, \ldots \rangle$ be the sequence of functions with domain $I = [0, 1]$ defined by $f_n(x) = nx(1-x)^n$.

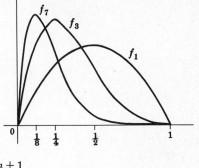

(i)   Show that $\langle f_n \rangle$ converges pointwise to the constant function $g(x) = 0$.

(ii) Show that $\langle f_n \rangle$ does not converge uniformly to $g(x) = 0$.

(iii) Show that, in this case,
$$\lim_{n \to \infty} \int_0^1 f_n(x)\, dx \; = \; \int_0^1 \left[\lim_{n \to \infty} f_n(x) \right] dx$$

29. Let $\langle f_1, f_2, \ldots \rangle$ be the sequence in $\mathcal{F}(\mathbf{R}, \mathbf{R})$ defined by $f_n(x) = \dfrac{n+1}{n}\, x$.

(i)   Show that $\langle f_n \rangle$ converges uniformly on compacta to the function $g(x) = x$.

(ii) Show that $\langle f_n \rangle$ does not converge uniformly to $g(x) = x$.

30. Let $\langle f_1, f_2, \ldots \rangle$ be a sequence of (Riemann) integrable functions on $I = [0, 1]$. The sequence $\langle f_n \rangle$ is said to *converge in the mean* to the function g if
$$\lim_{n \to \infty} \int_0^1 |f_n(x) - g(x)|^2\, dx \; = \; 0$$

(i)   Show that if $\langle f_n \rangle$ converges uniformly to g, then $\langle f_n \rangle$ converges in the mean to g.

(ii) Show, by a counterexample, that convergence in the mean does not necessarily imply pointwise convergence.

THE FUNCTION SPACE $C[0, 1]$

31. Show that $C[a, b]$ is isometric and hence homeomorphic to $C[0, 1]$.

32. Prove: Let $\langle f_n \rangle$ converge to g in $C[0, 1]$ and let $x_n \to x_0$. Then $\lim_{n \to \infty} f_n(x_n) = g(x_0)$.

33. Let p be a polygonal arc in $C[0, 1]$ and let $\delta > 0$. Show that there exists a sawtooth function g with magnitude less than $\frac{1}{2}\delta$, i.e. $\|g\| < \frac{1}{2}\delta$, such that $p + g$ does not belong to A_m (see Problem 14).

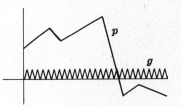

34. Let $\langle f_n \rangle$ be a Cauchy sequence in $C[0, 1]$ and let $\langle f_n \rangle$ converge pointwise to g. Then $\langle f_n \rangle$ converges uniformly to g.

UNIFORM CONTINUITY

35. Show that $f(x) = 1/x$ is not uniformly continuous on the open interval $(0, 1)$.

36. Define uniform continuity for a function $f : X \to Y$ where X and Y are arbitrary metric spaces.

37. Prove: Let f be a continuous function from a compact metric space X into a metric space Y. Then f is uniformly continuous.

FUNCTIONALS ON NORMED SPACES

38. Let f be a bounded linear functional on a normed space X. Show that
$$\sup\{|f(x)|/\|x\| : x \neq 0\} \; = \; \inf\{M : M \text{ is a bound for } f\}$$

39. Show that if f is a continuous linear functional on X then f is bounded.

40. Prove: The dual space X^* of any normed space X is complete.

Properties of the Real Numbers

FIELD AXIOMS

The set of real numbers, denoted by **R**, plays a dominant role in mathematics and, in particular, in analysis. In fact, many concepts in topology are abstractions of properties of sets of real numbers. The set **R** can be characterized by the statement that **R** is a *complete, Archimedian ordered field*. In this appendix we investigate the order relation in **R** which is used in defining the usual topology on **R** (see Chapter 4). We now state the field axioms of **R** which, with their consequences, are assumed throughout the text.

Definition: A set F of two or more elements, together with two operations called addition $(+)$ and multiplication $(\cdot)$, is a field if it satisfies the following axioms:

[A_1] Closure: $a, b \in F \;\Rightarrow\; a + b \in F$

[A_2] Associative Law: $a, b, c \in F \;\Rightarrow\; (a+b)+c = a+(b+c)$

[A_3] (Additive) Identity: $\exists\, 0 \in F$ such that $0 + a = a + 0 = a,\; \forall a \in F$

[A_4] (Additive) Inverse: $a \in F \;\Rightarrow\; \exists\, -a \in F$ such that $a + (-a) = (-a) + a = 0$

[A_5] Commutative Law: $a, b \in F \;\Rightarrow\; a + b = b + a$

[M_1] Closure: $a, b \in F \;\Rightarrow\; a \cdot b \in F$

[M_2] Associative Law: $a, b, c \in F \;\Rightarrow\; (a \cdot b) \cdot c = a \cdot (b \cdot c)$

[M_3] (Multiplicative) Identity: $\exists\, 1 \in F,\; 1 \neq 0$ such that $1 \cdot a = a \cdot 1 = a,\; \forall a \in F$

[M_4] (Multiplicative) Inverse: $a \in F,\; a \neq 0 \;\Rightarrow\; \exists\, a^{-1} \in F$ such that $a \cdot a^{-1} = a^{-1} \cdot a = 1$

[M_5] Commutative Law: $a, b \in F \;\Rightarrow\; a \cdot b = b \cdot a$

[D_1] Left Distributive Law: $a, b, c \in F \;\Rightarrow\; a \cdot (b + c) = a \cdot b + a \cdot c$

[D_2] Right Distributive Law: $a, b, c \in F \;\Rightarrow\; (b + c) \cdot a = b \cdot a + c \cdot a$

Here $\exists$ reads "there exists", $\forall$ reads "for every", and $\Rightarrow$ reads "implies".

The following algebraic properties of the real numbers follow directly from the field axioms.

Proposition A.1: Let F be a field. Then:

    (i)   The identity elements 0 and 1 are unique.

    (ii)   The following cancellation laws hold:

        (1) $a + b = a + c \;\Rightarrow\; b = c$,   (2) $a \cdot b = a \cdot c,\; a \neq 0 \;\Rightarrow\; b = c$

    (iii)   The inverse elements $-a$ and a^{-1} are unique.

    (iv)   For every $a, b \in F$,

        (1) $a \cdot 0 = 0$,   (2) $a \cdot (-b) = (-a) \cdot b = -(a \cdot b)$,   (3) $(-a) \cdot (-b) = a \cdot b$

Subtraction and division (by a non-zero element) are defined in a field as follows:

$$b - a \equiv b + (-a) \quad \text{and} \quad \frac{b}{a} \equiv b \cdot a^{-1}$$

Remark: A non-empty set together with two operations which satisfy all the axioms of a field except possibly [M_3], [M_4] and [M_5] is called a *ring*. The set $\mathbf{Z}$ of integers under addition and multiplication, for example, is a ring but not a field.

REAL LINE

We assume the reader is familiar with the geometric representation of $\mathbf{R}$ by means of points on a straight line as in the figure below. Notice that a point, called the origin, is chosen to represent 0 and another point, usually to the right of 0, is chosen to represent 1. Then there is a natural way to pair off the points on the line and the real numbers, i.e. each point will represent a unique real number and each real number will be represented by a unique point. For this reason we refer to $\mathbf{R}$ as the *real line* and use the words point and number interchangeably.

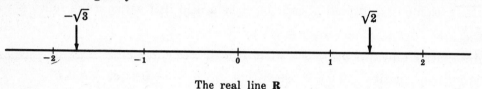

The real line $\mathbf{R}$

SUBSETS OF R

The symbols $\mathbf{Z}$ and $\mathbf{N}$ are used to denote the following subsets of $\mathbf{R}$:

$$\mathbf{Z} = \{\ldots, -3, -2, -1, 0, 1, 2, 3, \ldots\}, \quad \mathbf{N} = \{1, 2, 3, 4, \ldots\}$$

The elements in $\mathbf{Z}$ are called *rational integers* or, simply, *integers*; and the elements in $\mathbf{N}$ are called *positive integers* or *natural numbers*.

The symbol $\mathbf{Q}$ is used to denote the set of *rational numbers*. The rational numbers are those real numbers which can be expressed as the ratio of two integers provided the second is non-zero:

$$\mathbf{Q} = \{x \in \mathbf{R} : x = p/q;\ p, q \in \mathbf{Z},\ q \neq 0\}$$

Now each integer is also a rational number since, e.g., $-5 = 5/-1$; hence $\mathbf{Z}$ is a subset of $\mathbf{Q}$. In fact we have the following hierarchy of sets:

$$\mathbf{N} \subset \mathbf{Z} \subset \mathbf{Q} \subset \mathbf{R}$$

The *irrational numbers* are those real numbers which are not rational; thus $\mathbf{Q}^c$, the complement (relative to $\mathbf{R}$) of the set $\mathbf{Q}$ of rational numbers, denotes the set of irrational numbers.

POSITIVE NUMBERS

Those numbers to the right of 0 on the real line $\mathbf{R}$, i.e. on the same side as 1, are the *positive numbers*; those numbers to the left of 0 are the *negative numbers*. The following axioms completely characterize the set of positive numbers:

[P_1]   If $a \in \mathbf{R}$, then exactly one of the following is true: a is positive; $a = 0$; $-a$ is positive.

[P_2]   If $a, b \in \mathbf{R}$ are positive, then their sum $a + b$ and their product $a \cdot b$ are also positive.

It follows that a is positive if and only if $-a$ is negative.

Example 1.1: We show, using only $[P_1]$ and $[P_2]$, that the real number 1 is positive. By $[P_1]$, either 1 or -1 is positive. Assume that -1 is positive and so, by $[P_2]$, the product $(-1)(-1) = 1$ is also positive. But this contradicts $[P_1]$ which states that 1 and -1 cannot both be positive. Hence the assumption that -1 is positive is false, and 1 is positive.

Example 1.2: The real number -2 is negative. For, by Example 1.1, 1 is positive and so, by $[P_2]$, the sum $1+1 = 2$ is positive. Therefore, by $[P_1]$, -2 is not positive, i.e. -2 is negative.

Example 1.3: We show that the product $a \cdot b$ of a positive number a and a negative number b is negative. For if b is negative then, by $[P_1]$, $-b$ is positive and so, by $[P_2]$, the product $a \cdot (-b)$ is also positive. But $a \cdot (-b) = -(a \cdot b)$. Thus $-(a \cdot b)$ is positive and so, by $[P_1]$, $a \cdot b$ is negative.

ORDER

We define an order relation in **R**, using the concept of positiveness.

Definition: The real number a is *less than* the real number b, written $a < b$, if the difference $b - a$ is positive.

Geometrically speaking, if $a < b$ then the point a on the real line lies to the left of the point b.

The following notation is also used:

$b > a$, read b is greater than a,   means $a < b$

$a \leqq b$, read a is less than or equal to b,   means $a < b$ or $a = b$

$b \geqq a$, read b is greater than or equal to a, means $a \leqq b$

Example 2.1: $2 < 5$; $-6 \leqq -3$; $4 \leqq 4$; $5 > -8$

Example 2.2: A real number x is positive iff $x > 0$, and x is negative iff $x < 0$.

Example 2.3: The notation $2 < x < 7$ means $2 < x$ and also $x < 7$; hence x will lie between 2 and 7 on the real line.

The axioms $[P_1]$ and $[P_2]$ which define the positive real numbers are used to prove the following theorem.

Theorem A.2: Let a, b and c be real numbers.   Then:

(i)   either $a < b$, $a = b$ or $b < a$;

(ii) if $a < b$ and $b < c$, then $a < c$;

(iii) if $a < b$, then $a + c < b + c$;

(iv) if $a < b$ and c is positive, then $ac < bc$; and

(v)   if $a < b$ and c is negative, then $ac > bc$.

Corollary A.3: The set **R** of real numbers is totally ordered by the relation $a \leqq b$.

ABSOLUTE VALUE

The absolute value of a real number x, denoted by $|x|$, is defined by

$$|x| = \begin{cases} x & \text{if } x \geqq 0 \\ -x & \text{if } x < 0 \end{cases}$$

Observe that the absolute value of any number is always non-negative, i.e. $|x| \geqq 0$ for every $x \in \mathbf{R}$.

Geometrically speaking, the absolute value of x is the distance between the point x on the real line and the origin, i.e. the point 0. Furthermore, the distance between any two points $a, b \in \mathbf{R}$ is $|a - b| = |b - a|$.

Example 3.1: $|-2| = 2, \quad |7| = 7, \quad |-\pi| = \pi, \quad |-\sqrt{2}| = \sqrt{2}$

Example 3.2: $|3 - 8| = |-5| = 5 \quad \text{and} \quad |8 - 3| = |5| = 5$

Example 3.3:   The statement $|x| < 5$ can be interpreted to mean that the distance between x and the origin is less than 5; hence x must lie between -5 and 5 on the real line. In other words,

$$|x| < 5 \quad \text{and} \quad -5 < x < 5$$

have identical meaning and, similarly,

$$|x| \leq 5 \quad \text{and} \quad -5 \leq x \leq 5$$

have identical meaning.

The graph of the function $f(x) = |x|$, i.e. the absolute value function, lies entirely in the upper half plane since $f(x) \geq 0$ for every $x \in \mathbf{R}$ (see diagram below).

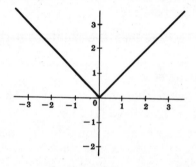

Graph of $f(x) = |x|$

The central facts about the absolute value function are the following:

Proposition A.4:   Let a, b and c be real numbers. Then:

   (i) $|a| \geq 0$, and $|a| = 0$ iff $a = 0$;

   (ii) $|ab| = |a||b|$;

   (iii) $|a + b| \leq |a| + |b|$;

   (iv) $|a - b| \geq ||a| - |b||$;   and

   (v) $|a - c| \leq |a - b| + |b - c|$.

LEAST UPPER BOUND AXIOM

Chapter 14 discusses the concept of completeness for general metric spaces. For the real line $\mathbf{R}$, we may use the definition: $\mathbf{R}$ is *complete* means that $\mathbf{R}$ satisfies the following axiom:

[LUB]   (Least Upper Bound Axiom):   If A is a set of real numbers bounded from above, then A has a least upper bound, i.e. sup (A) exists.

Example 4.1:   The set $\mathbf{Q}$ of rational numbers does not satisfy the Least Upper Bound Axiom. For let

$$A = \{q \in \mathbf{Q} : q > 0, \; q^2 < 2\}$$

i.e., A consists of those rational numbers which are greater than 0 and less than $\sqrt{2}$. Now A is bounded from above, e.g. 5 is an upper bound for A. But A does not have a least upper bound, i.e. there exists no rational number m such that $m = \sup(A)$. Observe that m cannot be $\sqrt{2}$ since $\sqrt{2}$ does not belong to $\mathbf{Q}$.

We use the Least Upper Bound Axiom to prove that **R** is Archimedean ordered:

Theorem (Archimedean Order Axiom) A.5:   The set $\mathbf{N} = \{1, 2, 3, \ldots\}$ of positive integers is not bounded from above.

In other words, there exists no real number which is greater than every positive integer. One consequence of this theorem is:

Corollary A.6:   There is a rational number between any two distinct real numbers.

NESTED INTERVAL PROPERTY

The *nested interval property* of **R**, contained in the next theorem, is an important consequence of the Least Upper Bound Axiom, i.e. the completeness of **R**.

Theorem (Nested Interval Property) A.7:   Let $I_1 = [a_1, b_1]$, $I_2 = [a_2, b_2]$, $\ldots$ be a sequence of nested closed (bounded) intervals, i.e. $I_1 \supset I_2 \supset \ldots$. Then there exists at least one point common to every interval, i.e.

$$\cap_{i=1}^{\infty} I_i \neq \varnothing$$

It is necessary that the intervals in the theorem be closed and bounded, or else the theorem is not true as seen by the following two examples.

> **Example 5.1:**   Let $A_1, A_2, \ldots$ be the following sequence of open-closed intervals:
>
> $$A_1 = (0, 1], \quad A_2 = (0, 1/2], \quad \ldots, \quad A_k = (0, 1/k], \quad \ldots$$
>
> Now the sequence of intervals is nested, i.e. each interval contains the succeeding interval: $A_1 \supset A_2 \supset \cdots$. But the intersection of the intervals is empty, i.e.,
>
> $$A_1 \cap A_2 \cap \cdots \cap A_k \cap \cdots = \varnothing$$
>
> Thus there exists no point common to every interval.

> **Example 5.2:**   Let $A_1, A_2, \ldots$ be the following sequence of closed infinite intervals:
>
> $$A_1 = [1, \infty), \quad A_2 = [2, \infty), \quad \ldots, \quad A_k = [k, \infty), \quad \ldots$$
>
> Now $A_1 \supset A_2 \supset \cdots$, i.e. the sequence of intervals is nested. But there exists no point common to every interval, i.e.,
>
> $$A_1 \cap A_2 \cap \cdots \cap A_k \cap \cdots = \varnothing$$

Solved Problems

FIELD AXIOMS

1. Prove Proposition A.1(iv):   For every $a, b \in F$,

$$(1)\ \ a0 = 0, \quad (2)\ \ a(-b) = (-a)b = -ab, \quad (3)\ \ (-a)(-b) = ab$$

Solution:

(1) $a0 = a(0 + 0) = a0 + a0$.   Adding $-a0$ to both sides gives $0 = a0$.

(2) $0 = a0 = a(b + (-b)) = ab + a(-b)$.   Hence $a(-b)$ is the negative of ab, that is, $a(-b) = -ab$. Similarly, $(-a)b = -ab$.

(3) $0 = (-a)0 = (-a)(b + (-b)) = (-a)b + (-a)(-b) = -ab + (-a)(-b)$.   Adding ab to both sides gives $ab = (-a)(-b)$.

2. Show that multiplication distributes over subtraction in a field F, i.e. $a(b-c) = ab - ac$.

    Solution: $a(b-c) = a(b + (-c)) = ab + a(-c) = ab + (-ac) = ab - ac$

3. Show that a field F has no zero divisors, i.e. $ab = 0 \Rightarrow a = 0$ or $b = 0$.

    Solution:

         Suppose $ab = 0$ and $a \neq 0$. Then a^{-1} exists and so $b = 1b = (a^{-1}a)b = a^{-1}(ab) = a^{-1}0 = 0$.

INEQUALITIES AND POSITIVE NUMBERS

4. Rewrite so that x is alone between the inequality signs:

    (i) $3 < 2x - 5 < 7$,   (ii) $-7 < -2x + 3 < 5$.

    Solution:

         We use Theorem A.2:

    (i)   By (iii), we can add 5 to each side of $3 < 2x - 5 < 7$ to get $8 < 2x < 12$. By (iv), we can multiply each side by $\frac{1}{2}$ to obtain $4 < x < 6$.

    (ii)   Add -3 to each side to get $-10 < -2x < 2$. By (v), we can multiply each side by $-\frac{1}{2}$ and reverse the inequalities to obtain $-1 < x < 5$.

5. Prove that $\frac{1}{2}$ is a positive number.

    Solution:

         By $[\mathbf{P_1}]$, either $-\frac{1}{2}$ is positive or $\frac{1}{2}$ is positive. Suppose $-\frac{1}{2}$ is positive and so, by $[\mathbf{P_2}]$, $(-\frac{1}{2}) + (-\frac{1}{2}) = -1$ is also positive. But by Example 1.1, 1 is positive and not -1. Thus we have a contradiction, and so $\frac{1}{2}$ is positive.

6. Prove Theorem A.2(ii): If $a < b$ and $b < c$, then $a < c$.

    Solution:

         By definition, $a < b$ means $b - a$ is positive; and $b < c$ means $c - b$ is positive. Now, by $[\mathbf{P_2}]$, the sum $(b-a) + (c-b) = c - a$ is positive and so, by definition, $a < c$.

7. Prove Theorem A.2(v): If $a < b$ and c is negative, then $ac > bc$.

    Solution:

         By definition, $a < b$ means $b - a$ is positive. By $[\mathbf{P_1}]$, if c is negative then $-c$ is positive, and so, by $[\mathbf{P_2}]$, the product $(b-a)(-c) = ac - bc$ is also positive. Hence, by definition, $bc < ac$ or, equivalently, $ac > bc$.

8. Determine all real numbers x such that $(x-1)(x+2) < 0$.

    Solution:

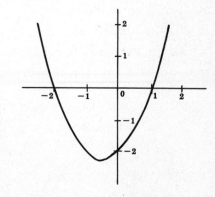

         We must find all values of x such that $y = (x-1)(x+2)$ is negative. Since the product of two numbers is negative iff one is positive and the other is negative, y is negative if (i) $x - 1 < 0$ and $x + 2 > 0$, or (ii) $x - 1 > 0$ and $x + 2 < 0$. If $x - 1 > 0$ and $x + 2 < 0$, then $x > 1$ and $x < -2$, which is impossible. Thus y is negative iff $x - 1 < 0$ and $x + 2 > 0$, or $x < 1$ and $x > -2$, that is, if $-2 < x < 1$.

         Observe that the graph of $y = (x-1)(x+2)$ crosses the x-axis at $x = 1$ and $x = -2$ (as shown on the right). Furthermore, the graph lies below the x-axis iff y is negative, that is, iff $-2 < x < 1$.

ABSOLUTE VALUES

9. Evaluate: (i) $|1-3| + |-7|$,   (ii) $|-1-4| - 3 - |3-5|$,   (iii) $||-2| - |-6||$.

Solution:

(i) $|1-3| + |-7| = |-2| + |-7| = 2 + 7 = 9$

(ii) $|-1-4| - 3 - |3-5| = |-5| - 3 - |-2| = 5 - 3 - 2 = 0$

(iii) $||-2| - |-6|| = |2-6| = |-4| = 4$

10. Rewrite without the absolute value sign: (i) $|x-2| < 5$,   (ii) $|2x+3| < 7$.

Solution:

(i) $-5 < x - 2 < 5$ or $-3 < x < 7$

(ii) $-7 < 2x + 3 < 7$ or $-10 < 2x < 4$ or $-5 < x < 2$

11. Rewrite using the absolute value sign: (i) $-2 < x < 6$,   (ii) $4 < x < 10$.

Solution:

First rewrite each inequality so that a number and its negative appear at the ends of the inequality:

(i)   Add -2 to each side of $-2 < x < 6$ to obtain $-4 < x - 2 < 4$ which is equivalent to $|x - 2| < 4$.

(ii)   Add -7 to each side of $4 < x < 10$ to obtain $-3 < x - 7 < 3$ which is equivalent to $|x - 7| < 3$.

12. Prove Proposition A.4(iii): $|a + b| \leq |a| + |b|$.

Solution:

Method 1.

Since $|a| = \pm a$, $-|a| \leq a \leq |a|$; also $-|b| \leq b \leq |b|$.   Then, adding,

$$-(|a| + |b|) \leq a + b \leq |a| + |b|$$

Therefore,
$$|a + b| \leq ||a| + |b|| = |a| + |b|$$

since $|a| + |b| \geq 0$.

Method 2.

Now $ab \leq |ab| = |a||b|$ implies $2ab \leq 2|a||b|$, and so

$$(a + b)^2 = a^2 + 2ab + b^2 \leq a^2 + 2|a||b| + b^2 = |a|^2 + 2|a||b| + |b|^2 = (|a| + |b|)^2$$

But $\sqrt{(a+b)^2} = |a + b|$ and so, by the square root of the above, $|a + b| \leq |a| + |b|$.

13. Prove Proposition A.4(v): $|a - c| \leq |a - b| + |b - c|$.

Solution: $|a - c| = |(a - b) + (b - c)| \leq |a - b| + |b - c|$

LEAST UPPER BOUND AXIOM

14. Prove Theorem (Archimedean Order Axiom) A.5:   The subset $\mathbf{N} = \{1, 2, 3, \ldots\}$ of $\mathbf{R}$ is not bounded from above.

Solution:

Suppose $\mathbf{N}$ is bounded from above.   By the Least Upper Bound Axiom, $\sup(\mathbf{N})$ exists, say $b = \sup(\mathbf{N})$.   Then $b - 1$ is not an upper bound for $\mathbf{N}$ and so

$$\exists\ n_0 \in \mathbf{N} \quad \text{such that} \quad b - 1 < n_0 \ \text{ or } \ b < n_0 + 1$$

But $n_0 \in \mathbf{N}$ implies $n_0 + 1 \in \mathbf{N}$, and so b is not an upper bound for $\mathbf{N}$, a contradiction. Hence $\mathbf{N}$ is not bounded from above.

15. Prove: Let a and b be positive real numbers. Then there exists a positive integer $n_0 \in \mathbf{N}$ such that $b < n_0 a$. In other words, some multiple of a is greater than b.

Solution:

Suppose n_0 does not exist, that is, $na < b$ for every $n \in \mathbf{N}$. Then, since a is positive, $n < b/a$ for every $n \in \mathbf{N}$, and so b/a is an upper bound for $\mathbf{N}$. This contradicts Theorem A.5 (Problem 14), and so n_0 does exist.

16. Prove: If a is a positive real number, i.e. $0 < a$, then there exists a positive integer $n_0 \in \mathbf{N}$ such that $0 < 1/n_0 < a$.

Solution:

Suppose n_0 does not exist, i.e. $a \leq 1/n$ for every $n \in \mathbf{N}$. Then, multiplying both sides by the positive number n/a, we have $n \leq 1/a$ for every $n \in \mathbf{N}$. Hence $\mathbf{N}$ is bounded by $1/a$, an impossibility. Consequently, n_0 does exist.

17. Prove Corollary A.6: There is a rational number q between any two distinct real numbers a and b.

Solution:

One of the real numbers, say a, is less than the other, i.e. $a < b$. If a is negative and b is positive, then the rational number 0 lies between them, i.e. $a < 0 < b$. We now prove the corollary for the case where a and b are both positive; the case where a and b are negative is proven similarly, and the case where a or b is zero follows from Problem 16.

Now $a < b$ means $b - a$ is positive and so, by the preceding problem,

$$\exists \; n_0 \in \mathbf{N} \quad \text{such that} \quad 0 < 1/n_0 < b - a \quad \text{or} \quad a + (1/n_0) < b$$

We claim that there is an integral multiple of n_0 which lies between a and b. Notice that $1/n_0 < b$ since $1/n_0 < a + (1/n_0) < b$. By Problem 15, some multiple of $1/n_0$ is greater than b. Let m_0 be the least positive integer such that $m_0/n_0 \geq b$; hence $(m_0 - 1)/n_0 < b$. We claim that

$$a < \frac{m_0 - 1}{n_0} < b$$

Otherwise $\quad \dfrac{m_0 - 1}{n_0} \leq a \quad$ and so $\quad \dfrac{m_0 - 1}{n_0} + \dfrac{1}{n_0} = \dfrac{m_0}{n_0} \leq a + \dfrac{1}{n_0} < b$

which contradicts the definition of m_0. Thus $(m_0 - 1)/n_0$ is a rational number between a and b.

NESTED INTERVAL PROPERTY

18. Prove Theorem A.7 (Nested Interval Property): Let $I_1 = [a_1, b_1]$, $I_2 = [a_2, b_2]$, $\ldots$ be a sequence of nested closed (bounded) intervals, i.e. $I_1 \supset I_2 \supset \cdots$. Then there exists at least one point common to every interval.

Solution:

Now $I_1 \supset I_2 \supset \cdots$ implies that $a_1 \leq a_2 \leq \cdots$ and $\cdots \leq b_2 \leq b_1$. We claim that

$$a_m < b_n \quad \text{for every} \quad m, n \in \mathbf{N}$$

for, $m > n$ implies $a_m < b_m \leq b_n$ and $m \leq n$ implies $a_m \leq a_n < b_n$. Thus each b_n is an upper bound for the set $A = \{a_1, a_2, \ldots\}$ of left end points. By the Least Upper Bound Axiom of $\mathbf{R}$, $\sup(A)$ exists; say, $p = \sup(A)$. Now $p \leq b_n$, for every $n \in \mathbf{N}$, since each b_n is an upper bound for A and p is the least upper bound. Furthermore, $a_n \leq p$ for every $n \in \mathbf{N}$, since p is an upper bound for $A = \{a_1, a_2, \ldots\}$. But

$$a_n \leq p \leq b_n \quad \Rightarrow \quad p \in I_n = [a_n, b_n]$$

Hence p is common to every interval.

19. Suppose, in the preceding problem, that the lengths of the intervals tend to zero, i.e. $\lim\limits_{n \to \infty} (b_n - a_n) = 0$. Show that there would then exist exactly one point common to every interval. Recall that $\lim\limits_{n \to \infty} (b_n - a_n) = 0$ means that, for every $\epsilon > 0$,

$$\exists \; n_0 \in \mathbf{N} \quad \text{such that} \quad n > n_0 \; \Rightarrow \; (b_n - a_n) < \epsilon$$

Solution:

   Suppose p_1 and p_2 belong to every interval. If $p_1 \neq p_2$, then $|p_1 - p_2| = \delta > 0$. Since $\lim_{n \to \infty} (b_n - a_n) = 0$, there exists an interval $I_{n_0} = [a_{n_0}, b_{n_0}]$ such that the length of I_{n_0} is less than the distance $|p_1 - p_2| = \delta$ between p_1 and p_2. Accordingly, p_1 and p_2 cannot both belong to I_{n_0}, a contradiction. Thus $p_1 = p_2$, i.e. only one point can belong to every interval.

Supplementary Problems

FIELD AXIOMS

20. Show that the Right Distributive Law $[D_2]$ is a consequence of the Left Distributive Law $[D_1]$ and the Commutative Law $[M_5]$.

21. Show that the set $\mathbf{Q}$ of rational numbers under addition and multiplication is a field.

22. Show that the following set A of real numbers under addition and multiplication is a field:
$$A = \{a + b\sqrt{2} : a, b \text{ rational}\}$$

23. Show that the set $A = \{\ldots, -4, -2, 0, 2, 4, \ldots\}$ of even integers under addition and multiplication satisfies all the axioms of a field except $[M_3]$, $[M_4]$ and $[M_5]$, that is, is a ring.

INEQUALITIES AND POSITIVE NUMBERS

24. Rewrite so that x is alone between the inequality signs:
    (i) $4 < -2x < 10$, (ii) $-1 < 2x - 3 < 5$, (iii) $-3 < 5 - 2x < 7$.

25. Prove: The product of any two negative numbers is positive.

26. Prove Theorem A.2(iii): If $a < b$, then $a + c < b + c$.

27. Prove Theorem A.2(iv): If $a < b$ and c is positive, then $ac < bc$.

28. Prove Corollary A.3: The set $\mathbf{R}$ of real numbers is totally ordered by the relation $a \leqq b$.

29. Prove: If $a < b$ and c is positive, then: (i) $\dfrac{a}{c} < \dfrac{b}{c}$, (ii) $\dfrac{c}{b} < \dfrac{c}{a}$.

30. Prove: $\sqrt{ab} \leqq (a + b)/2$. More generally, prove $\sqrt[n]{a_1 a_2 \cdots a_n} \leqq (a_1 + a_2 + \cdots + a_n)/n$.

31. Prove: Let a and b be real numbers such that $a < b + \epsilon$ for every $\epsilon > 0$. Then $a \leqq b$.

32. Determine all real values of x such that: (i) $x^3 + x^2 - 6x > 0$, (ii) $(x - 1)(x + 3)^2 \leqq 0$.

ABSOLUTE VALUES

33. Evaluate: (i) $|-2| + |1 - 4|$, (ii) $|3 - 8| - |1 - 9|$, (iii) $||-4| - |2 - 7||$.

34. Rewrite, using the absolute value sign: (i) $-3 < x < 9$, (ii) $2 \leqq x \leqq 8$, (iii) $-7 < x < -1$.

35. Prove: (i) $|-a| = |a|$, (ii) $a^2 = |a|^2$, (iii) $|a| = \sqrt{a^2}$, (iv) $|x| < a$ iff $-a < x < a$.

36. Prove Proposition A.4(ii): $|ab| = |a|\,|b|$.

37. Prove Proposition A.4(iv): $|\,|a| - |b|\,| \leq |a-b|$.

LEAST UPPER BOUND AXIOM

38. Prove: Let A be a set of real numbers bounded from below. Then A has a greatest lower bound, i.e. $\inf(A)$ exists.

39. Prove: (i) Let $x \in \mathbf{R}$ such that $x^2 < 2$; then $\exists\, n \in \mathbf{N}$ such that $(x + 1/n)^2 < 2$.

         (ii) Let $x \in \mathbf{R}$ such that $x^2 > 2$; then $\exists\, n \in \mathbf{N}$ such that $(x - 1/n)^2 > 2$.

40. Prove: There exists a real number $a \in \mathbf{R}$ such that $a^2 = 2$.

41. Prove: Between any two positive real numbers lies a number of the form r^2, where r is rational.

42. Prove: Between any two real numbers there is an irrational number.

Answers to Supplementary Problems

24. (i) $-5 < x < -2$ (ii) $1 < x < 4$ (iii) $-1 < x < 4$

32. (i) $-3 < x < 0$ or $x > 2$, i.e. $x \in (-3, 0) \cup (2, \infty)$ (ii) $x \leq 1$

33. (i) 5   (ii) -3 (iii) 1

34. (i) $|x - 3| < 6$ (ii) $|x - 5| \leq 3$ (iii) $|x + 4| < 3$

INDEX

235

Index of Symbols

$(X, \mathcal{T})$	topological space, 66		$\text{ext}(A)$	exterior of A, 70
(X, d)	metric space, 114		e_x	evaluation mapping, 207
$\|\cdots\|$	norm, 118		$\mathcal{F}(X, Y)$	class of functions from X into Y, 207
$\preceq$	coarser, 71		$F(A, B)$	class of functions from A into B, 211
$\Rightarrow$	implies, 7		iff	if and only if
$\exists$	there exists, 7		$\inf(A)$	infimum of A, 36
$\forall$	for all, 7		$\text{int}(A)$	interior of A, 70
$\setminus$	(e.g. $A \setminus B$), difference, 3		$\mathbf{N}$	the set of positive integers, 2, 226
A'	derived set of A, 67		$\mathcal{N}_p$	neighborhood system of p, 70
$\bar{A}$ or A^-	closure of A, 68		$\mathcal{P}(A)$	power set of A, 3
$\mathring{A}$ or A°	interior of A, 70		$\mathbf{Q}$	the set of rational numbers, 2, 226
A^c	complement of A, 3		$\mathbf{R}$	the set of real numbers, 2, 225
$\prod \{A_i : i \in I\}$ or $\prod_{i \in I} A_i$ or $\prod_i A_i$	Cartesian product, 19		$\mathbf{R}^m$	Euclidean m-space, 117
π_i	ith projection function, 19		$\mathbf{R}^\infty$	l_2-space, 117
$\langle \ldots, \ldots \rangle$	ordered pair, 5		$S(p, \delta)$	open sphere, 113
$\emptyset$	empty set, 2		s.t.	such that, 7
$b(A)$	boundary of A, 70		$\sup(A)$	supremum of A, 36
$C[0, 1]$	continuous functions on $[0, 1]$, 111, 210		$\mathcal{T}$	topology, 66
$d(A)$	diameter of A, 112		$\mathcal{T}_A$	relative topology on A, 72
$d(a, b)$	distance from a to b, 111		$\mathcal{U}$	usual topology, 66
$\mathcal{D}$	discrete topology, 66		$\mathbf{Z}$	the set of integers, 2, 226

Schaum's Outlines
and the Power of Computers...
The Ultimate Solution!

Now Available! An electronic, interactive version of *Theory and Problems of Electric Circuits* from the **Schaum's Outline Series.**

MathSoft, Inc. has joined with McGraw-Hill to offer you an electronic version of the *Theory and Problems of Electric Circuits* from the **Schaum's Outline Series.** Designed for students, educators, and professionals, this resource provides comprehensive interactive on-screen access to the entire Table of Contents including over 390 solved problems using Mathcad technical calculation software for PC Windows and Macintosh.

When used with Mathcad, this "live" electronic book makes your problem solving easier with quick power to do a wide range of technical calculations. Enter your calculations, add graphs, math and explanatory text anywhere on the page and you're done – Mathcad does the calculating work for you. Print your results in presentation-quality output for truly informative documents, complete with equations in real math notation. As with all of Mathcad's Electronic Books, *Electric Circuits* will save you even more time by giving you hundreds of interactive formulas and explanations you can immediately use in your own work.

Topics in *Electric Circuits* cover all the material in the **Schaum's Outline** including circuit diagramming and analysis, current voltage and power relations with related solution techniques, and DC and AC circuit analysis, including transient analysis and Fourier Transforms. All topics are treated with "live" math, so you can experiment with all parameters and equations in the book or in your documents.

To obtain the latest prices and terms and to order Mathcad and the electronic version of *Theory and Problems of Electric Circuits* from the **Schaum's Outline Series**, call 1-800-628-4223 or 617-577-1017.

Schaum's Outlines and Solved Problems Books
in the
BIOLOGICAL SCIENCES

*SCHAUM OFFERS IN SOLVED-PROBLEM AND QUESTION-AND-ANSWER FORMAT
THESE UNBEATABLE TOOLS FOR SELF-IMPROVEMENT.*

❀ Fried **BIOLOGY** ORDER CODE 022401-3/$12.95
(including 888 solved problems)

❀ Jessop **ZOOLOGY** ORDER CODE 032551-0/$13.95
(including 1050 solved problems)

❀ Kuchel et al. **BIOCHEMISTRY** order code 035579-7/$13.95
(including 830 solved problems)

❀ Meislich et al. **ORGANIC CHEMISTRY, 2/ed** ORDER CODE 041458-0/$13.95
(including 1806 solved problems)

❀ Stansfield **GENETICS, 3/ed** ORDER CODE 060877-6/$12.95
(including 209 solved problems)

❀ Van de Graaff/Rhees **HUMAN ANATOMY AND PHYSIOLOGY** ORDER CODE 066884-1/$12.95
(including 1470 solved problems)

❀ Bernstein **3000 SOLVED PROBLEMS IN BIOLOGY** ORDER CODE 005022-8/$16.95

❀ Meislich et al. **3000 SOLVED PROBLEMS IN ORGANIC CHEMISTRY** ORDER CODE 056424-8/$22.95

Each book teaches the subject thoroughly through Schaum's pioneering solved-problem
format and can be used as a supplement to any textbook. If you want to excel in
any of these subjects, these books will help and they belong on your shelf.

Schaum's Outlines have been used by more than 25,000,000 student's worldwide!

PLEASE ASK FOR THEM AT YOUR LOCAL BOOKSTORE OR USE THE COUPON BELOW TO ORDER.

SCHAUM'S SOLVED PROBLEMS SERIES

- ■ **Learn the best strategies for solving tough problems in step-by-step detail**
- ■ **Prepare effectively for exams and save time in doing homework problems**
- ■ **Use the indexes to quickly locate the types of problems you need the most help solving**
- ■ **Save these books for reference in other courses and even for your professional library**

To order, please check the appropriate box(es) and complete the following coupon.

❑ **3000 SOLVED PROBLEMS IN BIOLOGY**
ORDER CODE 005022-8/**$16.95 406 pp.**

❑ **3000 SOLVED PROBLEMS IN CALCULUS**
ORDER CODE 041523-4/**$19.95 442 pp.**

❑ **3000 SOLVED PROBLEMS IN CHEMISTRY**
ORDER CODE 023684-4/**$20.95 624 pp.**

❑ **2500 SOLVED PROBLEMS IN COLLEGE ALGEBRA & TRIGONOMETRY**
ORDER CODE 055373-4/**$14.95 608 pp.**

❑ **2500 SOLVED PROBLEMS IN DIFFERENTIAL EQUATIONS**
ORDER CODE 007979-x/**$19.95 448 pp.**

❑ **2000 SOLVED PROBLEMS IN DISCRETE MATHEMATICS**
ORDER CODE 038031-7/**$16.95 412 pp.**

❑ **3000 SOLVED PROBLEMS IN ELECTRIC CIRCUITS**
ORDER CODE 045936-3/**$21.95 746 pp.**

❑ **2000 SOLVED PROBLEMS IN ELECTROMAGNETICS**
ORDER CODE 045902-9/**$18.95 480 pp.**

❑ **2000 SOLVED PROBLEMS IN ELECTRONICS**
ORDER CODE 010284-8/**$19.95 640 pp.**

❑ **2500 SOLVED PROBLEMS IN FLUID MECHANICS & HYDRAULICS**
ORDER CODE 019784-9/**$21.95 800 pp.**

❑ **1000 SOLVED PROBLEMS IN HEAT TRANSFER**
ORDER CODE 050204-8/**$19.95 750 pp.**

❑ **3000 SOLVED PROBLEMS IN LINEAR ALGEBRA**
ORDER CODE 038023-6/**$19.95 750 pp.**

❑ **2000 SOLVED PROBLEMS IN Mechanical Engineering THERMODYNAMICS**
ORDER CODE 037863-0/**$19.95 406 pp.**

❑ **2000 SOLVED PROBLEMS IN NUMERICAL ANALYSIS**
ORDER CODE 055233-9/**$20.95 704 pp.**

❑ **3000 SOLVED PROBLEMS IN ORGANIC CHEMISTRY**
ORDER CODE 056424-8/**$22.95 688 pp.**

❑ **2000 SOLVED PROBLEMS IN PHYSICAL CHEMISTRY**
ORDER CODE 041716-4/**$21.95 448 pp.**

❑ **3000 SOLVED PROBLEMS IN PHYSICS**
ORDER CODE 025734-5/**$20.95 752 pp.**

❑ **3000 SOLVED PROBLEMS IN PRECALCULUS**
ORDER CODE 055365-3/**$16.95 385 pp.**

❑ **800 SOLVED PROBLEMS IN VECTOR MECHANICS FOR ENGINEERS
Vol I: STATICS**
ORDER CODE 056582-1/**$20.95 800 pp.**

❑ **700 SOLVED PROBLEMS IN VECTOR MECHANICS FOR ENGINEERS
Vol II: DYNAMICS**
ORDER CODE 056687-9/**$20.95 672 pp.**

**ASK FOR THE SCHAUM'S SOLVED PROBLEMS SERIES AT YOUR LOCAL BOOKSTORE
OR CHECK THE APPROPRIATE BOX(ES) ON THE PRECEDING PAGE
AND MAIL WITH THIS COUPON TO:**

McGRAW-HILL, INC.
ORDER PROCESSING S-1
PRINCETON ROAD
HIGHTSTOWN, NJ 08520

OR CALL
1-800-338-3987

NAME (PLEASE PRINT LEGIBLY OR TYPE)

ADDRESS (NO P.O. BOXES)

CITY   STATE   ZIP

ENCLOSED IS   ☐ A CHECK   ☐ MASTERCARD   ☐ VISA   ☐ AMEX   (✓ ONE)

ACCOUNT # _____   EXP. DATE _____

SIGNATURE _____

MAKE CHECKS PAYABLE TO MCGRAW-HILL, INC. PLEASE INCLUDE LOCAL SALES TAX AND **$1.25** SHIPPING/HANDLING
PRICES SUBJECT TO CHANGE WITHOUT NOTICE AND MAY VARY OUTSIDE THE U.S.   FOR THIS
INFORMATION, WRITE TO THE ADDRESS ABOVE OR CALL THE **800** NUMBER.